W9-BVH-576

LEVEL 4

LEVEL 4

VIRUS HUNTERS OF THE CDC

Joseph B. McCormick, M.D.
Susan Fisher-Hoch, M.D.
With Leslie Alan Horvitz

STERLING
New York

To
Jonathan Mann
&
MaryLou Clements-Mann

STERLING
New York

An Imprint of Sterling Publishing
1166 Avenue of the Americas
New York, NY 10036

STERLING and the distinctive Sterling logo are
registered trademarks of Sterling Publishing Co., Inc.

© 1996 by Joseph B. McCormick, M.D.
New material © 1999 by Joseph B. McCormick, M.D.

All rights reserved. No part of this publication may be reproduced,
stored in a retrieval system, or transmitted in any form or by any means
(including electronic, mechanical, photocopying, recording, or otherwise)
without prior written permission from the publisher.

ISBN 978-1-4549-1665-9

Distributed in Canada by Sterling Publishing
C/o Canadian Manda Group, 664 Annette Street
Toronto, Ontario, Canada M6S 2C8

For information about custom editions, special sales, and premium and corporate purchases,
please contact Sterling Special Sales at 800-805-5489 or specialsales@sterlingpublishing.com.

Manufactured in the United States of America

2 4 6 8 10 9 7 5 3 1

www.sterlingpublishing.com

"Biosafety Level 4 is required for work with dangerous and exotic agents which pose a high individual risk of life-threatening disease. Members of the laboratory staff have specific and thorough training in handling extremely hazardous infectious agents, and they understand the primary and secondary containment functions of the standard and special practices, the containment equipment, and the laboratory design characteristics. They are supervised by competent scientists who are trained and experienced in working with these agents. Access to the laboratory is strictly controlled by the laboratory director. The facility is either in a separate building or in a controlled area within a building, which is completely isolated from all other areas of the building. A specific facility operations manual is prepared or adopted.

"Within work areas of the facility, all activities are confined to Class III biological safety cabinets or Class I or Class II biological safety cabinets used along with one-piece positive pressure personnel suits ventilated by a life support system. The maximum containment laboratory has special engineering and design features to prevent microorganisms from being disseminated into the environment."

—Standard definition of "Level 4," from CDC/NIH, *Biosafety in Microbiological and Biomedical Laboratories*, 3d ed. (Washington: U.S. Government Printing Office, 1993)

TABLE OF CONTENTS

PREFACE

This book is a personal story of our experiences hunting Level-4 organisms and other emerging infectious diseases. Level 4 is the highest degree of laboratory containment for isolation and experimentation on microbiological organisms. Level-4 organisms produce lethal illnesses in humans and, for the most part, have no treatment and no prevention. Most are viruses, and among the most infamous are Ebola and Lassa. This elite group of organisms exerts a special fascination and gives those of us who work with them unparalleled experiences.

Our intent is therefore to give the reader a personal encounter with Level-4 organisms in various situations: in patients, in nature, in the laboratory. We hope that through our stories the reader in the safety of his or her home may vicariously face and care for a seriously ill patient with viral hemorrhagic fever; experience the search for virus reservoirs in remote places; sweat over the search for a new virus in the laboratory; work in a space suit; or look down an electron microscope. We hope to share with you the agony of uncertainty in the face of an epidemic of fatal disease and the triumph of containing that epidemic. Finally, we invite the reader to join us on the hunt in the jungles, villages, cities, and even deserts for those who are infected, and to visit with us the lairs of the viruses. Ultimately, through these personal accounts of real people and places, we also hope to bring about a greater understanding of the viruses and their impact on humans.

The journeys and accounts we share with you are our own; they are not meant to be more than that. All the characters are real, and all the incidents we describe truly happened. We have changed only the names of patients, in order to preserve medical confidentiality. The way we recount the events reflects the view we were afforded as they occurred; we obtained this evidence through our own eyes, our own ears, and our own emotions, and it is filtered to some extent—since we are human—through our prejudices and previous experiences.

With underlying emotions ranging from horror and pity to joy and humor evoked by the events we witnessed, we hope to share both the human tragedies of those who died and the triumphs of those who survived. We had our successes and our failures, but as time goes on it becomes more apparent that wars we thought we were winning against these tiny monsters are far from over. Indeed, each time we achieve a victory, the battleground shifts. There is a new outbreak, a new virus, a new enigma.

But let us not deceive ourselves: the driving force behind the emergence and reemergence of these Level-4 organisms is our own species. The microbes are not lurking, searching for human prey. It is we who interfere with them. They reside successfully, and often silently, in biological balance with their natural hosts. Only when humans invade their environment, intrude in their sphere, do we become infected. That is when the trouble starts. Viruses are not emerging; overpopulation and expansion of human habitation into their hiding places are forcing them into the open, providing them with new opportunities for epidemic fame and media attention. Humankind serves no purpose to a hemorrhagic fever virus and is not required for its long-term survival. We are dead-end hosts, and the virus dies with us.

Furthermore, with few exceptions viral hemorrhagic fevers are diseases of the poorest of the poor. The chain of events that turn the single case of a man infected—from having cut down some virgin forest or killed a wild animal—into an outbreak is a modern phenomenon. Western medical practices are spreading throughout the world, and are life-saving to many. Better medical care for all must be supported and encouraged. However, in many places invasive medicine and surgery is not well practiced and is even less well regulated. Surgery practiced in remote areas, under poor conditions, and by staff with limited education is commonplace. It has proved fatal to surgeons and other hospital staff unfortunate enough to operate in error on a patient with one of these viruses, and to patients unfortunate enough to be sharing the same hospital ward and sometimes the same syringes and needles. This is the genesis of the epidemics we hear about on television today.

Use of needles, syringes, and drips is exploding in developing nations, where these devices are popularly believed to provide instant relief from the myriad miseries of the poor. Poverty, ignorance, and sometimes greed lead to reuse of nonreusable sharp instruments, without sterilization, providing an opportunity for a Level-4 virus to hitch a ride. Ebola spreads in this way.

But Level-4 viruses are not the big-time killers. The real threat is from the viruses that really are exclusive human pathogens. We are their natural host, and they are without doubt out to get us. AIDS and hepatitis C have flourished in the explosive population growth and urbanization of the last couple of decades, particularly in developing countries. Their transmission depends on human behavior, and we have not hesitated to help them, by blood transfusions and sometimes even by reusing needles. Large numbers of people with no risk factors other than unsafe medical care are now dying of these terrible, slow diseases.

Our experience started with the hemorrhagic fevers and has evolved slowly but naturally into the study of the broader issue of emerging diseases, which now must also include violence and environmental health. We have come to see the commonality of the roots of emerging infections with the roots of malnutrition, urban violence, toxic poisoning, and extermination of biological species. All result from the consequences of overpopulation, poverty, and urbanization.

We ask the reader of our tales to keep in mind the role that we humans play in emerging infections. We must get at the root causes if we are ever to do more than delay the crisis. This includes dealing with overpopulation and poverty. We hope our experiences will open the eyes of both the curious and those motivated to understand how to make our world a better, safer place. Remember that in the world of viruses, we are the invaders.

JOSEPH B. MCCORMICK, M.D.

SUSAN FISHER-HOCH, M.D.

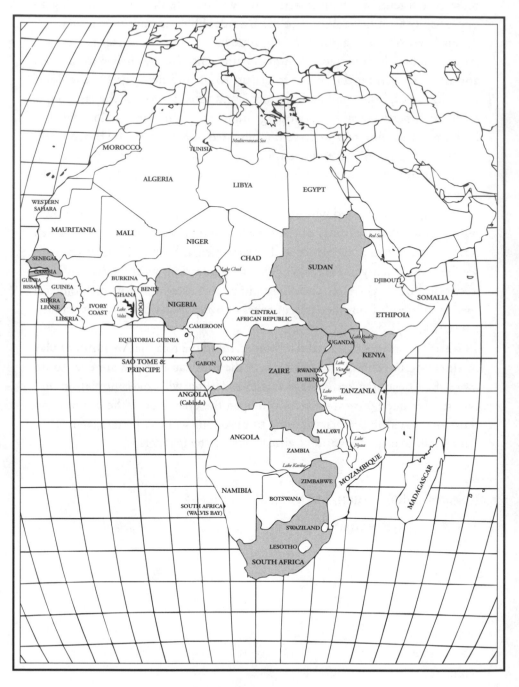

A F R I C A

Nzara, 1979

Whether Roy Baron and I touched down in Nzara, it was close to dusk. Night brought with it no relief; the heat was every bit as brutal as it had been in Juba before our plane took off. The humidity of southern Sudan sucked every drop of sweat out of us. There was simply no way to cool down.

Because the pilots could only fly by sight, they were compelled to stay overnight at the same government rest house where my colleague and I were being lodged. They weren't pleased. The thought of spending any time in an area rife with a lethal infection had them unnerved. But there was nothing else they could do, not if they wanted to make it back to Khartoum in one piece.

The delay worked in my favor, providing me with a terrific opportunity: this way they would be able to take something back to Khartoum for me and get it to the U.S. embassy, which could then ship it back to the States. They wouldn't know what was in the package I'd be giving them. If they did, they'd certainly refuse it—for (if our hunch was correct) the package would contain the very thing that they were most anxious to flee: the Ebola virus.

Not that they would be in any danger of infection. I knew how to pack specimens carefully, so that the air crew wouldn't run the risk of exposure. When I told them that I wanted them to take a package back to Khartoum, they were actually quite agreeable. We needed to get samples back to our lab at the Centers for Disease Control (CDC) in Atlanta as soon as possible so that we could determine the cause of the outbreak. Today there are other faster and more sensitive tests, some of which can even be done in the field, but in 1979 the only way to know for certain that the disease was in fact Ebola was to identify specific antibodies or to isolate the virus in cultured cells or tissues. Clinical observation alone wasn't sufficient. In this part of the world, where the average lifespan is less than fifty years, infectious diseases are ubiquitous. Many different infections can look like Ebola in the early stages of the illness. Symptoms like high fever, headache, abdominal pains, and sore throat can signal the onset of the flu as well as Ebola. Telltale signs of bleeding suggest Ebola but, even then, establishing a correct diagnosis may be tricky. And if what we had turned out to be Ebola, there was another question that needed to be answered: Was it the same strain as in 1976 or was it yet another virus, something that we'd never seen before?

We decided to examine the patients at the makeshift hospital in Yambio—a town located not far from Nzara—that same night. As soon as Roy and I could deposit our gear, we gathered together all the instruments we would need to draw serum samples from the patients. Obtaining these samples was only half the battle; we had to preserve them, too. First we would have to take the blood from the patients, then separate the red blood cells from the yellow serum containing the virus, freeze the serum samples in the dry ice we'd brought with us all the way from Khartoum, and finally package them for shipping.

When we arrived at the hospital, all we found was a dimly lit mud-walled building with a thatched roof. There were no windows. A few people were clustered outside the entrance—families of the patients dying inside. Their faces were filled with anguish. Roy and I donned protective suits, which looked like nothing so much as U.S. Air Force jumpsuits—only these jumpsuits were made out of white plastic-coated paper. Then we put on our full-face respirators. Not only were they unbearably hot and cumbersome to wear, they invariably had the effect of scaring the patients to death. That is, if Ebola didn't kill them first.

Inside the hut we were greeted by a truly macabre spectacle. In the uncertain light of two small kerosene lanterns, we could make out about a dozen patients, all adults, lying on mats on the bare earth. Some of them were thrashing about, delirious, in a futile attempt to escape the disease consuming their bodies. Others lay still, the rattle in their throats an unmistakable sign that they would soon be dead. The terrible heat of the equatorial night held us in a merciless grip. The jumpsuits and respirators made it worse. Sweat was pouring off us. We could barely breathe.

To examine each patient it was necessary for me to get down on my knees, lantern in hand, while Roy helped organize the blood samples as soon as I collected them. He had never worked in a developing country before, much less witnessed the devastation of Ebola, so this must have been a big shock for him.

Frank bleeding is not common in the early stages of Ebola, but minor manifestations, such as bleeding in the eye, are often present. I had to look into the whites of the eyes and inspect the interior of the nose and the gums for evidence of bleeding. Petechiae, which are tiny points of bleeding into the skin, are a significant clue, but they're hard to see in dark-skinned Africans, even under optimal conditions. By the dismal light of a kerosene lantern, they were impossible to detect, except in the whites of the eyes or the roof of the mouth and throat. A fine rash is also associated with Ebola in some cases, but again, I could scarcely expect to see it in such bad light. Yet one glimpse at these people's throats provided me with convincing evidence. Two or three days into the infection, following an incubation period of five days, Ebola can produce a throat so swollen and painful that a victim of the disease can't even swallow his own saliva. When you peer down such a throat, you see what could be mistaken for raw hamburger. A yellowish pus-like substance may ooze from the tonsils.

Bleeding from the rectum is another sign of Ebola. But unless the bleeding is significant, which it often is not, you have difficulty being sure. Either you ask a family member or else you have to get a stool sample and test it for blood, but for logistical as well as cultural reasons, this sort of test is often difficult to do.

It was like practicing medicine the way it had been done in the Dark Ages. We had no X-ray machines, no blood counts, no blood cultures, no diagnostic capacity except our own training and experience. Nevertheless, we pressed on, patient by patient.

Each patient presented a different challenge. Three of them were delirious and uncooperative. In order to examine them, I had to find either a nurse or a family member to help me and to hold their arm still while I took blood. When we were done, I was fairly certain that at least seven patients had Ebola. About the others I was less convinced. But it was our obligation to separate the ones we were sure were infected from those who might be suffering from other ailments, and then try to treat them as best we could.

At the end of about three hours we were done. It was nearly 11 P.M. I'd been up for some twenty hours, and I was exhausted. But my job wasn't over; I still needed to separate the sera from the red blood cells. The lab technicians at the CDC would be upset—and rightly so—if they received samples of serum filled with lysed red blood cells, which would interfere with the accuracy of their tests. Without electricity, I had to resort to improvisation. I'd had enough foresight to bring along an antiquated hand-cranked centrifuge, but because it contained only two buckets, it could handle only two samples at a time. I had thirteen to separate. Since I had to spin each pair of tubes for about ten minutes to get a reasonable separation, it would take me at least a solid hour of cranking to get the sera separated. Then I would have to separate each one into aliquots, label them, and pack them in dry ice. It didn't take long for me to discover that, in my depleted state, I was simply incapable of cranking the centrifuge for ten minutes without taking a break.

I was working in a room equipped with a small, shaky wooden table. I had to work alone; certainly, I couldn't let anyone else in with me. A tube might accidentally break, or else in my fatigue, I might splash myself. Why subject anyone else to the risk? I was wearing a simple surgical mask (it was just too hot to use a respirator) and my protective paper suit. Since I wear glasses, I didn't see any need to put on goggles. In spite of my exhaustion, I was as careful as I could be, but that didn't mean that I was safe. I might have infected myself at almost any time without realizing it. I wouldn't know for sure until the incubation period passed—after which coming down with a fever and the wracking body pains would characterize early Ebola infection. It took me nearly five hours to finish all thirteen serum specimens. It was now almost five in the morning, and I could hear the first sun birds in the acacia trees above the tall grass beginning to chirp to the first light.

I was beyond exhaustion, and the beautiful African dawn seemed slightly out of focus in the early mists. The rest house, where I tried to get some sleep, consisted of a few rooms with sagging wire-frame beds supporting thin cotton mattresses. There were no sheets. Though it was cooler here than it had been in the hospital, the suffocating heat was still oppressive. Getting to sleep was difficult. And it didn't last for long. Somehow I managed to get up again around 7 A.M. to deliver my precious box filled with the arduously separated sera to the pilots before they could take off for Khartoum. And this was only the prelude. Our formal investigation began the next day.

Our job was clear: we had to determine the extent of the epidemic, identify all the cases, and isolate the victims so that they couldn't transmit the infection to others. Over the next few days and weeks, we conducted an intensive search for people stricken with Ebola. At the same time, we set up a small lab to test for antibodies. This way we wouldn't have to wait for test results to come back from Atlanta.

Two nights after our arrival in town, I was back in the makeshift hospital, examining a suspected Ebola patient. On this occasion I decided to forego the respirator, which was simply too uncomfortable. The patient was an elderly woman who'd been brought in from an area where there had been a confirmed case of Ebola. She was delirious and spiking a high fever. Just before she was admitted, we were told, she'd suffered a seizure. Ebola patients sometimes have seizures, particularly in the end stage of the disease. Although I failed to detect any signs of bleeding, there was no question that she was very ill.

I knelt down to take the blood sample. Because she was thrashing around, I would ordinarily have had someone hold her arm, but she was old and frail. I thought that I could handle her by myself. Gripping her left arm tightly, I prepared to place the needle in her vein. I inserted the needle and then began to pull back the syringe to ensure that I was in the vein. Suddenly she gave a violent lurch, displaying far more strength than I thought possible from such an ill lady. The unexpected motion caused the needle to slip. My glove was punctured. The next thing I knew, there was a small drop of blood on the glove. My blood.

A moment later, I registered the sting of the needle. I saw now that it had broken the skin at the base of my thumbnail.

I swore under my breath.

How could I have been so careless? I had bled over three hundred victims of Lassa fever and never come close to pricking myself. My first instinct was to pull off my glove and cry out, but what good would that have done? Though I rinsed off the glove with disinfectant, I knew the damage had been done. So the only thing I could do was finish bleeding the woman and continue with my work. I can't say that I was calm, but I wasn't in a panic, either. Still, I had a nauseating feeling. I knew, more than most people, that when you get stuck by a potentially contaminated needle in the midst of a deadly epidemic—like the one I'd earlier investigated in Zaire—the odds for survival aren't very good.

Actually, I'd have to say that the fatality rate was about 100 percent.

Of course, now I was in the Sudan. It was possible that this particular strain of the virus might be less virulent, though the data were unclear. I also knew that a British researcher named Geoff Platt had stuck himself with a needle full of Ebola near Salisbury, England, while injecting mice in the Porton Down "hot lab" there after the 1976 Zaire epidemic. He became critically ill within days. He'd pricked his thumb just as I had, but he hadn't even drawn blood! Like me, he'd applied disinfectant to the wound immediately afterward. Later, a colleague asked him why he hadn't simply cut his thumb off. So I couldn't take much reassurance from precedent. Nor could I put any hope in a cure—there was no cure, there was no vaccine, there was no treatment.

Or almost no treatment. There was one possibility: it might not be worth much, but it was all I had. In 1979, it was thought that convalescent plasma—that is to say, plasma taken from patients who'd recovered from Ebola—might have some therapeutic benefit. But the data from the few experiments carried out so far in England—most of which were poorly designed—didn't give much basis for confidence. However, I had taken several units of plasma, which we'd collected in 1976 in anticipation of just this kind of situation, with me to the Sudan. The plasma wasn't in the best condition, but at least it had been run through a filter so that most of the green gunk that had collected in it had been removed. Well, I thought, I'll let Roy administer some of it to me and hope it works. What else could I do?

The contingency plans we'd made for this kind of emergency called for me to be shipped out in a plane specially equipped with a field isolator. The plane in question, however, was sitting on a runway in Europe, while the isolator was stored away in the U.S. Naval Medical Research Unit in Cairo. And even if I could get out, I would in effect be shutting down the investigation. No one else on the team had my experience; they couldn't continue on their own. This meant that almost all of our efforts so far would have been for nothing. What would happen if I stayed on—and what would happen if I left? I began to play out the different scenarios in my mind.

One: The old woman died. In that event, I was probably incubating Ebola. But I figured at least there would still be sufficient time after her death for me to be evacuated.

Two: She might have Ebola but survive. If then we tested her blood and found antibodies, we'd have incontrovertible evidence that I was infected, in which case I'd have to take immediate action—whatever that might be.

Three: She might not have Ebola at all. In that case, if I went ahead and arranged for transport from Cairo, I would feel ridiculous. After all, I reasoned, since plasma was the only treatment available (as questionable as it was), I wouldn't be any better off incubating it in the U.S. than I was here in the Sudan. In any event, there wouldn't be enough time for me to return to the States before I took sick. The incubation period for Ebola is at most only a few days after having been exposed the way I was.

Then I had to consider another question: If I stayed and I did have Ebola, would I be putting anyone else's life at risk? In the early stages of the disease, it is not easily transmissible unless someone is exposed to infected blood. That settled it. I would stay and ride it out.

That evening, Roy set up a line and gave me the plasma. We washed this down with well over half a bottle of whiskey. I liked to think that the alcohol might have some therapeutic effect. If this was a fantasy, at least I took some consolation from it. The bottle empty, there was nothing else for me to do but go to the rest house, to my cot and paper-thin mattress.

From that point on, I kept careful watch on the poor old lady, looking in on her at least twice a day, checking her vital signs and drawing her blood to see if she'd developed antibodies. I slept fitfully when I slept at all. Though I went about with my normal routine, she was constantly on my mind. Whatever happened to her was likely to happen to me. Her fate had become my own.

CHAPTER ONE

Zaire, 1983/1965

T ears of anger and frustration welled up in my eyes. I was standing at the end of a squalid bed in the Mama Yemo Hospital, Kinshasa, Zaire, watching a woman die. She could only be about twenty-five, I guessed. Her practically motionless form was stretched out on a stained mattress. There were no sheets to cover her nakedness. I swatted at the flies, which buzzed relentlessly around my ears. The ward held about thirty or forty women, among whom several were in a similar condition to the woman before me. She had lost her hair, her face was sallow, and her eyes sunken. Her lips were studded with raw sores. Her tongue must have caused her great pain; when I examined it more closely, I could see that it was encrusted with the yeast infection we now know to be so common in people afflicted with terminal AIDS. Her skin stretched tautly over her bones like an unpainted canvas over a frame. In places it was sown with the livid, bulging blotches of Kaposi's sarcoma, a cancer of the blood vessels in the skin, widely seen in AIDS patients. Over other parts of her body, bed sores had excavated extensive suppurating craters. Although she was of average height, she could hardly have weighed more than fifty or sixty pounds.

There was no one to attend to her needs—no family, no friends. This was a situation alien to Zairian culture as I knew it. It is the African tradition for the family to gather around the dying person to ensure that the transition from life to death is as gentle as possible. Where circumstances permit, it is important that this transition take place in comforting and culturally as well as personally meaningful surroundings, so that the dying soul can be assured of a welcome by the ancestors. I had seen people in Africa dying of diabetes and tuberculosis and leprosy, even Lassa fever. Always, they were surrounded by their families, whose presence gave them succor. Indeed, it was most common for patients with terminal disease to be taken away by the family to die at home rather than in the hospital.

What I saw before me now was new, a human tragedy unparalleled in my experience of modern Africa: a young woman, dying alone, abandoned.

How had it come to this?

What catastrophic, culture-wrenching changes could possibly have caused this to happen?

I had certainly seen my share of misery and poverty and disease while teaching school in Zaire. I had seen people dying of rabies and smallpox. I had watched small children succumb to malaria while their mothers looked on, helpless, grieving. But never anything like this. In Africa, people do not die alone. Or so I had thought.

As accustomed as I was to seeing the terminally sick, the plight of this young woman overwhelmed me. She epitomized for me the emergence in Africa of the discordant, chaotic modern world, where none of the old rules applied. I struggled to understand the reason for this devastation. What kind of disease could so dramatically and cruelly transform traditional culture? Of all the fatal diseases I'd seen in Africa, I realized that AIDS had created a cultural watershed. All the experience I'd acquired trying to fight disease in remote regions of the world now had to be reevaluated. I had no antidote to offer, no solace to give, and precious little hope to draw on.

Suddenly, she groaned and lifted her eyes toward me. In spite of her pain, she greeted me politely: *"Moyo Wanji"*—Hello, Sir.

The small flicker of a smile she managed to give me suggested that she might have enough energy to answer a few questions.

Where did she come from? I asked.

"Wembo Nyama," she replied.

I knew the town. I had arrived there in 1965 at the beginning of the school year, fresh and eager after four years of college and a year of French instruction in Brussels. I'd just turned twenty-one, and I was about to become the assistant principal of a high school where I was to teach science and mathematics.

Nearly twenty years seemed to dissolve.

If anyone told me when I was growing up in rural Indiana that I would one day end up teaching science to impoverished students in rural Zaire . . . well, there was no one in rural Indiana to tell me any such thing. As for myself, I doubt I could even have located Zaire—then called the Belgian Congo—on a map.

Raised as I was on a farm, far from any major city, I led a fairly sheltered, very simple life. My mother wanted to see me receive a good education, but no one in my family had ever gone to college. In my entire high-school class, there were only five students who even bothered to apply. But I was fortunate in having befriended a couple, a pastor named Jim Colvert and his wife, Sue, who more or less took me under their wings. Because Sue had gone to Southern Florida State College, a small liberal arts institution, she encouraged me to apply there. Even then, I wondered how I could hope to go. We had no money. But the Methodist Church offered me a partial scholarship.

Fearful, doubtful, and excited, I left for Florida in the fall of 1960 on a Greyhound coach, with a footlocker, a suitcase, and about fifty dollars in cash.

College was not so much an answer for me as it was a question. It made me more eager to expand my horizons. I knew I loved science—I took every course in physics, chemistry, and biology that I could—and for any American youngster interested in the sciences, the times couldn't have been better. The Soviet Union had launched Sputnik and now threatened to dominate the U.S. in space. The race was on to catch up, and there was suddenly a great deal of money available for science education. I applied for a National Science Foundation grant that would allow me to go on and study physics. No sooner did I receive the grant than I thought: *Is this what I really want?*

There was a whole world out there that I knew nothing about, and like so many other people my age, I was idealistic. I felt that I could make a difference, given the chance. The Peace Corps, which was only a few years old at the time, seemed like a logical choice for someone curious and looking for adventure. That they were prepared to send me to a far-off corner of the globe really appealed to me, but I wanted to teach science—and teach it in the language of the people I'd be teaching it to. All the Peace Corps could offer was a chance to teach English.

It was at this point that the Methodist Church again came to the rescue. The Church was seeking recent graduates, young, unmarried, and willing to teach in Zaire. The new recruits would be replacing teachers either killed or forced to leave the country during the turbulent era that followed independence in the early 1960s. When I learned that the Church was willing to let me teach science and would in addition provide me with a year of French instruction in Brussels beforehand, my mind was made up.

Two months after my graduation, I was on a ship bound for Belgium with the same footlocker and suitcase I had taken to college. Belgium exposed me to a culture and way of life that I never knew existed, but Europe couldn't begin to prepare me for life in the countryside of Africa.

In the 1960s, Kinshasa, the capital of Zaire, was a city that worked—a marked contrast to what it's like these days. Traffic coasted down wide, well-maintained boulevards flanked by palm trees. At night, the city was brilliant with light, and when you turned on the taps, water would flow out. It was all marvelous and new for me, a long way from Indiana. It was while I was in Kinshasa that I first heard of Wembo Nyama, a small town located in Kasai province in the east of Zaire. That was where I'd be teaching. It was now five years since Patrice Lumumba, an avowed Marxist, had helped to lead the country to independence. In fact, Lumumba had attended the boarding school I was being sent to—and had been expelled for "misbehavior." By 1965, Lumumba was slain and an army colonel named Joseph Mobotu was in power. And, to this day, there he remains, presiding in splendid isolation over a divided, demoralized, and poverty-stricken nation. In the conflict that swept over Zaire after its difficult birth, the school was shut down and remained closed for three years. Just one year before my arrival, two missionaries were killed there. The school's reopening meant that the young people of Wembo Nyama would have a second chance at an education. It was a revival that I was pleased to be a part of.

But when I arrived in Wembo Nyama, I discovered that my responsibilities weren't limited to teaching. I was also put in charge of the living arrangements of the students. As many as twelve to fifteen students slept in each dormitory room, which was only twenty feet square. I realized that the students needed more space. In the absence of bricks and mortar, I expanded our school's facilities using whatever materials I could lay my hands on, chiefly baked mud. I also had to forage for food to feed my charges. This gave me a great lesson in self-reliance. The school had no food, and the local markets could not provide sufficiently for two hundred students. I became adept at conducting scavenging expeditions into the countryside to procure whatever was available. These were hard times for Zaire (they would only get harder), and people had enough trouble finding food for themselves, much less providing for our students. The task of foraging required art, diplomacy, and no little determination. The legacy of the war complicated the situation. Many locals had spent the last few years hiding out in the forests, leaving their fields unplanted. When all else failed, we'd hire locals to hunt wild game for us. I had to feed the kids.

In addition to teaching, constructing dormitories, and foraging, I also took on work at the local hospital. Heavily damaged during the civil war, it was in desperate need of repair. Growing up on a farm, I'd developed a variety of useful skills that came in handy now. I was able to fix a generator, for instance, and even succeeded in getting a portable X-ray machine functioning—just about. No high-quality films here, but it worked. At least you could tell if a femur was broken.

Being the hospital's electrical handyman gained me some new colleagues. Late one night, I was pulled out of bed and summoned to the operating theater. I squinted through darkness at a group of surgeons huddled over a gurney. They were holding flashlights, trying to finish emergency abdominal surgery. The generator had failed. I stood transfixed by an image that would remain with me for the rest of my life: the flashlight-lit form of a woman with most of her intestine lying outside of her abdomen.

"Come on!" somebody said.

I wiped my hand across my face, took a breath, and went to the generator. I found a circuit shorting out and fixed it.

Soon after this, I started making the rounds with physicians. I was now in a position to see firsthand the astonishing variety of diseases that afflicted the people of Wembo Nyama: rabies, smallpox, tuberculosis, cholera, malaria, to name a few. Malaria was especially hard on young children. On one occasion, a baby—he was no more than eight or nine months old—was admitted with malaria and severe anemia. The pediatrician, Ray Isley, said the boy was in heart failure. He showed me the blood sample he had just taken to test for the parasites. The boy's blood was so thin that it was pink rather than red. Ray began to transfuse fresh blood into the child.

This will make it better, I said to myself.

I stood over his bed, watching in silence, first assuming—then hoping—that the boy would recover.

But the anemia caused by the malaria had progressed too far, and so this child became the first baby I ever saw die.

It was a terrible experience: the thin, pink blood; the still, small body. But it *was* an experience. And it started me thinking: Could a Ray Isley working in a country like Zaire make much of a difference? For every patient he saved, how many, like this little boy, did he reach too late with too little?

How, I wondered, could the work of a single physician be multiplied over an entire people who were engulfed in so much misery and disease? Though I didn't know it at the time, with this question, I had just introduced myself to the field of public health.

At the end of my first year of teaching in Zaire, I decided to apply to medical school in the States. I prevailed on Ray Isley to administer the exam that would qualify me for admission to medical school. Applying from Zaire was anything but routine, however. First of all, I had practically no information about different schools and scarce resources to draw on for advice. Having never considered medical school before, I didn't know anyone to turn to, except for the physicians in Wembo Nyama. One of them had been trained at the University of Minnesota and the other at the University of Kansas. I also had a close friend, Pete Peterson, who'd graduated from Duke Divinity School; it was he who suggested that I apply to Duke Medical School. In the end, I wrote to several top schools, including Stanford, Harvard, Yale, Indiana, and Duke. Indiana assumed I was a foreigner. Harvard, Yale, Stanford, and Duke all wrote back to say that due to the fact that they had no alumni available to interview me in Zaire, they wouldn't be able to consider my candidacy.

Over the last several months, I'd struck up a deep friendship with another teacher, named Shannon. We'd drawn so close that we began to talk about becoming engaged. (We didn't only get engaged; in 1968, we got married.) It so happened that she had an uncle who had graduated from Duke Medical School. When I showed her the letter from Duke, she wrote a strong note to Syd Osterhout, Duke's dean of admissions, expressing her disappointment over their rejection. Nor did she neglect to mention that her uncle was an alumnus. To my surprise, Dean Osterhout wrote back to say that Duke had reconsidered and would "keep my file open" until I returned in June for an interview.

But neither Duke nor I had reckoned on the vicissitudes of Zairian politics. To suppress a revolution in the early sixties in the breakaway province of Katanga, Mobutu had hired Belgian mercenaries to fight on his behalf. But now the mercenaries themselves had become a renegade force. That they hadn't been paid for some time did nothing to bolster their loyalty to the regime. In late April and early May 1967, mercenaries concentrated in and around Kisangani (formerly Stanleyville), decided to revolt, and they took over some key buildings, including the radio station. Mobutu responded in a characteristic manner, by placing the white population in the entire country under house arrest. He ordered his troops to fan out to schools, hospitals, and other institutions and "guard" the white people. No white person was to leave his compound for any reason. Our school was no exception to the rule. One day, Zairian troops swept over the small airfield near the school and placed empty metal drums on the strip so that no plane could land or take off. Then they deployed themselves throughout the school and the dormitories to insure that we couldn't leave.

Nonetheless, we were allowed to go about our business and continue teaching. We got on surprisingly well with the soldiers, sharing such essentials as food and comic books with them. But under no circumstances were they about to allow us to leave our compound. To make certain that we could not even communicate with the outside world, they confiscated our radios. They even seized electric razors and any other gadget that they suspected might be a radio in disguise. As the weeks wore on, I began to realize that I was in jeopardy of losing whatever chance I had of getting into medical school. Without a radio, I couldn't even let Duke know that I was unavoidably detained.

June came and went. I was in despair. At last, late in July, after ten weeks of house arrest, I tried to reason with the soldiers, asking that that they permit me to go into town.

What was the harm in that? I prodded.

Nothing doing.

Reasoning escalated into arguments, which always ended one-sidedly in no. That was when I got into my Land Rover and started to drive toward the road. As I approached Mobutu's troops, they waved their rifles menacingly. I smiled like an idiot and kept driving. I tried to look serene, but I was scared as hell. All I could do was hope that they wouldn't decide to make an example of me by blowing my brains out. I fought the impulse to duck down in order to avoid the bullets I was sure would soon start flying my way.

But nothing happened.

I drove to the Kananga airstrip—to lobby for a precious seat on a DC-3. I was always afraid that someone would notice that a "blanc" was wandering around loose in the city, but to my relief, no one seemed to care. I spent three days trying to book my seat. It was a sort of lottery, random, without rules. "Go to the airport and see if you can get a seat," people would tell me. Dutifully, each day, I set out to the airport with my suitcase and bag.

I finally managed to convince the man in charge at the airport to put me on a flight. It was now the last week of July.

More delays at Kinshasa meant that I didn't arrive in New York until the end of the first week of August. I drew my pay from Mission headquarters, and then set off for Durham, North Carolina, home of Duke University. The next morning, I turned up at Syd Osterhout's office, all 116 pounds of me, wearing a faded polo shirt and equally faded khaki pants, and introduced myself. I tried to explain to him why I'd been unable to come earlier. He seemed to be amused. Either my story was so good that it didn't matter whether it was true or not, or else he really did believe me. At any rate, he made me feel welcome and said that he would see about setting up some interviews for me. All of the people who interviewed me seemed more interested in hearing about my adventures than in asking me any complicated medical questions. Then I was told that, within two weeks, a decision would be made. If I was accepted, I would have just a little over a week to prepare for the start of school. I took a bus to Indiana, to my mother's house. She was horrified by the sight of her deeply tanned, very thin son, the result of three sun-baked years on a rural Zairian diet.

CHAPTER TWO

Of Epidemiology and Potato Salad

A letter from Duke. I was in!

Since the acceptance came a scant ten days before the start of the semester, my immersion in basic medicine was immediate. It would also prove total. Yet I soon realized that I was intrigued by subjects that weren't being taught at Duke Medical School. Midway through my second year, I sought out the new, charismatic professor of pediatrics, Sam Katz.

"I've had experience in Africa," I told him, "and I am interested in working in developing countries. Do you have any suggestions for my elective period in basic sciences where I could use that experience?"

"Why don't you spend a year with my old friend Tom Weller in Boston at the School of Public Health? He's doing the kind of things you're interested in."

I flew to Boston and called on Dr. Thomas Weller, who was friendly, if more than a trifle reserved. Well into middle age, his face nevertheless retained its boyishness. As chairman of the Tropical Medicine Department at the Harvard School of Public Health, he enjoyed an international reputation as a virologist and had earned a Nobel Prize for isolating the polio virus. Although he'd never before taken on a mere medical student at the School of Public Health, he was willing to give me a chance. I didn't hesitate to move to Boston for my junior year.

Dr. Weller long admired the CDC for its expertise in public health and epidemiology; in fact, five of my classmates were already working for the agency. He suggested that I might consider joining CDC myself. I returned to Duke for my senior year and applied to CDC just as I was about to begin my internship. Sam Katz had instilled in me an interest in children's medicine, so I interned in pediatrics at Children's Hospital of Philadelphia under Dr. C. Everett Koop. An imposing man with a full beard, Dr. Koop resembled a learned rabbi. He was a superb teacher and a great surgeon, but what struck me most about him was the way in which he engaged the parents of his patients. Most of his surgical cases involved very young children who were suffering from a variety of serious afflictions, usually birth defects. Dr. Koop had an uncanny ability of being able to explain the nature of the child's problem clearly and directly. He made sure that the parents understood the risks; and while he was never unduly sanguine, he tried to give them as much hope as he realistically could. The relationship he established with the families of his patients was an inspiration to me.

More than any other clinical discipline, pediatrics embraces prevention and public health. Think of the importance of immunizations in the lives of young children. Far from representing a detour, my training in pediatrics was one more step in the career I was building for myself in public health. With my pediatrics residency completed in 1973, I took the next step and began my life at CDC.

I arrived at CDC's Atlanta headquarters in July 1973, just in time to begin the courses for incoming EIS (Epidemic Intelligence Service) officers. I would be replacing an outgoing officer named David Fraser, who was returning to the University of Pennsylvania to complete his fellowship in Infectious Diseases. I was assigned to the Special Pathogens Branch of the Division of Bacterial Diseases and was enrolled in the EIS course. That vital preparation is supposed to last a month, but, barely a week into the course, while I was attending a lecture, Roger Feldman, chief of Special Pathogens, walked in to seek me out. Now, Roger is a big man with a deep rumbling voice—a difficult guy to ignore. He located me, gave me a brisk tap on the shoulder, and said, "I'm going to send you to Parker, Arizona, to an Indian reservation. There's a report that they have an epidemic of sore throats there. It could be streptococcal disease, but we're not sure."

"When do I leave?" I asked, trying to suppress my excitement at such an early opportunity to get into the field, even if the affliction was nothing more exotic than sore throats.

"You need to leave this afternoon," Roger said.

It was then nearly ten o'clock in the morning.

Each year, one or two students would be pulled out of the EIS class to undertake an assignment because an emergency situation had arisen that demanded the presence of an investigator—any investigator—in the field immediately. Experience, if you didn't have it already, was something you acquired as you hit the ground running.

I was elated; couldn't believe my luck. My first epidemic sounded fantastic. An epidemic of sore throats—in the middle of summer! All I was told was that the victims had attended a Fourth of July picnic. At this stage of my professional preparation, I didn't even know that it is possible to get a sore throat from a food-borne infection.

I discovered that, in the enduring tradition of EIS, the first thing you have to do is become an instant expert. You have to get your hands on everything you can find on the subject and read it, for the most part en route to the site of the outbreak. Of course, you have to find real experts to brief you, too. However obscure a disease is, there is usually someone at CDC who knows about it. Mostly, though, you have to rely on yourself. You need to have an instinct for finding and absorbing information. Then you have to know how to use it sensibly. Although your immediate supervisor is someone who has done it before, and who knows most of the tricks, each epidemic is a little different, with its own twists and dilemmas, and it is up to you to solve them, but, more than that, to learn something new.

Once you've gathered what you need to become an instant expert, your next step is to scramble around to collect the materials you'll need for the investigation: swabs, vials, syringes, sterilized packets of silicon gel for collecting cultures of strep, and so on. In the confusion, it's all you can do to remember to take a few pairs of clean socks and extra underwear along with you.

On no account can you leave without an EPI 1 in your hand. This piece of paper constitutes your marching orders. Among other things, it confirms that the relevant state or local health agency has requested the assistance of CDC. As a federal agency, CDC must obtain the state's permission to initiate an investigation in its jurisdiction. The EPI 1 also specifies the individuals in the state health department you need to contact as soon as you get to where you're going. Once you reach your destination, you must first establish lines of communication with CDC in Atlanta, so that you'll have someone available twenty-four hours a day to answer questions and help you make decisions. The essence of the training is learning on the job, backed by extensive experienced support and supervision.

As a raw EIS officer, you're always wondering about whether you're adequately prepared. Questions constantly weigh upon your mind. Can I find the source of this disease? What can I do to stop this epidemic? Will I manage to get the correct data so I can completely define the problem and come up with the solution to it? Will I obtain cooperation from the state and local people I have to work with?

A representative from the Arizona State Health Department was on hand to meet me in Phoenix. He told me that he'd be accompanying me to Parker, which was located about a hundred miles away, not far from the California border, and that he'd provide me with any assistance he could. Parker was a small town, but it was important because it served as a commercial hub for several nearby Indian reservations. The night of my arrival, I met with the physician who ran the small clinic on the reservation. He filled me in as much as he could. His story was straightforward: it began with a large Fourth of July picnic. Naturally, there was plenty of good food and beer. Then, a few days later, many of the picnickers—but not all—came down with severe streptococcal sore throats. The physician had seen many of the patients and had deduced that the one thing they had in common was that picnic. My job was to figure out why attending a picnic had put them at such risk, and then determine what I could do to stop any further cases from occurring. In theory, it sounded simple enough. But in practice?

I would soon find out.

In the articles I'd read on board the plane, I came across a report of an outbreak that had reminded me of this one. About a decade before, several people had developed sore throats after eating food contaminated with a certain strain of the streptococcus bacteria. This bacterium is the most common cause of sore throats. What made this particular infection serious, however, was that individuals, especially children, often go on to develop grave complications. Strep has been known to lead to rheumatic heart disease, kidney failure, severe skin conditions, and arthritis. Though these complications had rarely been reported in association with an epidemic, the possibility was worrying. I would have to move fast.

An outbreak investigation is very much like the investigation of a crime. It consists of detective work, following hunches, and carefully collecting evidence. In epidemiology, however, the criminal is the bug. Find the bug. And then find how it got to its human hosts. The bug's motive? Making a lot more bugs, I guess.

But it's not just bugs that you're dealing with. You have to deal with people, especially the victims. It requires some effort to explain to them what you're doing, and then to convince them to cooperate. Fortunately, that wasn't a problem in Parker. Here, people were clearly concerned about the outbreak, and not the least concerned were those individuals who had been responsible for organizing the event. This was my first time on a Native American reservation, and it was turning into an important learning experience. To elicit information, I needed to work through a certain hierarchy of authority to insure that I didn't offend any of the leaders or elders. I was fortunate in having already learned the importance of this in African villages.

I decided that this investigation needed to be done according to the book—what is called a case control study. This is a scientific method used by epidemiologists to discover the most important differences between those people who did become sick and those who did not. If you can determine these differences—particularly when it comes to food-borne outbreaks—you are usually close to pinpointing the cause or the route of infection. So I divided people who had been to the picnic into two groups: "cases" (people who had had sore throats) and "controls" (people who had not had sore throats).

Okay, now that I had my subjects, I had to figure out what to ask them. So I prepared a questionnaire. It had to be carefully worded, in order to elicit answers that were clearly in the category of yes or no. Basically, I was fishing for clues. Nonetheless, there was a certain logic to the questions. Some of them were obvious: Did you go to the picnic? Did you eat this dish, that dish, or the other dish? Did you drink this beverage or that one?

The investigator must be very careful in the way questions are phrased, so that they help people think clearly and insure accurate answers. It's easy to reach the wrong conclusions in investigations like the one I was conducting in Arizona. People tend to forget, or they may give a false answer in an effort to say what they believe is expected of them. There are certain things you tell the doctor, and other things you don't think he wants you to say. At the same time, I had to take specimens. This can only be done with the consent of the subjects, and it often required me to exercise all my powers of persuasion. In this particular outbreak, the obvious specimen was a throat swab. Once I was able to collect my swabs, I put them in special transport packets with silica gel in them. The gel would keep the bacterium alive until it reached the labs at CDC. There the specimens would be spread on special plates of agar jelly, which is rich with nutrients. If the strep was there, it would grow into round, grayish piles of organisms surrounded by transparent halos in the agar. Subsequent tests would tell us exactly what kind of strep we were dealing with.

I set about going from house to house, talking to people, marking my questionnaire, and sticking swabs down people's throats. I had no trouble confirming that the picnic had been the one common denominator. However, my case control study allowed me to pick up another crucial clue. All those who became sick had eaten one particular dish at the picnic: potato salad.

Now I had to find that potato salad—if there was any left.

My door-to-door inquiries turned up information that there were leftovers from the picnic. Next question: Who had them? This required a survey—with surprisingly intimate questions. Then I had to ask permission to go rummaging in people's fridges and freezers. After locating samples of several of the picnic dishes in a freezer belonging to the community center at the reservation, I very carefully packed up this precious evidence and dispatched it back to CDC to see if the lab could culture the guilty streptococcus from it.

After spending a week in Parker, I was ready to return to Atlanta for the second half of the investigation, which would continue in the lab. In the days before the personal computer, to obtain the statistical results of the questionnaire and other investigation inquiries, I had to enter all my data on punch cards. On the sixth floor of the CDC was an IBM machine known as a card sorter. It operated exclusively on a Yes/No answer system and sorted cards based on where holes had been punched in them. Although the machine was able to process each operation rather rapidly, it required a huge number of operations to arrive at an answer. Holes are punched in cards if the answer was yes, and left alone if the answer was no. The yes cards would then be separated into one pile, and the no cards in another. But if there are several levels to sort for, then the process of determining an answer grows progressively more difficult. If, for example, I wanted to find out whether all of the women who'd eaten potato salad had become ill compared to the men who had eaten it, I would have to sort the cards for all people who ate potato salad, then for all of the females and males. Then I would have to sort those cards for the females who ate potato salad and were ill. Cards would end up strewn all over the place. The whole business was very tedious. Nowadays, we put the information directly into a computer, enter a few instructions, and obtain the answers in a few seconds.

The lab had the final word: the potato salad harbored the culprit. Definitely strep. Apparently, after the potato salad was prepared, it had been left in large containers in the refrigerator, but because it took some time for the cold to penetrate the container, the center of the food remained warm for several hours—the perfect spot for bacteria contamination to live. It appeared that whoever had mixed the salad had been infected with strep. Because this person had not taken adequate food preparation precautions, the strep had got into the salad. The bacteria were just as happy about being in the warm center of the container, even in a refrigerator, as they would later be resting in the back of their host's throat.

So we prepared a report, an EPI II, to the health authorities in Parker, in which we described earlier recommendations to dispose of the frozen potato salad, and to be certain that individuals were appropriately treated with penicillin if suspected to harbor the organism in their throats. These measures were enough: the outbreak stopped, and no further cases occurred. I returned to my EIS course, but it was the Parker outbreak that provided me with the most practical foundation of my training in the scientific methods of epidemiology. It wouldn't be long before I would find myself searching for rats in Nigeria, surveying Sudanese villagers for incidences of Lassa, and asking patients about injections they'd received from general practitioners in Pakistan. No matter what situations I would face in the future, the basic methods at my disposal were those I learned tracking down suspect potato salad in a small southwestern town.

CHAPTER THREE

Suffer the Children

More than a thousand people were crowded into São Paulo's Emilio Rebas Hospital, which had beds for no more than five hundred. They were lying in the halls, stretched out on mattresses on the floors, and crammed into every nook and cranny that could be found. All of them were sick with meningococcal disease. Some rooms held only children.

Entering these rooms, you looked, and looked again, hoping that what you saw wasn't real. But it was real: the children here had lost their hands, or arms and legs, even parts of their noses and ears to the disease.

Meningitis is a bacterial infection of the fluid around the spinal cord. The bacterium is named meningococcus; hence the disease is called meningococcal meningitis. Its symptoms include headache, fever, nausea, and vomiting. In severe cases, it can cause convulsions in the victim, particularly in children. Sometimes it can result in coma. Since the bacterium is also carried in the bloodstream, it can produce shock and hemorrhage. The reason meningitis may lead to the loss of appendages is that it tends to block blood vessels with clots that cut off the blood supply to the extremities. The affected body parts turn black and gangrenous. Even though this horrible condition occurs in only about 10 percent of the survivors of the disease—and in the U.S., for some reason, is a much rarer complication—the number of children affected in a country the size of Brazil was still quite large. Plastic surgeons and physical therapists were desperate to figure out what could be done to help them.

The disease is terrifyingly swift. First come the splotches in the skin, and then the skin may actually start to blacken and slough off—a finger or toe, or even a nose or an ear lobe. It is very much like severe frostbite.

Meanwhile, many other children were in danger of dying from renal failure, another complication of the disease. There was only one way to save this group. It was a technique called peritoneal dialysis. I had never seen so many children undergoing this type of dialysis in my life. Peritoneal dialysis relieves the kidneys when they are incapable of carrying on their normal function of cleansing the blood of waste products and excess water. It works by pumping fluid through tubes inserted into the peritoneum, a thin membrane that covers the whole surface of the gut. Permeable, the peritoneum allows fluid and other substances to pass through. The introduction of fluid into the peritoneum essentially washes out the toxins in the bloodstream and expels them through the tubes. Obviously, the fluid has to be replenished and removed on a regular basis. Because it is such a cumbersome process, peritoneal dialysis is practical for only a few days. The hope is that this period will give enough time for the kidneys to recover. If peritoneal dialysis is continued for a protracted period, the tubes create a high risk of infection.

The bacterium that causes meningococcal meningitis is a small, round organism that turns red under the light microscope when treated with the standard Gram's stain. These little red cocci usually appear in pairs, which is why they are named diplococci, or double cocci. There are several distinct types of meningococci, but the most important are types A, B, and C. The Brazilian outbreak was type A. It is a respiratory disease, meaning that the bacteria are spread through the air. As a result, it is easily transmissible from one person to another, carried by nasal secretions and respiratory droplets. What makes meningitis especially diabolical is that for every patient who has the disease, there may be up to ten others who harbor the bacterium without ever getting sick. So, any time an epidemic like this erupts, crowding and close contact favor rapid spread, even among apparently healthy individuals.

Just as had happened during the Black Death in fourteenth-century Europe, many of Brazil's wealthiest people simply fled the country and remained abroad until the epidemic had run its course. Others, who could not afford to leave, took refuge inside their homes and refused to let their children go to school. The epidemic further exacerbated the rift between the rich and the poor, in large part because the rich looked upon the disease as the scourge of the lower classes. In their panic, they'd fire their servants rather than keep them in their houses, believing that this step minimized their risk of becoming infected. The impoverished masses of Brazil were thereby made even poorer. Thus, in every way, they bore the brunt of the epidemic.

It was this unprecedented outbreak of meningitis that brought me to São Paulo in 1974. I wasn't exactly new to epidemics—I was now in my second year as an EIS officer—but I had never confronted one so big. My specific assignment was to work with the Pan American Health Organization (PAHO) and the Brazilian government to assess the extent of the outbreak in the larger cities, and to help develop a strategy for control. In São Paulo, the number of cases was estimated to be twenty thousand. By the time I arrived, there were reports of cases turning up in Rio de Janeiro, Belo Horizonte, and Brasilia. In all Brazil, the total number was believed to be in excess of 120,000. It wasn't only the big cities that were affected. Even hospitals and clinics serving less densely populated communities were filling with victims. Not surprisingly, everyone experiencing the mildest headache or fever would run to the doctor, fearing that they had fallen prey to meningitis. Physicians and medical dispensers were being overwhelmed with the worried, and that only made their jobs that much harder.

Conditions in Brazil were conducive to the spread of the disease. The bacterium flourished in a setting burdened by overpopulation and cursed by devastating poverty. The poverty I'd seen in Zaire was completely different from what prevailed here. In Africa, there is at least some kind of subsistence economy always at work: barring prolonged droughts, people are usually able to grow enough to feed themselves. But I'd never seen anything like São Paulo. This was poverty on an unimaginable scale, an epidemic in its own right, which had transformed whole parts of Rio and São Paulo into hideous slums, fertile breeding grounds for violence as well as for lethal pathogens. I was told that the population in the slums of São Paulo, known as *favelas*, which sit on the hills ringing the city, grew at the rate of half a million a year. Most of the slum dwellers came from rural areas, where the economy could no longer sustain them, lured by the prospect of work. With the rapid explosion of the population in Brazil came cutthroat competition for jobs and resources. The same scenario obtained in Rio, and to a lesser degree in Belo Horizonte, Belém, San Salvador, and Fortaleza. In fact, the world over was urbanizing at an unprecedented rate.

In the *favelas*, space was at a premium, and people would stake a claim to whatever territory they could defend. The typical *favela* was a warren of hovels constructed of tin, burlap, and cardboard, all held in place, more or less, by wire and string. Everywhere you looked there were ragged children, covered with dirt and infested with scabies and other parasites, playing in the mud. It was impossible to distinguish the boundaries of anyone's home. As far as I could make out, there was no such thing as personal space. Privacy was beyond the imagination of these urban poor. The floors were dirt, furnishings confined to a few mats, some broken-down chairs, and maybe a ragged carpet or a shabby wooden bed. Even asleep, everyone was jammed together. Sanitary facilities, of course, were practically nonexistent, so that when it rained, the *favelas* were turned into oceans of mud and rank water reeking with sewage and detritus.

The terrible epidemic was also compounded by the ineffectiveness of PAHO officials, who had little idea how to cope with the emergency, and demonstrated even less motivation to do so. In those days, PAHO officials were from other Latin American countries, and they spoke only Spanish, never having learned any Portuguese, Brazil's native language. This resulted in a poor relationship with the Brazilians, which seriously hampered PAHO's ability to deal with the epidemic, and they often found themselves cut out of the decision-making process. (We can only hope that this has now been remedied.) Since I had no wish to be marginalized as well, I learned Portuguese as quickly as I could. If I were to get anything done at all, I would have to figure out how to maintain a good relationship with PAHO, while forming a real alliance with the Brazilians.

There was one bright ray of hope in this situation. Meningococcus is one of the very few bacteria that remain highly sensitive to penicillin, a cheap and effective antibiotic. However, there was still the problem that, in the early stages, you were never quite sure who had meningitis and who did not. If you got it, you got it fast, and you would need treatment very quickly. These days, with a population gripped by fear, every illness that turned up began to look as if it could be meningitis.

In Belo Horizonte, the capital of the Brazilian state of Minas Gerais, I went to work with the state health laboratory to help set up a diagnostic capability. Ambition, endurance, and good intentions are important to this work, but improvisation is essential. While we did pull some kind of lab together, we lacked an incubator, a key piece of equipment when you're breeding bacterial cultures. A resourceful CDC technician named George Gorman transformed an old wooden cabinet into an incubator by installing a lightbulb and a thermostat. If the incubator was our greenhouse, we still needed "pots" to grow the bacteria that was taken from suspected cases. We used candle jars, which get their name from the very simple fact that they are jars containing candles. The candles burn up oxygen, leaving an atmosphere rich in carbon dioxide inside the jar. Meningococcal bacteria hunger for just such an environment. Once they get it, they grow, producing colonies large enough to identify. A colony is put under a microscope and treated with a Gram's stain. If it is positive for meningococci, the typical red diplococci show up.

We were lucky, though, because a vaccine had just become available. Developed principally by Emil Gotschlich at Rockefeller University in the late 1960s, it was a polysaccharide vaccine and was effective against meningococcus types A and C. (Polysaccharides are the molecules that make up the heavy sugar coat of the meningococcus bacterium. The human immune system responds to the sugar coat as if it were the whole bacterium. That immune response will insure protection if the individual is later exposed to the bacterium itself.) Serendipitously, because the outbreak in Brazil was predominantly type A, with some residual type C, the vaccine was likely to be effective there.

The vaccine campaign of the Brazilian government was a Herculean effort and a resounding success. In the two years of the epidemic, health authorities succeeded in vaccinating between sixty and seventy million people, approximately 70 percent of the total population in 1974. That so many lives were saved, and that so many children were spared cruel disfigurement, can only be considered cause for rejoicing. The tragic note, however, is that the vaccination campaign was not begun soon enough. As many as five to ten thousand people lost their lives by the time we started.

Just before I completed my work in Brazil in 1976, I received a call from my advisor at CDC, Dr. Bill Feoge. Bill, who was later to become the agency's director, is a tall lanky figure who brought to his work a messianic fervor. He was a Pied Piper to public health people, including me. He is a man possessed of the very qualities guaranteed to make certain politicians and bureaucrats view him with mistrust, which is to say that he is dedicated, honest, and straightforward. So when he called me to say that the agency was interested in establishing a field station in Sierra Leone to study a new disease, I was inclined to listen very carefully.

"I know that you've had experience working in Africa," he said, "and I'd like to know whether you'd consider heading up a program to study a new disease in West Africa."

I asked what the disease was called.

"It's called Lassa fever."

My new boss, Karl Johnson, chief of Special Pathogens at CDC, occupied an office at CDC so small that when I came to occupy it myself years later, Bobby Brown, a primate veterinarian colleague, pronounced it well under regulation size for a chimpanzee. When I met Karl, he'd only recently returned from Panama, where he'd been attached to the Middle American Research Unit. He had spent much of his time tracking down a new arenavirus called Machupo, carried by small rats, that causes devastating hemorrhagic fevers in human beings. The cause of Lassa fever, it turned out, had just been shown to be an arenavirus. At this time Karl was a couple of years shy of his fiftieth birthday. His was an imposing, if rather eccentric, presence. Nearly six feet tall, with a shock of thick black hair spattered with gray, Karl sported a full, erratically trimmed beard that made him look more like Che Guevara than a sober-minded scientist. I'm sure that he preferred the image of a revolutionary to that of a medical researcher. His style of dress was determinedly casual as well. He was especially partial to embroidered shirts from Central America, with low pockets. Though he was soft-spoken and cultivated a studied air of relaxation, Karl was actually full of nervous energy. He could scarcely put one cigarette out without lighting another. (In those days, smoking enjoyed a certain popularity among many epidemiologists.) Once I'd come to know him well, I discovered that he loved parties and delighted in sitting up all night talking with friends while killing a bottle of bourbon. His warmth, intellect, and charm attracted a lot of admirers.

It didn't take long for Karl to realize that I didn't know anything about viruses. But I did know Africa, I did know my way around a laboratory—and I did know computers. That was something Karl appreciated. It was already becoming clear that the computer would be an increasingly valuable tool for analyzing epidemiological and laboratory data. And by the end of March 1976, Karl and I were on a plane headed for Sierra Leone.

Sandwiched between Guinea and Liberia, Sierra Leone is about the size of South Carolina and has a population of close to three million. The land was once covered almost entirely by primeval rain forest, but what I saw was mostly secondary bush. Slash and burn agricultural practices had denuded the trees and created one of the poorest countries on earth. The eleven major tribes of Sierra Leone share a Lingua Franca called Krio, which you can hear spoken everywhere in the capital of Freetown and in many places elsewhere in the country. Krio is a version of Pidgin English peculiar to Sierra Leone, but similar to the language spoken in many former British colonies up and down the West African coast. It is most closely related to a language spoken on a tiny island off the South Carolina coast, populated by descendants of escaped slaves. Krio is a charming but rather strange tongue, which borrows heavily from French and Portuguese, as well as from several other African languages and local dialects. The result is a pungent linguistic stew characterized by several delightful expressions such as "How de go-de-go?" meaning, more or less, "How are you doing?" Or, if you want to state that something is happening, or that you are in possession of an object, you use the word "deh." If, on the other hand, you don't have the object, you use the phrase "No deh." For example, "Cold beer, no deh" tells you that the fridge is out of kerosene and you'd better brace yourself for a warm brew.

Reflecting the attitude of many people who speak it, Krio is a language that exists entirely in the present. If you want to talk about something happening in the past or in the future, you have to devise elaborate verbal constructions.

What counts is what happens today. But if they had a live-for-today attitude, the people of Sierra Leone also had to reckon with the prospect of dying tomorrow of Lassa fever. The disease is endemic to the country. You could safely say that Sierra Leone is the home of Lassa fever, even though the name Lassa comes from the Nigerian town where the virus was first isolated. Typically, victims of the disease experience fever, headache, and a terrible sore throat and suffer from vomiting and diarrhea in addition to extensive pains throughout the body. Shock and hemorrhage finally carry victims off. Lassa virus is a close relative of Bolivian Hemorrhagic Fever, which Karl had been pursuing in South America. This time, the culprit of transmission—the reservoir—was an African rat: *Mastomys natalensis.*

There was no question that we had our work cut out for us. When we were introduced to the secretary of health of Sierra Leone, his first question was: "What is the CDC? Does that stand for the Colonial Development Company?"

Not exactly a splendid start.

After we explained our purpose, the secretary wanted to know what the annual budget of CDC was.

"Well," Karl replied, "it's about $120 million."

His jaw dropped. He couldn't believe it. The annual budget for the entire nation of Sierra Leone wasn't that big. Then he sat back, studying us thoughtfully. He decided it was well worth his while to be collaborating on a project with such a lucratively endowed partner.

Fortunately for us, the officials we met with at the U.S. Embassy knew what the CDC was. The Americans wanted Lassa fever stopped; some Peace Corps volunteers had contracted the disease in Sierra Leone and in neighboring Liberia. None had died, but one had become permanently deaf. Our job was now to find the best place to set up shop. The embassy offered us a vehicle, which we used to drive to the town of Bo, about 160 miles northeast of Freetown. After inspecting the hospital there, we headed north to Panguma, where a CDC team, made up of Tom Monath, Kent Campbell, and David Fraser, among others, had carried out an extensive study of Lassa fever back in 1972. Panguma is a small, sleepy town of about three thousand people situated at the base of an imposing crest of rain forest–covered hills. The hospital was operated by Irish nuns, who had acquired a great reputation for skill and dedication. The hospital here was also in much better shape than the dreary one we'd seen in Bo, with better beds (all with thick mattresses) and brightly colored rooms filled with light. The sisters had been expecting us. They were somewhat guarded, though, since they were aware that their hospital had already figured in the annals of Lassa fever just four years before. They weren't particularly eager for their institution to become permanently linked in the mind of the public with the disease, which isn't to say that they didn't show us the utmost hospitality, offering us a plentiful lunch of rice, local greens, and our choice of range-fed chicken, beef, or goat. Without question, the best part of the meal was the Star Beer.

"Cold beer, deh." Very cold, too. I could see that the sisters knew how to live.

Over lunch, the sisters provided us with a running account of all the problems they had to contend with trying to manage a hospital out here. They needed clean water. We could imagine how hard that was to come by. And they needed electricity twenty-four hours a day. Almost impossible. And where were they to find trained people? None of this was new to us. These problems were the same wherever you went in rural Africa.

What about Lassa fever? we finally asked.

Yes, the nuns acknowledged, cases of Lassa fever continued to occur. It was practically routine.

Karl and I pressed on to a town called Segbwema, about twenty-five miles southeast from Panguma. Segbwema was where Lassa fever—then unnamed—had first been described in the scientific literature of 1952. What we saw in the local hospital convinced us that Lassa fever had never left Segbwema.

After leaving Segbwema, Karl and I concluded that our best strategy was to set up a central operation in the northern village of Kenema, a provincial capital, where we could be assured of enough electricity to run a laboratory, but near to both Panguma and Segbwema, where we could work with patients. First we returned to Atlanta to set up support for the project. With that accomplished, I traveled back to Kenema alone. I had my hands full dealing with logistics: Where was I going to live, how was I going to obtain the right equipment and find personnel who could assist me? We had clearly found plenty of patients in Panguma and Segbwema, but how would we search for an effective treatment?

After a month on the project, when I was just settling in, I received a telegram from Karl. It contained news of a disease even more dramatic and lethal than Lassa.

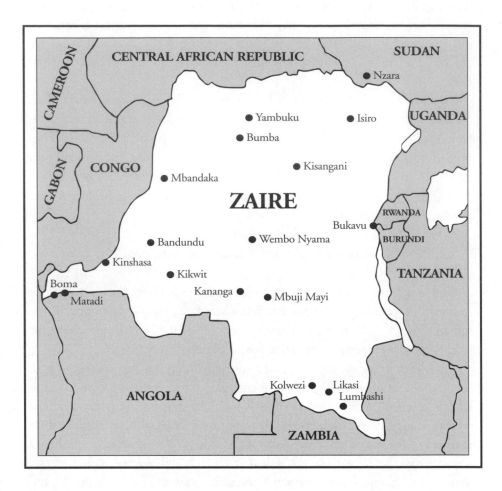

CHAPTER FOUR

The Death of a Nurse from Yambuku

Yambuku. It was a name that would eventually evoke terror around the world, but the day that Karl's telegram reached me, I had never heard of it. The telegram described an outbreak of an unknown hemorrhagic fever virus in Zaire. Several people were already dead, and new cases were continuing. Because Karl was still in Atlanta, he was somewhat at a disadvantage. He could only take a guess at what the disease was: "It could be Lassa, or yellow fever, or Crimean Congo. Or it could be Marburg."

Whatever it was, one thing was clear: it was a rapid infection, devastating in its effects, accompanied by bleeding from the nose and gums and, occasionally, massive bleeding from other parts of the body. People suffered from such excruciating diarrhea that they soon became dehydrated, causing their skin to stretch like paper and their eyes to recede in the sockets. Most victims died within a few days. All attempts at therapy had failed. An arsenal of antibiotics accomplished nothing. Vitamins didn't work. Intravenous fluids, intended to replace the fluids the patients had lost, also had no effect. Flowing out of membranes the disease had made permeable, the intravenous fluids actually drowned the patient—internally. In many instances, the victims couldn't even be treated, effectively or otherwise, because they lived in such a remote and inaccessible area.

Lassa fever, the disease that had brought me to Sierra Leone, was a viral hemorrhagic fever. First identified in Nigeria in 1969, it had spread to Liberia and then to Sierra Leone, but it had never been seen outside of West Africa. In Zaire, for example, it could produce symptoms similar to those Karl was describing: fevers, hemorrhaging, swelling, shock, and convulsions. Marburg was another distinct possibility. Also known as Green Monkey Disease, Marburg is characterized by high fevers, rashes, bloody vomiting, and severe diarrhea. It earned its name after it had claimed the lives of several lab workers in Marburg, Germany (then, later, in Belgrade, Yugoslavia). Under an electron microscope, Marburg doesn't look anything like other human or animal viruses. Human viruses are usually small and round, or slightly oval organisms. Marburg, however, was long and snakelike, with strange loops and twists. It was so peculiar-looking that some people theorized that it must have come from outer space. Of the thirty-one people stricken with it, seven had died.

But, at this point, Marburg had occurred only once, nearly ten years before, in 1967. Then it vanished, for a time, as mysteriously as it had appeared. It was believed that the disease originated with green monkeys imported from Uganda and that the human victims had all come into direct contact with the animals' blood and tissues. Later, a researcher from CDC went to Uganda to investigate, seeking to discover a possible source of the virus and whether the monkeys were in fact spreading the infection. The investigation yielded nothing conclusive.

Both Marburg and Lassa had a high mortality, but not so high as was apparent in this outbreak in Zaire. Usually only about 15 to 30 percent of those infected died. Nonetheless, it was still possible that what was happening in Yambuku was Marburg. But suppose it wasn't Marburg at all, or any of the other candidates Karl had named. Suppose this was a disease that the world had never experienced before.

What then?

"If we do receive authorization to go ahead with an investigation," Karl wrote in his cable, "would you be willing to join it?"

I wanted nothing more. I'd taught in Zaire. I'd spent years in villages that sounded just like Yambuku. I knew French, which was widely spoken in the country, as well as two regional dialects. Besides, I had never done anything remotely like this. My experience up to this point had been mainly with bacteriological diseases like streptococcus anthrax, leprosy, and meningitis. I'd only recently become involved in the investigation of hemorrhagic fevers, but fortunately timing was not critical in Sierra Leone. Lassa fever would always be there. I could safely break off for the project if the Zaire expedition went ahead.

For the time being, there was nothing to do but wait. The Zairian government had yet to give the go-ahead for an investigation. This was not unusual. Governments are seldom quick to acknowledge that their people are imperiled by an epidemic they can't control. Such an admission can devastate the tourist industry and play havoc with the economy. Only, in Zaire's case, the economy was already in tatters, the victim of a misguided effort to nationalize private industry. As for the tourist trade, Zaire had little to lose. Hardly any tourists came anymore. Sightseers generally aren't interested in impoverished people, crumbling roads, and empty store shelves. Nor are they particularly drawn by stories of teachers and other educated nationals, as well as missionaries, being slaughtered by renegade soldiers. Zaire was at peace now, but it was a peace of the dead and the dying.

Karl promised to keep me updated on the situation in Yambuku. In the meantime, assuming that approval would eventually come from the government in Kinshasa, I went ahead with my preparations to leave Sierra Leone. Field investigations tend to be improvisational affairs. When it comes to strange epidemics in the heart of Africa, you can be sure that there won't be a limitless source of funds to draw upon. People don't go to sleep at night in the U.S. worrying about whether they're going to wake up with Lassa fever. Funding at an agency like the Centers for Disease Control is often directly proportional to the degree of alarm a particular disease arouses in the population. The closer to home it is, the more you may be sure that Congress is prepared to appropriate funds to investigate it. It wasn't until Ebola appeared to pose a threat to Fairfax County, Virginia, that this actually happened. But that event was still far in the future.

Lacking adequate funds, Karl and I had constructed a Level 4 "field isolator" (a portable lab) before I left for Sierra Leone. The lab was designed to make it possible for us to work with infectious material inside an airtight chamber without risk of exposure. It consisted simply of a box with ports in the side walls fitted with large black latex gloves in which you inserted your hands and arms. You couldn't exactly call it high tech. Nonetheless, the engineer we worked with had done an ingenious job. Mounted on a piece of 4-by-8 plywood that supported steel rods at each corner, the lab was enclosed in a plastic envelope that could be collapsed for easy transport. The envelope itself was fitted with an external blower, which sucked air out, creating negative pressure within. Negative pressure ensured that, even if the plastic wrapping sprang a leak, no infectious material in the chamber would escape. Instead, the air was blown out through special HEPA (high-efficiency particle) filters, which can trap particles as small as a virus. The trouble was that the gloves were awkward; as we later realized, the greatest risk didn't come from accidentally breathing in these viruses, but from sticking your finger with an infected needle or other sharp instrument. Much later, once we understood where the actual danger lay, we started to work with specimens from patients with Lassa fever in Sierra Leone on an open bench, where we could protect our hands more easily. At this point, however, we had no idea of what type of virus we would be working with or how it was transmitted. There was a very real possibility it could be airborne.

On October 19, while I was still carrying out my preparations to go to Zaire,

Karl Johnson and Patricia Webb at CDC; Ernie Bowen in Porton Down, England; and Stefan Pattyn and Guido van der Groen in Antwerp all succeeded in isolating a virus from the blood of a patient dying in Yambuku. The unnamed virus, they said, bore a resemblance to Marburg, being filamentous and twisted into strange loops, yet it would not react with the Marburg diagnostic reagents. It seemed that it must be a relative of Marburg, but something much deadlier. Deadlier than Marburg—and almost any other infection known to mankind, this virus from Zaire grew with alarming speed, blasting apart tissue culture cells in the laboratory.

Though no one knew it at the time, the virus was already making its way south toward the capital of Zaire. It was being carried in the blood of a Belgian nun, Sister Myriam, who had worked at the hospital in Yambuku. Admitted to Ngaliema Hospital in Kinshasa, she was cared for by a nurse named Mayinga, an African. Before long, Mayinga began to experience the early symptoms, too: fever, headache, malaise. Having seen the disease and heard the terrible stories from up country, she was driven by a fear any of us could understand. Distraught and scared, she began to wander aimlessly around Kinshasa, making the rounds of doctors and emergency rooms without daring to face the terrible truth, waiting to be told that it was nothing, just malaria, not to worry. As she moved from waiting room to outpatient clinic, she risked exposing those around her to the virus. This she understood little as she struggled with symptoms that gave her no rest. She grew worse and worse.

One wonders what she would have noticed first. The headache? The fever? The increasing difficulty in swallowing and the intense soreness in the back of her throat? Terrified of the truth, she continued to deny to herself that she had the virus, but desperately sought reassurance, an everyday explanation. Malaria. It must be malaria. People had malaria all the time. And they got better.

But Mayinga did not have malaria.

Finally, she could go no farther and took refuge at Ngiliema Hospital, the same place where she'd contracted the disease. She immediately became the center of a flurry of activity, as physicians rushed to give her plasma derived from patients who'd recovered from Marburg. They were acting in the forlorn hope that it might have a therapeutic benefit. Anyone known to have come into contact with her in the hospital or anywhere else in the city was quarantined. Ironically, the virus that was extracted from her blood would turn out to be a kind of bequest to humanity, since it has been the source of virtually all knowledge we've subsequently obtained about this previously unknown virus and its effects on the human body.

On the same day that scientists were identifying the virus in Atlanta and Europe, October 19, I walked out of the building that served as my up-country home in Sierra Leone and saw a truck pull up. The driver approached me and asked if I knew who Dr. Joseph McCormick was. I identified myself.

"Then this is for you," he said, handing me a message from the U.S. Embassy. It was the news that I'd been waiting for: WHO (the World Health Organization) had been given permission by the Zairian government to launch an investigation. My instructions were to leave immediately for Kinshasa. I sent a message back to the embassy with the driver, requesting assistance in arranging a flight for me. The adventure was about to begin.

Karl, of course, was one step ahead of me. He and a colleague named Joel Breman, a newly hired CDC epidemiologist, were already in Zaire. On board the flight from Atlanta to Kinshasa, Joel and Karl happened to meet a man who would play a crucial role in the investigation: Bill Close, director of Mama Yemo, the largest public hospital in Zaire and, probably, in all of Africa. Bill was about five and a half feet tall, with a large round head. Somewhat portly in body, he was energetic, idealistic, and humane in spirit. When I first met him, he smoked constantly. He was anything but relaxed. He spoke French like a Parisian and, like a Parisian, was able to imbue his words with an acidic air of impatience carried off by a certain "presence." His was a utopian cause—a movement called Moral Rearmament—that had inspired him to go to Zaire in the early 1960s, shortly after independence. Few other people would have relocated their families and settled them in a Third World country in the middle of a war. But Bill wasn't like most other people. For Bill, Zaire was an unprecedented opportunity to exercise his ideals and to do something exciting. He still lives this way today.

Bill's idea of excitement was to travel into the middle of a war zone—which was what the mineral-rich province of Katanga had turned into at the time—and then to see what kind of help he could offer. He wasn't taking sides in the bloody insurrection—a prudent choice on his part. He was interested only in treating the wounded. Undaunted by the danger, Bill continued to operate on the sick and injured, sometimes performing surgery on soldiers at gunpoint. All in all, I suppose it was possible that he got more excitement than he'd bargained for. The conflict did give him the chance to meet a man who would later turn out to be a useful friend to have, an ambitious colonel named Mobutu Sese Seko. Mobutu would later go on to take power (which he still holds) in Zaire, dominating the country early on with shrewdness, later to be combined with ruthlessness and greed. Nevertheless, as events later would prove, the association wasn't just advantageous to Bill. The WHO investigators benefited from it as well. With Bill's help, Karl and Joel realized they were in a far better position to get things done than they would have been otherwise. In a country where nothing was achieved quickly—or at all—Mobutu could cut through the red tape with a few words, for failure to comply with his wishes came at a heavy price.

Once they reached Kinshasa, some members of WHO, led by Karl, remained in the capital, while Joel took charge of a smaller advance team that was to travel into the Yambuku area hundreds of miles to the north. Within twenty-four hours of their arrival in Kinshasa, still jet-lagged, Joel and his colleagues from France and Zaire boarded another plane headed into the interior.

And then they disappeared. All contact with them was lost.

At this point, just as Karl was about to set up operations in Kinshasa, the news arrived that Mayinga had died. Now there was no hiding the truth: the unnamed disease had come. Few people believed that she would be the only victim.

But I knew nothing of all this. I was struggling to get to Kinshasa from Sierra Leone, and this was no simple task. Once I got to Freetown—I risked a Sierra Leone Airways Fokker flight rather than endure eight hours of punishing travel over the road—there was the matter of whether there would be any flight out of Freetown at all. Getting around Africa is at best a game of chance punctuated by mainly unwelcome surprises. Nowadays, we often detour back to Europe to travel between two African capitals. It is easier, infinitely more comfortable, and safer. In this instance, I didn't have that option; I was in too much of a hurry. Airline schedules in Africa are notoriously unreliable. The two airlines—the Nigerian and Ghanaian—considered to be among the most dependable on the continent, showed up only when it suited them and with scant attention to timetables. It wasn't unusual to wait for a day in some dingy airport terminal for a flight to appear. To complicate matters further, very few African countries maintained diplomatic missions with Sierra Leone, which made getting the requisite visas all but impossible.

Given the fact that a direct flight to Kinshasa was out of the question, I was told I would have to stop first in Abidjan, Ivory Coast, and then in Douala, Cameroon. I had visas for neither country. Somehow I would have to find lodging for the night in each place and see to the safety of my portable lab. I would have to rely on my ingenuity and the good graces of the customs officials.

The customs officer in Abidjan gave me a funny look. What business did I have in the Ivory Coast? he demanded. I tried to explain: I just needed to find a place for myself and my lab to spend the night. Then why, he wanted to know, didn't I have a visa? I tried to weave a plausible story for his benefit, telling him that there was no way that I could have obtained a visa in Freetown. After all, the Ivory Coast didn't maintain an embassy there. I went on to explain that there was an outbreak of disease in Zaire that required my immediate presence. Was this so hard to understand? It became clear that a "gratuity" was required to put him in a compliant mood. But as a matter of principle I wasn't prepared to offer him one, even though he was certainly within his legal rights. I doubt if someone from the Ivory Coast would have expected to be able to talk his way through a U.S. immigration check in this way. Nevertheless, as a representative of the U.S. government, I was never supposed to take the easy and, indeed, customary way out of problems by a greasing of palms. This being Africa, however, I assumed that if I kept pressing him, I would eventually wear him down, and, at least get a bed for the night. In the end, he did relent. But he still wasn't completely satisfied that everything was on the level.

"Give me your passport for safekeeping," he insisted.

Well, I had no intention of surrendering my passport. U.S. passports are valuable commodities, highly prized. Parting with mine would have been a very stupid thing to do. So I handed him my yellow WHO vaccination card instead. The functionary studied it for a moment, seeming to ponder what to do with it. Finally, deciding that it looked sufficiently official, and knowing I would need it for further travel in West Africa, he shrugged and waved me through.

In Cameroon I was obliged to repeat the process all over again—and yet again, when I finally arrived in the capital of Zaire, on October 23. The airport was just as I remembered from my younger days in that country: a place of bedlam and graft, where travelers moved like fugitives in an atmosphere made dark and menacing by well-armed and ill-trained soldiers. Although I had no visa for Zaire, I did have a copy of the WHO invitation asking me to join the investigation. Not that it mattered. This time, I met with no objection once I made my purpose clear. Everyone at the airport was familiar with the outbreak—the more so since the death of Mayinga—and they weren't disposed to quibble over formalities. The only snag came when I tried to get the crate containing the isolator through customs.

"Qu'est-ce-que vous pouvez faire pour moi aujourd'hui?" the customs official asked when he saw the crate: What can you do for me today?

The implication was unmistakable. I informed him that I didn't think that he could make much use of the lab, but if he wanted to guard it overnight until I brought a representative from WHO back to retrieve it, he was more than welcome to the job. He looked a bit crestfallen, but he understood that his dreams of an especially profitable day and a couple of extra Simba beers had just been dashed. He gave up.

Emerging from customs, weary but expectant, I was greeted by a man from Karl's team. "Bad news," he said, once I got into the car.

"The epidemic is already here. People are in a panic. Ngaliema Hospital is under quarantine." Then he turned and gave me a wan smile. "Welcome to Kinshasa, Dr. McCormick."

CHAPTER FIVE

The Battle Commences

Thirty minutes after getting through customs, I was riding through the teeming streets of Kinshasa. I was struck by how much the place had deteriorated since I had last been there. Though I detected no overt signs of panic, I was sure that everyone I passed was aware of what was happening. They might not have known the name Mayinga, but they all must have heard rumors about the "nurse from Yambuku" who had died. And that knowledge had to be frightening. Because if they believed that virtually anyone in the city could be infected, then that automatically made every stranger a suspect, presumed guilty until proven innocent. Trouble was, there was no way to prove innocence. One person was dead of the virus in Kinshasa today, but tomorrow it could be fifteen or twenty. No one knew anything for certain. Everyone was afraid.

The WHO team had taken up residence in a part of the Belgian government cooperative mission in Kinshasa called Fometro, which stood for Fonds Medical Tropical. It was a sort of guest house, warehouse, and motor pool rolled into one. It also served as the center for the mission's medical programs, which included managing Mama Yemo Hospital. While no cases of the new virus had so far been identified at Mama Yemo, Bill Close and his staff were taking no chances. At two thousand beds—the hospital housed a large population of patients, with a hundred new births added to it every day—any outbreak of the virus there could have been catastrophic. Bill organized a quarantine system intended to insure that any instances of the infection could be contained. New admissions were carefully screened for any signs of the disease and immediately removed to a special ward, where they were carefully monitored.

While there were still no confirmed cases at Mama Yemo, the same could not be said of Ngaliema Hospital. With one patient infected and another dead, the quarantine was especially strict, even onerous. One ward, called Pavilion 2, had been designated as a quarantine area for the patients, although movement within the hospital itself was not restricted; a second ward, Pavilion 5, was set aside for members of the staff who had been exposed. No fewer than thirty-seven people had been identified by officials as having come into contact with Mayinga. A South African doctor named Margarita Isaacson had taken charge of the quarantine process. She was a small, highly energetic woman who wore large spectacles, which covered half of her face. Reputed at one time to have been an Israeli paratrooper, she seemed completely at home giving orders, which she expected to be obeyed without question. She'd assumed control in a highly charged atmosphere. The fear in the hospital was palpable. There were questions even in the eyes of patients: *Do you have it? Do I? Am I already dying?*

Under such circumstances, the iron-willed Dr. Isaacson did an extraordinary job of keeping panic at bay. In the best of times, getting anything done in a city like Kinshasa is difficult enough, but maintaining a quarantine was a logistical nightmare. In addition to keeping an eye on the quarantined patients, she also had to make sure that they were fed and their families kept apprised of their status. But what made her exceptional was her determination to insure that those families never got past the door. In Africa this was unheard of. When a patient is admitted to a hospital, his family practically moves in with him. A patient, after all, cannot depend on the hospital for food. Indeed, he usually can't even depend on the hospital for basic nursing care. Such responsibilities routinely fall to the family. So for Dr. Isaacson to forbid them to visit was unprecedented.

Every morning, the families would camp outside the hospital entrance and wait until someone in a position of responsibility would appear. Then the questions would begin. Why should the hospital prevent them from visiting their loved ones? Who was going to see to it that they got enough to eat? What would happen if the family member became worse and there was no one there to comfort them? Dr. Isaacson met with the families herself. She assured them that she sympathized. "But there's nothing I can do. You can't see them until there's no further risk of exposure. Do you want to get sick, too?"

No, they did not. The matter seemed to be settled, a certain understanding seemed to have been reached; but the next day, the families would be back, seeking word about their loved ones. Once more Dr. Isaacson would address them and let them know that there was nothing to be done. The quarantine was absolute.

I hadn't been in Kinshasa for more than twenty-four hours before I was asked to attend my first meeting with members of the WHO team and officials from the Ministry of Health, all of whom took their direction from Karl. He called most of the shots, as well he should have. He was more scientifically savvy than anyone else on the team. Moreover, his experience in leading epidemiological investigations in Latin America had made him especially sensitive to foreign cultures. Without that sensitivity, he could never have orchestrated the management of an outbreak like the one in Yambuku.

While Karl was the acknowledged leader, the team meetings were usually chaired by the Minister of Health, Dr. Ngwete Kikhela, a small, round-faced man, quite well-spoken, who'd acquired his training in public health in Belgium and Canada. While Dr. Ngwete spoke some English, it wasn't sufficient to allow him to converse easily with his foreign colleagues. And when it came to infectious diseases or the management of a complex outbreak, he was completely at sea. Although he was eager to cooperate, it was clear that he looked to the WHO team to figure out how to resolve the crisis in his country. That, after all, was our job.

Yet there was always a good deal of jockeying for power among the other members of the team, especially the Belgians. As the former colonial power, they believed—probably correctly—that they knew more about Zaire's politics and culture than the rest of us. So they took it upon themselves to assume the role of the nation's protectors. Their problem was that they simply didn't have the ability to conduct an investigation like this. They were reduced to venting their frustrations and stewing along with some of their French and even a few of their American colleagues when decisions were made with which they disagreed. In any event, the virus wasn't lying low, waiting for us to settle our petty differences. We were, after all, on a war footing, consuming every scrap of information we could get in an attempt to keep abreast of what was happening up country. We were aware that new cases of the virus were continuing to occur. But at what frequency? How many people were in quarantine? Was everything being done to prevent more infections?

With everything else that was going on, we had another problem to contend with: What had happened to the team led by Joel Breman that had been dispatched to the site of the epidemic? They were already supposed to be in Yambuku, having been placed on a plane just hours after they arrived in Kinshasa. We were worried about them. Anything could be happening up there. All we knew was this: that after leaving Kinshasa, the team had flown into Bumba, a port city about eighty miles south of Yambuku. But that was five days ago, and there still was no word from them. Astonishingly, no one even had any idea how to get in touch with them. After giving it some thought, I realized that there might be a relatively simple solution to the problem. From my experience in Zaire, I figured that the best way to get in touch with anyone in the country was to use the missionaries. Missionaries in Africa operate their own efficient "bush telegraph," and Zaire was no exception. After making a survey in Kinshasa, I learned that there was a field station not too far from Yambuku operated by a North American missionary organization. More importantly, this mission had a radio, used to keep in daily contact with colleagues in Kinshasa. The next morning found me seated by the radio in the Kinshasa mission office, waiting for Yambuku to call in. As soon as I had the missionary on the other end, I explained our problem. Would it be possible, I asked, for the mission to send someone in search of Joel Breman and the rest of his crew, and maybe get them to the radio? Failing that, could they find out what was happening to them?

Yes, I was assured, someone would radio back that evening.

Twelve hours later, I was back at the mission center. There was good news. Joel's team had been located, and everyone was in good health. I was told that Dr. Breman himself would be calling the following morning. I didn't know Joel well at the time, but we were acquainted with each other. A big bear of a man, Joel stood over six feet tall and spoke in a mellifluous baritone that put you in mind of a DJ on a late-night radio station. He craved nothing more than a daunting challenge, which was why he'd abandoned his job as a public-health official in Michigan to join the investigation in terrain far less hospitable than, say, the Upper Peninsula. He spoke French fluently, albeit with a decided drawl. Having spent many years in former French West Africa trying to curb smallpox, he'd gained a good understanding of African culture. When we were finally able to communicate directly, Joel told me what had happened to his team after leaving Kinshasa.

"The Zairian Air Force dropped us off—literally," he said, referring to their arrival at Bumba. "After the plane landed on the airstrip, the pilots refused to turn off the engines. They just opened up the backdoor of the C-130, told us to take our things and scram. As soon as we were out, they took off."

As they'd started out toward Yambuku, he went on, the investigators stopped in the villages along the way to inquire as to whether anyone had fallen ill. Though the team had confirmed no new cases in their travels, it was apparent that the villagers were aware of the danger. People came running out, crowding and jostling around their car on the way to Yambuku, all of them petrified with fear of the disease. Many villages had instituted their own quarantine: Strangers weren't being permitted to enter, and even residents who'd been away for any length of time were put under watch for signs of the illness. After having to cope with bouts of smallpox over the years, the villagers had learned—the hard way—about the necessity of imposing such restrictions.

When Joel's team arrived in Yambuku, they discovered a town plunged into chaos. The hospital was deserted, the staff having been decimated by the virus. Zairians who thought they'd been exposed had fled to their villages. Those who remained behind were waiting in terror to find out whether they were infected. This story is well described in Bill Close's novel *Ebola*. Despite fictionalization, this account of the Yambuku outbreak is, in fact, the best researched and thus most accurate account of all the events that took place there in 1976. Joel voiced to me his suspicion that the hospital in Yambuku, far from being a refuge for the sick, was actually contributing to the spread of the disease. The absence of sterile techniques—and especially the reuse of hypodermic needles— was doubtless one of the principal vehicles of transmission. "We're still hearing of new cases," Joel said, "but the problem is we can't determine whether they're definitely this particular virus. With the hospital shut down, people aren't coming into Yambuku. They're staying put in their villages, so it's difficult to get a good picture of the situation."

Before signing off, I told him we would contact the Zairian military and try to set up a flight to Bumba to pick them up. Finding anyone to fly into Bumba, though, wasn't simple. As far as the Zairian military officials were concerned, the virus was airborne, and they'd only have to take one breath in Bumba and drop dead. Worse, there was no way that I could assure them that every member of Joel's team was free from infection. So we turned to Bill to help us.

Even though I'd just met Bill, I had a considerable confidence in him. He struck me as someone who was a genius when it came to organization. How else to account for the efficiency with which he ran Mama Yemo? In a country that was riddled with corruption, he distinguished himself by refusing to put up with laziness or incompetence. Moreover, he was an optimist. The same sense of idealism that had led him in the first place to volunteer to practice medicine in the middle of a bloody insurrection hadn't deserted him here. Others would have given up under the strain and frustration. So I cannot say that I was surprised when Bill came back from his meeting with Mobutu to tell us that a plane would be made available to fly north to retrieve our investigators. Even so, again the pilots refused to get out of the aircraft when they arrived in Bumba, and once Breman's team had boarded, they insisted that the investigators remain as far away from the cockpit as possible.

On my fourth day in Kinshasa, we were alerted to more bad news. This time it wasn't from Kinshasa or Yambuku. The virus had struck apparently before the Yambuku outbreak in southern Sudan, about five hundred miles northeast of Yambuku. The outbreak there sounded exactly like the one that had killed several people in Yambuku. Were the two epidemics related? Since the epidemic in the Sudan seemed to have started before the one in Zaire, we could only conclude that, if there was a connection, the infection must have originated in the Sudan. In that case, it would have traveled southwest over the border, following a route into Zaire that had changed little since Dr. Livingstone had explored the area in the 1870s. Someone would have to get as close to the Sudanese border as possible to see whether a link existed between the outbreaks. Then I thought: Why shouldn't it be me?

Karl didn't object to my decision. Nor did anyone else on the team. That was probably because no one else wanted to go. The part of the world I would be traveling to was as remote as any on the face of the planet. Few people outside of the region had any idea of what was to be found there. Thinking it would help, I got hold of a Michelin map of the area. Later, I came to realize that it must have been drawn up by sublime optimists. The map was more an exercise in faith than in cartography. At that, the advice the map offered wasn't exactly reassuring: "Where there is no clear trail, a guide and navigation instruments are essential. It is most unwise for a single vehicle to journey on its own." That was nice. And then there was this intriguing piece of information: "The delineation of international boundaries must not be considered authoritative." Coming from a prestigious mapmaker, this was surely an abject admission of cartographical defeat.

While I was undertaking my exploration of the border, the main team would focus on Yambuku, searching for additional cases. If all went well, investigators might succeed in limiting the spread of infection, too. At the same time, another member of the WHO team, Belgian sleeping sickness expert Dr. Simon van Nieuwenhove, planned to set out along a more southerly route in the province of Haute Zaire, to those parts better served by such public transport as the railroad. Here Ebola might be expected to travel at greater speed. The virus would nevertheless encounter two major obstacles. First, to get to that region at all, it would have to pass through the remote area to which I was headed; second, the pace of the railroad challenged the snail for slowness, and its erratic reputation was legendary. We once sent a barrel of diesel oil by rail to supplement our fuel supply. It never arrived.

The daily meetings at Fometro took on added urgency the closer we came to our departure for the north. We had a variety of troublesome issues to wrestle with. How would we go about trapping animals and insects that might turn out to be the vehicles for spreading the infection? How should the outbreak in Yambuku be investigated? What kind of procedures should we be developing to use for our field lab? There was one other thing. It was always in the back of our minds. What were we going to do if one of our own team came down with the disease?

Our days were consumed in frenzied efforts to obtain supplies and equipment. We needed vials, syringes, rubber gloves, and other materials. We needed Land Rovers and fuel. There would be no fuel in the north. We would have to take our own along in steel drums. Economically, Zaire was a disaster. Ever since its independence in 1960, when it was known as the Belgian Congo, Zaire had struggled to stand on its own. But by 1976, things had gone from bad to worse, in large part because, three years previously, the government had nationalized all privately owned businesses, sending the economy into collapse. Like most of rural Africa, the interior wasn't electrified, yet by 1976 it was practically impossible to purchase something as basic as a candle. Conditions were so dire that, in many parts of the countryside, beer was unobtainable. Beer serves as a barometer of how an African economy is faring. When the beer goes, you know that things have hit rock bottom.

Although conditions were somewhat better in Kinshasa, we still encountered difficulty finding the supplies we needed. Bill Close and his colleagues at Mama Yemo, a couple of British army vets, proved resourceful in helping us obtain what we needed. In addition, the U.S. Embassy provided us with vacuum-packed combat rations in tins you couldn't open without a key. They were practically antiques. The ones I was given were dated 1945 or earlier. Maybe, I thought, 1945 would prove a particularly good vintage for C rations.

Another one of our major concerns was establishing some means of communication among the teams, so that we could avoid a repetition of the experience we'd gone through with Joel Breman. Once again, the long arm of the missionaries served us well. Sure enough, a Belgian Catholic mission in Kinshasa had a system it was willing to let me use. It was quite something—a Rube Goldberg device consisting of a small single-sideband receiver operating off a twelve-volt battery. The antenna was mounted on the back of the Land Rover. A mechanism resembling a bicycle pump allowed it to be extended almost to a height of forty feet. In addition, a pair of wires was suspended from its top to catch faint radio signals. The radio itself plugged into the Land Rover's battery. To my delight, it actually worked. At least it did when I tested it in Kinshasa. What I couldn't predict was whether it would work equally well when I reached my destination. Wherever that might be.

A couple of days before we were ready to leave Kinshasa, news came that the epidemic appeared to be waning in Yambuku, but no one could be certain whether the danger was over. For all we knew the disease might still be raging in the bush. This report made it even more imperative that we get a second team into Yambuku as soon as possible before the trail grew cold or the virus became so widespread in the villages that it would be impossible to stop.

Finally, the day came for us to leave for the north. It was October 30, exactly one week after my arrival in Kinshasa. Early in the morning, the members of the three teams—the Yambuku team and the two smaller teams that would be led by Simon and myself—collected at N'djili military airfield. There we waited, watching while airport workers loaded the three Land Rovers, forty drums of diesel fuel, several boxes of "vintage" combat rations, and other essential supplies on board the C-130. As I watched, I was seized by a familiar anxiety, a mixture of fear and excited anticipation. I'd never done anything quite like this before. I had no idea what to expect, but I couldn't wait to get started.

Unfortunately, I would have to.

As soon as we were strapped into the jump seats along one side of the craft, we confidently waited to get under way. Nothing happened.

The engines didn't start. The plane didn't move. We waited some more. Still, nothing. Then the cause of the delay became clear. Out on the tarmac, an air force commander was talking to the pilot, informing him that the plane couldn't leave unless some of his superior's friends—or maybe they were his relatives—came on board with us. He also insisted that we take on some extra supplies he wanted delivered to Bumba. No one was about to defy him and, in fact, this kind of thing was quite customary. Anyone who had any power in Zaire suffered few compunctions about exercising it. Indeed, anyone who had power and *failed* to use it was like as not to lose it forever.

Once we were finally airborne, it took us two hours to get to Bumba. As soon as we set down on the red-clay runway, scores of people came running out to see what was happening. Most of them were children who watched, open-mouthed, as the cargo was offloaded. Not much ever happened in Bumba. The arrival of a transport plane was a big event.

The second leg of our flight took us to Kisangani. It was late afternoon when we arrived. Since the city is so close to the equator, darkness falls suddenly, just after six. This meant that we barely had time to get the two Land Rovers off the C-130 along with the supplies and seek accommodation for the night. I was able to find lodging at the local Catholic mission. One of the first things I observed when I walked in was a panel displaying photos of missionaries who had previously worked at the mission. All of them were dead, killed by renegade Zairian soldiers ten years earlier. The little gallery served as a grim reminder that it wasn't only disease that made this region so dangerous.

The virus couldn't remain anonymous forever. The honor of giving it a name fell to Karl. Although Yambuku would seem to suggest itself, Karl didn't think it sounded quite right, perhaps because he didn't want to stigmatize the town any further. After studying a map of the area, he noticed a river that ran close to Yambuku. He decided to appropriate its name for his purpose. The name of the river was Ebola.

CHAPTER SIX

The Ebola Trail

When it turned out that no one in Kisingani knew anything about an outbreak of disease, I decided to leave the others and set out alone for Isiro about a hundred and twenty miles to the northeast. Isiro is the largest town between Kisangani and the Sudanese border. I had only my driver for company. Considering the driver's personality and temperament, I would have preferred to be alone. He was laconic to the point of being mute. His silences had something belligerent about them, though, and he gave every impression that driving me around was a terrible imposition. It soon became apparent that he wasn't a very good driver, either. He seemed to think that he could turn our trip into the Grand Prix, taking rutted dirt roads at speeds that would, I thought, be sure to bring us to ruin. But should I try to reprimand him, or even give him a few words of caution, he would glare at me as if to say: *What business is it of yours how I drive?* The problem was that I didn't have a choice about him. He was a young expatriate in his early twenties who had grown up in a mission family in southern Zaire. He'd been assigned to me by the team. I was told that he was familiar with the terrain and local culture and that I could rely upon his judgment. I had my doubts. Recognizing, though, that the Michelin maps were unlikely to be of much use, I resigned myself to making the best of the situation. I just had to hope that I survived the journey intact.

It was the rainy season, so conditions along the roads were even worse than they would ordinarily have been. The rains transform the soil into rich red mud, whose substantial appearance lures the unsuspecting driver into a quagmire. The soil is primarily laterite, a clay with a low-grade iron ore that gives it a rust color. When wet, laterite roads are as slick as a frozen lake, with a surface so treacherous that tires normally capable of handling heavy mud became functionally bald. Under these circumstances, we did well to average ten to fifteen miles an hour. Compared to other roads in this part of the country, the road to Isiro was a major highway. At least it was clearly delineated and had two ruts to accommodate the wheels of my Land Rover.

Shrouded in a perpetual mist, the elephant grass that dominated so much of the savanna assumed the look of a half-remembered dream. As soon as they heard our approach, monkeys, antelopes, and baboons would suddenly scatter and melt into the murk. The piercing cries of the animals reached us from far away. Birds swooped overhead, shadowy forms that soon vanished in the vast grayness of the sky. Then the landscape would fall strangely silent. For long periods of time, we would hear nothing except for the sound of the Land Rover's engine and the unrelenting tattoo of rain against its roof.

Medical facilities barely existed in this part of the world, nor were there many physicians. The best that people could expect was a dispenser, who would supply them with primitive remedies. The dispensers, I believed, were in the best position to know whether any outbreak of disease had occurred in their districts. But I also spoke to village chiefs and teachers—anyone, in fact, who might have any information. While people were usually eager enough to be of help, I still found it difficult to get a sense of what was really happening. I had to use local interpreters, so there was always the possibility of something important getting lost in translation. And then there was a more general problem: I was dealing with an illiterate population, which tended to attribute special powers to foreigners. I could never be certain whether those I questioned were conveying accurate information or just saying something that they thought I wanted to hear.

"Do you know of any case of someone coming down with fever and bleeding?" I'd ask.

"Oh yes, sure," they would reply. "That kind of thing occurs."

But then they would quickly add that nothing like this had occurred any time recently. And, of course, I had no way of knowing for certain whether any disease they described was actually Ebola. There are, after all, many diseases in Africa, and people fall sick and die all the time, including young, healthy people. Even a well-trained physician would have difficulty diagnosing a particular complaint in such primitive conditions, let alone administering an effective treatment. I was so intent on locating people who might know about Ebola that it wasn't until late in the day that I realized that we hadn't had anything to eat since leaving Kisingani. Maybe that explained the dark looks that my driver kept throwing my way. When I suggested that we stop at the next village marketplace and see whether we could find some food, the driver exhibited the most enthusiasm I'd seen from him all day.

The next village we came to, though, offered nothing in the way of food. There was no marketplace to speak of, and the shops were empty. Wherever we went, we found the same situation. I'd been prepared for spartan conditions in this region, but nothing this bad. It was obvious that people ate what they grew and that there was nothing left over to sell. Until now, I'd put off trying the C-rations. Suppose this thirty-year-old food turned out to be spoiled and inedible. That would probably mean that all the tins were bad. And then what were we going to do for food?

We stopped by the side of the road, but didn't get out. The rain was coming down harder, nearly obliterating the view through the windshield. I brought out two tins. With unconcealed skepticism, the driver watched me open them. "It'll be all right," I told him without conviction. I kept the chicken and gave him the turkey meal. My tin also contained cheese, peanuts, and soup. I gave the cheese a try, chewing it carefully. Not bad, I thought, not bad at all. I tried the chicken. It was actually quite tasty. I nodded to the driver, who so far hadn't touched his newly opened tin. "It's okay, go ahead." Taking no assurance from my words, he had an experimental taste. It was obvious that he approved. Maybe he would grant me more credibility in the future, I thought.

Water was a different story. My predicament wouldn't have been unfamiliar to the narrator of the *Rime of the Ancient Mariner*. Water was everywhere. It never ceased to come down from the sky. But there wasn't a drop to drink. Any water we could draw out of a village water hole or even a well was sure to be contaminated, so all I could do was to use iodine tablets to purify it. The taste of iodine is so unpalatable that it erased any doubt in my mind about the water being safe to drink. No microorganism, I thought, could possibly survive in such a foul-tasting liquid and still have enough strength to make me sick.

By the time we reached Isiro, dusk had fallen. Anxious to test my radio, I pumped up my antenna, strung out the wires, and connected them. Then I turned the radio on and played with the dial, straining to pick out a local broadcast. But all I could obtain were conversations between missionaries, lonely voices seeking contact in the dark. Then I tried to see whether I could reach anyone. All the while I was conscious of my driver's eyes on me. I ignored him and continued what I was doing. "This is Dr. Joe McCormick with the World Health Organization team in Zaire," I said, speaking into the mike. "Can anyone read me?"

Nothing.

I readjusted the antenna and tried again. Still, nothing. A burst of static. No matter what I did, I failed to get any response. And to think how pleased I'd been to acquire my radio, how confident I was that I'd be easily able to communicate with the outside world. I felt as if I were truly lost. No one knew where I was, and if I couldn't get the radio to work, they never *would* know. I went to sleep that night trying to keep myself from giving into despair and loneliness. Worse was the gnawing fear that I might end up traveling for hundreds of miles through desolate terrain and still find no evidence of disease. And while it was true that I knew that an epidemic was going on in the Sudan, I had no papers, no way of getting over to the other side of the border. I had to face the possibility of returning from my first search for a virus with nothing to show for it.

The next morning I ventured into Isiro, but I had no better luck. No one knew anything about a disease like Ebola. There was nothing to be done but head farther north, closer to the Sudanese border. Our next destination was a town called Dungu, a distance of fifty miles. The road we took was little more than a path. Ever since the departure of the Belgians in 1959–60, the whole northern area had been neglected, and none of the roads and bridges were maintained. We almost never caught sight of another vehicle. Why would anyone in his right mind want to drive up here, and where in the world would they be going?

I began to harbor serious doubts about whether there could be a connection between the two outbreaks of Ebola. In this part of the world, if anyone needs to travel, they either walk or ride a bicycle; their horizons are generally limited by the distances they can cover in a day. With travel between Zaire and the Sudan so difficult, it was unlikely that many people would have been in a position to transmit the disease from one location to the other. The disease has a short incubation period of only a few days, and it would have been an impossible walk or bike ride for an infected person. All my inquiries bore out my suspicion: no one I spoke to knew of any travel or trade between the two countries in this region. Later reports that the two outbreak sites were connected by trucking traffic were entirely unfounded. I was virtually alone, traveling the road connecting the two outbreaks. Simon was covering the area farther south, but anyone traveling there had to pass over the roads he was exploring.

The remoteness of this area was underscored by the response my arrival brought when I entered some of the villages. I probably couldn't have created more excitement if I'd come from outer space. Small children gathered around me, marveling at the extraordinary sight in their midst. Most of them had probably never seen a white face. My unusual appearance even succeeded in inspiring terror. Some just screamed and fled at my approach.

Interesting, but unsettling.

Once people got over the initial shock, however, I found that they were willing to speak to me. But these encounters took time. It is inappropriate to speak to anyone in Africa, especially in rural areas, without exchanging detailed assurances about the health of the families of both parties. Even asking for directions ran the risk of initiating a conversation lasting two hours. And from experience, I knew that you couldn't simply ask the first person you happened to meet. To do so would be unacceptable. You had to find the person in authority.

Just outside of Dungu I told my driver to stop so I could try the radio again. I was still irritated by my failure to get it to transmit.

"This is Dr. Joseph McCormick of the World Health Organization. Can anyone read me? Please come in."

I waited. Then I tried again. This time I thought I detected a voice. I fiddled with the dial. "This is Dr. Joseph McCormick. Can you read me?"

"Yes," I heard faintly, "I can read you, Doctor."

I'd succeeded in making contact with a missionary in Bunia, about one hundred miles southeast of our position. I asked him if he'd mind conveying a message for me to the WHO team headquarters in Kinshasa and let them know what progress—or lack of it—I'd made so far. The missionary assured me that he would be happy to do so through his mission. I was elated. My umbilical had not been cut after all.

At Dungu we found refuge at a local mission. Hospitality in such places is generous and sincere. A visitor makes for a rare source of entertainment, at least until after dinner. What electricity is available is supplied by private generators and then only during the few hours after dusk. Otherwise, you keep the same hours as the birds. After supper, everyone usually goes to bed, villages fall silent, and when the last of the cooking fires expire, darkness completely envelops the world. One of the great joys of rural Africa at night is walking outside and looking up at a sky that has suddenly filled with the light of a thousand stars. The only response is awe. You have a sense that the Africans who live in these villages enjoy a special intimacy with the universe. This sense of wonder may give meaning to lives otherwise so heavily burdened by poverty and disease.

Like most Africans, I moved as soon as it was light. This was a time to scour the villages around Dungu for evidence of Ebola-like disease. Again, I came up empty handed. Nor did I uncover any kind of commercial communication between this part of Zaire and southern Sudan. People in Dungu rarely traveled any distances, and why should they? There was nowhere to go, nothing to do.

After Dungu, I headed farther north, to a town called Aba, closer still to the border with Sudan. But, before we could go anywhere, we first had to cross the Uele River by ferry, a jury-rigged conveyance composed of four leaky dugout canoes with planks laid over them. It looked too shaky to convey my Land Rover, let alone a commercial truck weighing five tons. Even though this crossing was situated on the main route between Sudan and Yambuku, it was apparent that vehicles used the ferry only rarely, making it unlikely that it ever could have carried much traffic with Ebola virus. Nonetheless, we were good-naturedly assured that our Land Rover would make it safely to the other side. We were dubious, but lacking any alternative, we proceeded to drive it up onto the planks. The sudden weight caused the planks to wobble alarmingly, but, surprisingly, they held. Neither my driver nor I had any intention of remaining inside the vehicle while we crossed over to the other side. We weren't that crazy. So we stood in the canoes, watching anxiously as the crew paddled and pulled the fragile vessels with their unsteady burden to the opposite shore. All the while, other crewmen were busy bailing out the canoes with tin cans. At any moment, I expected us all to go toppling into the river. By some miracle, though, we gained the opposite shore. In Africa, I long ago learned, there is a solution for just about any problem—but often not in the one you were expecting.

The missionaries in Dungu told me that it was difficult to get to Aba by road, but they were wrong. It was, in fact, impossible. To reach Aba it was necessary to cross another river, but the ferry to do this was "unavailable." Actually, it was missing altogether, having broken loose from its moorings and floated down river during a recent storm. There was no bridge, either—not close by, certainly, maybe not anywhere. So we journeyed on and eventually came to the town of Doruma, which was as close as you could come to the border with Sudan. The two Italian fathers who ran the local mission welcomed me with great warmth. They hadn't had visitors in a long time, and they wanted to shower me with hospitality. We spoke to each other, using a curious mixture of French, Italian, and primitive Spanish, sprinkled with a few words of English.

That night, they served a typical meal of rice and beans with eggs, cooked in palm oil. The fathers brought out what they said was bread made specially in my honor. It was sour and bitter, but the taste didn't seem to bother them. The fathers apologized for having no beer. We drank water, which was, thankfully, boiled, so that I didn't have to endure the terrible flavor of iodine. At the conclusion of dinner, the fathers produced their store of local firewater for a small night cap. I was accustomed to palm wine, but this stuff must have been made from jet fuel and sulfuric acid, which was just about how it tasted. When one of the fathers lit his pipe, I half expected us all to go up in flames.

With my head spinning from the potent concoction, the fathers showed me to my lodging for the night: a storeroom furnished with a small camp cot. The cot was surrounded by boxes of gnarled potatoes, sacks of flour, and tins of dried porridge. The room smelled strongly of the contents of these boxes. Taking a peek, I saw that the flour was moldy. So that was how the bread got to taste so bad: aspergillus.

Exhausted, I immediately fell asleep.

But not for long.

Insistent squeals and rustling brought me wide awake. What was happening? Straining to see in the pitch darkness, I realized that I was surrounded by rats rummaging in the flour and potatoes. I decided that if I left them alone, they would do the same for me. Resigned, I fell back to sleep, although it could not be said that it was an especially restful sleep. The next morning, I politely suggested to the fathers that they might have quite a few rats on their hands. Since they acted unconcerned, I didn't press the issue. Maybe they regarded the rats as just another source of protein. Rodents are eaten widely in Africa and are even considered something of a delicacy.

Over a breakfast of coffee and more fungus-flavored bread, I asked them the same questions I'd put to everyone else along the route to Doruma. Did they know of any cases of Ebola? No, they said, there had been none in Doruma. What about the outbreak in the Sudan, just a few miles across the border? No, they hadn't heard of it. Then one of the fathers said, "Wouldn't you like to go up to Nzara and see what's happening there for yourself?"

"Of course. But I don't have any papers. I don't have a visa. So I really don't see how I could."

"Oh!" the father replied. "This is no problem. We'll find one of the big chiefs around here. He'll arrange everything, and he'll probably want to lend you one of his speakers who can accompany you." A "speaker" was a curious amalgam of legal expert, interpreter, and educated village person.

That is exactly what happened. That day, we toured the neighboring villages, eventually locating a chief who readily agreed to provide me with a speaker to accompany my driver and myself across the border. Just as the fathers had predicted, he also volunteered to write a letter in Zande, the language widely used in both northern Zaire and southern Sudan, requesting the appropriate authorities to allow us to cross the border. To make sure that the letter would look suitably official, the chief put his own personal stamp and seal on it. That evening, back at the mission compound, I decided to try using my radio to get a message back to my people in Kinshasa. I wanted to let them know that I would be crossing the Sudanese border the next day. I might not have found evidence of Ebola yet in my journey north, but that was about to change.

CHAPTER SEVEN

The Deserted Hospital

By the time we pulled out of Doruma, I'd been on the road for seven days. But at least now I had better company for the journey. The speaker turned out to be a genial young man and a school teacher who knew some French as well as the local languages, Lingala and Zande. He also had the advantage of being familiar with the area we'd be traveling through.

The main highway leading to the Sudan turned out to be just another path. There were few tire tracks visible—a telling indication that it went largely unused by vehicular traffic. The border, when we reached it, was deserted. The only way we knew that we were there was a makeshift barrier across the road, which consisted of a pole resting on two forked sticks. Obviously, it wasn't intended to keep anyone out. We lifted it up and crossed over into the Sudan. After a few miles, we came upon a broken-down truck. A man sitting by the side of the road told us that the axle had given out and that the driver had gone to Kinshasa for a new one—over a thousand miles away as the crow flies—much, much farther by road. This man had been hired to make sure that nothing happened to the truck while he was gone. He'd been doing this for five weeks now and evidently had little expectation of seeing the driver any time soon. But he didn't seem to mind. He lived in the area. It was a job. What else did he have to do?

About ten miles after entering what we believed to be Sudan, we stumbled on the official crossing. It was nothing more than a small compound with a few soldiers under the supervision of a commandant. A stocky, mustached figure, he came out to greet me and my party. He seemed a bit bemused by our appearance. Probably so few people came across the border from this direction that any activity would have surprised him. Well, I thought, now I'll find out whether the chief's letter will have any value. The worst that could happen was that he would order us to turn around and go back. But that would mean that I'd never get to carry out an investigation in the one place I knew Ebola to exist.

As his soldiers looked on with curiosity, the commandant put out his hand. I identified myself and handed him the letter. His eyes scanned the chief's words. The seal and the stamp seemed to impress him. Then he looked up at me. "Welcome to the Sudan," he said. "Please, come have some tea with us."

Gratified by his favorable response, the three of us followed him into the compound, which was little more than a few tin-roofed structures. The atmosphere was one of lassitude and inertia. It seemed that nothing ever happened here. The commandant instructed one of his soldiers to bring us cups of tepid tea. "What has brought you to the Sudan?" he asked.

I told him that we'd heard about an outbreak of a disease that causes fever and bleeding. I saw at once that he knew what I was talking about.

"Ah, yes, of course. You mean the disease in Nzara," he said. "But now this same disease is going on in Maridi, the district capital." He added that the outbreak had hit Maridi some weeks after it had exploded in Nzara. I was aware that the epidemic had spread to Maridi, but until now I hadn't any corroboration that it was still going on.

Then the commandant returned his gaze to me. "But I still don't understand. Why do you wish to go to Nzara?"

"I am a doctor," I explained. "It's my job to find the source of the disease."

It seemed that he'd never heard of anything like this.

"In Nzara," he said, "there is no need for doctors anymore. The people you have come to save, they are all dead."

Nzara, a town of about 3,000 people, is built up around a cotton factory that dates back several decades, to the time of British rule. Most of the inhabitants live in family compounds on the outskirts of town. The factory is Nzara's main source of income. It was also, for all I knew, the source of Ebola.

When I got to Nzara, I didn't have any idea what to expect. On the surface, nothing looked out of the ordinary. People were going about their business, betraying no sign of panic or fear, but when the speaker approached a man to ask for directions to the hospital, I noticed a distinct change in attitude. The man's face darkened. Then he made a brusque gesture, mumbled something to the speaker, and quickly turned away.

"Did he tell you where to find the hospital?" I asked the speaker.

"Yes, he says it is just up the street. But he says we shouldn't go there. He says it is a bad place. And he says that no one is there anyway."

"Did he say why?" I asked.

"I asked, but he wouldn't answer me."

More curious than afraid, I directed the driver to take us to the hospital. The driver didn't say anything, but I could tell that he wasn't happy about it. He didn't want to get anywhere close to the hospital.

It was a simple one-story brick building with scarred, dun-colored walls. There was no sign of life within. I tried the door. It was unlocked. Entering, I found myself in a darkened vestibule that yielded into a large empty room. There was a sour smell in the air, a mixture of dried blood and excretion. It was the only room in the hospital, so I had to assume that it was used by male and female patients alike. The beds were nothing more than broken metal springs set on frames. Undoubtedly, the patients had to supply their own mats—that is, assuming there were any patients. I called out. I heard only the dull echo of my own voice in reply.

Just as I was ready to leave, I registered the sound of footsteps. I turned to see a smiling, robust man approach me. His white tunic was stained.

"I am Dr. Mohammed," he declared.

After introducing myself, I asked what had happened to everyone. "They are all gone. The patients, the nurses, all of them are running away."

"But you stayed?"

"I am the doctor. Where else would I go?"

"Why did everyone flee?" I asked.

"They see what has happened to others here. So many deaths, so quickly. They think they will die, too. So they run away. I can't blame them."

I was talking to a captain who was prepared to go down with his ship.

While I had never been in the Sudan before, I knew enough about its tormented politics to understand how anomalous the doctor's position was. Like most physicians in the Sudan, he was a Muslim, undoubtedly from the north of the country, while his patients were Christians and animists. The ethnic and religious rivalries between the two halves of the country had triggered a civil war that continues intermittently to this day. I imagined that little love was lost between the doctor and the people he treated. The epidemic that had emptied his hospital couldn't have helped.

Dr. Mohammed went on to say that, altogether, thirteen people had become infected with a disease the likes of which he'd never seen before. "Seven of them died. There is nothing I can do for them." He sounded both enraged and bewildered. I asked if he knew whether any cases were still occurring. No, the doctor said, the epidemic had ended about five weeks before. "In Maridi, yes, there is a problem still. The government has imposed a *cordon sanitaire*. There is no trade with the city, no one can get in or out." (Later, I would learn that the *cordon sanitaire* was also preventing a team of WHO investigators from reaching Maridi; they were reduced to cooling their heels in the southern provincial capital of Juba.)

"What were the symptoms of this disease?" I asked.

"People cannot swallow, their bodies are full of pain, the blood vessels break in the eyes. Blood is coming from the gums, too. They are having a bad fever."

"Can you tell me about the first patient you saw with this disease?"

Dr. Mohammed proceeded to describe a man in his twenties who had worked at the cotton factory and lived in a family compound outside of town. He had been admitted into the small hospital with familiar symptoms: high fever, headache, sore throat, abdominal pain, diarrhea, and bloody stool. In six days he was dead. From what I could determine, it sounded as though this patient may have been the first (or "index") case of Ebola in the Sudan.

Before I left, Dr. Mohammed said, "There is one thing more I neglected to tell you. I sent one of my patients to Maridi. They have a better hospital there."

He must have noticed a change in my expression.

"What is it? What is wrong?"

"Nothing," I said.

I didn't want to make him feel worse than he already did. It had just occurred to me that by sending his patient to Maridi, Dr. Mohammed had helped spread the disease to that city.

Ironically, if the epidemic hadn't spread to Maridi, the epidemic in Nzara would probably never have become known to the outside world. Again and again, the viruses that emerge from the remote parts of the earth and assail the indigenous population only gain attention when they move out of a small area to affect larger numbers—or when they kill off wealthy people or foreigners, especially Americans. Diseased white Westerners are always a sure bet when it comes to attracting attention. If the right people aren't infected or dying, outbreaks that occur all the time in places like Nzara go unnoticed.

Although the people I interviewed in Nzara registered some concern about a possible recurrence, they exhibited no sign of panic. The passage of more than a month since the last case may have provided them with some reassurance that the worst was over. How could I be sure that these were definitely Ebola cases? I lacked the biological evidence to make an accurate assessment. I needed reagents and my portable laboratory. There weren't even any hospital records. All I had were gravestones and stories. For now, these would have to do.

In keeping with the custom of all African hospitals, the index case had been attended by his family during the course of his brief illness. Soon after his death, his brother had also fallen ill. He was luckier, though. He survived. Curiously, the dead man's wife was even more fortunate, because she never became ill at all. The serological evidence we later took from her confirmed that she'd escaped infection completely.

After talking to the doctor, I went to see the widow of the suspected index case. She was now living with her parents and her two children. A woman in her late teens or early twenties, she wore her head shaven as a sign of mourning, but she showed no emotion the whole time I was with her. Africa is fascinating: When someone dies or is just about to die, people gather outside the house and wail and beat their breasts. In some societies, mourners smear their skin with chalk dust. These rituals can go on for hours and hours. From a Westerner's point of view, such displays may seem overly dramatic, even artificial, but they are an integral part of the culture and not fake at all. For once the mourning is over, people adopt a stoicism that often contrasts markedly with behavior in the West. It is possible that the ritual provides a catharsis, a means for people to exorcise the grief and pain so that they can get on with their lives.

What I was trying to discover from the widow and others in town was where the infection might have come from. I could see, though, that the woman was nervous, unaccustomed to being interviewed. There was an added drawback. In this society, a wife generally didn't know much about her husband's activities once he left the house. I hoped to obtain a good history of how he'd spent his time before he fell ill, so that I could trace the source of the infection. Had the dead man been bitten by an insect? Had he been hunting? Did he eat something that might have been contaminated? Had he received any injections? In these circumstances, compiling a history like this was practically impossible, but I was able to gather some evidence that suggested how the virus was being transmitted. The hospitals were one possible route; between the reuse of needles and the absence of quarantine measures, other patients and the victims' families had been needlessly exposed to the disease. But I soon learned of a custom that also put people at terrible risk: in this culture, people felt the need to handle the dead body before burial. It is a loving act, a means of ratifying the importance of the deceased to the survivors. It is common for family members to come up to the casket and kiss the deceased. They also believe that the body must be completely cleansed. To do this adequately, it first must be thoroughly washed, then all matter must be expunged from it, including urine and feces. Because, in Ebola victims, the feces inevitably contain blood, this process serves only to spread the infection to the mourners. From what I later learned, this was how Ebola was finding so many victims in Maridi.

For the time being, the only thing I could do was to circulate through the

compounds and try to find people who might have information about the epidemic. Was there any evidence of a common source, I wondered, or was this
index case the only one? I interviewed the families of four victims, all of whom
had had some contact with the index case. But their accounts didn't provide me
with any definitive answers. Then I took a look at the cotton factory. Was this
the site where Ebola had first made its leap to human beings from its natural
reservoir, whatever that might be?

Though the managing director of the cotton factory tried to be helpful, meeting with him did not lead to any better understanding of the possible source of
the virus or the role the cotton factory might have played in spreading the disease.
There was one intriguing fact: The products of the cotton factory were shipped
out through Juba up the Nile to Khartoum. There was no export through Zaire,
and no way to export through Zaire. In fact the director was amazed at my question, responding with a laugh that it would be impossible to get any goods to
market through Zaire, and he assured me there had not been a market for his
goods in Zaire for many years. I proceeded to examine the cotton factory.

It consisted of several brick and wood buildings at the edge of Nzara. The
compound was enclosed by a rusting wire fence, and the grounds, which once
must have been planted with flowers and shrubs, were now overgrown with
uncut grass and weeds. The buildings were a colonial type built some five or six
decades earlier. Now they had mostly paneless windows, either open to the elements or covered with a hodgepodge of paper and wood. Any residual glass
panes contained cataracts of dust and grime. Inside, the building was dimly lit,
most of the bulbs dead or missing. The ceilings were high—a hallmark of the
colonial style—with ceiling fans that seemed mostly inoperative. There were two
striking features about the interior. One was the incredible clatter and whir of the
ancient spinning machines, shuttles, and looms. The machinery here looked like
relics from an industrial museum. The second striking aspect was the strange
odor in the air, an atmosphere that was a colloidal suspension of cotton fibers,
dust, and noise. But the smell—the smell added a dimension difficult to define.
It was a smell I vaguely recognized, but could not immediately name. I looked
deeper into the building, and, as my eyes turned on the high ceilings, I knew
immediately the source of that odor. The ceiling was suspended, and it was
grossly discolored gray to black in many areas. Some of the ceiling had decayed
completely. The discoloration, the smell—all of a sudden I knew.

Bats.

As in most such buildings throughout tropical Africa, the ceiling was fes-
tooned with bats. At night, these nocturnal creatures probably poured out of
the vents in the roof, a black, squeaking cloud, to find their prey of insects or

fruit. They returned in the morning to hook themselves upside down for a good day's sleep. As they rustled and snoozed, they excreted bat guano. This is what gave the ceiling its peculiar color. Where the ceiling had deteriorated, the guano spilled onto the factory floor, giving rise to the musty ammoniac smell that hitched a ride on the cotton and dust. I had lived long enough in Africa to know that bats were a harmless part of life. We used to get them in our house at night—through the thatch roof and mat ceiling—and we would try to catch them with our baseball mitts, without injuring them, and then release them outside. People here were used to bats, so that I knew no one working in the cotton factory was concerned about them. For my part, I couldn't help wondering: could the bats somehow be mixed up in the transmission of Ebola?

CHAPTER EIGHT

My Place Among the Dead

Could the bats' excretions be the ecological niche of the virus? If so, how did the virus perpetuate itself in the bats? We had to suppose the virus was relatively harmless to bats, but fatal to humans and other primates—not an impossible concept, given what we now know about species adaptation of viruses. But if bats were responsible for Ebola in the Sudan, or anywhere else, it would be difficult to prove. We had heard of no reports of bats coming into contact with Ebola patients in Zaire; but bats are ubiquitous in Africa, so that meant nothing.

Just because the factory seemed the most likely source of the infection in Nzara didn't make it so. In order to solve the riddle of the infection's origin, we would have to set up some fairly extensive studies to test the bats for the presence of Ebola. I was certainly not in a position at that moment to go about obtaining a large enough sample to see whether bats were infected in sufficient numbers to transmit the infection, if they were infected at all. The WHO Sudan team that was to follow a few days later did gather a few bats. However, their sampling was inadequate, and they were unable to isolate Ebola virus or even demonstrate past infection with Ebola virus in any of the bats. Given the apparent rarity of human infections from the natural reservoir, I tended to think that whatever the animal reservoir, infected individuals of the species were relatively uncommon. Karl Johnson and his teams were to test a host of animal species, including brown bats and fruit bats in 1978 when they conducted an extensive sweep of the jungle fauna searching for the Ebola reservoir. Later, another attempt was carried out by Gene Johnson of the United States Army Medical Research Institute of Infectious Diseases (USAMRIID), who conducted an extensive blitzkrieg investigation in a cave where the suspected bats lived. He had far more resources than we could have dared dream of at CDC for such a project. For all that, unfortunately, he came up with nothing.

Could the infection have crossed from the Sudan into Zaire? I very much doubted it. If bats were involved, it certainly would not be a single bat that could have carried it from one site to another. And, as I'd previously concluded, human traffic was not likely to have been a route of transmission either. Why would anyone travel all the way from Nzara to Yambuku, even if he were willing to brave long stretches of nearly impassable roads? There was no trade. The thread and cloth the Nzara factory produced was carried through Juba to Khartoum or to Nairobi. There was no way to export it through Zaire, and no one to buy it in Zaire.

Before leaving Nzara, I sat down and wrote a letter to Don Francis. (This is the same Don Francis who would later become principal advisor on HIV to the late Randy Shiltz, author of the bestseller *And the Band Played On*.) It was a kind of "Dr. Livingstone, I presume" note. Don was a friend who'd been involved in a smallpox investigation in the Sudan. From radio transmissions, I found out that he was now with the WHO team still stuck in Juba. I was certain, though, that sooner or later he would get to Nzara and would want to know what had happened there. In the note, I identified the index case and indicated where he'd been buried. I also provided Don with details of the outbreak and gave him instructions as to how he could go about finding the doctor and the survivors. All in all, I must confess, it was a very satisfying note to have written since it would confirm that I had got into Nzara before any other investigator. Later, he told me he was very surprised and amused to find it waiting for him.

After spending three days in Nzara, I decided to leave. The investigation was interesting enough, but there was no way that I was going to establish a link between the Sudan and Yambuku outbreaks by staying any longer. The only other thing I could have done was to collect specimens and trap animals that might be carrying the virus, but I had no facilities or equipment for mounting an investigation on such a scale, so I made up my mind to return to Yambuku in Zaire and then head farther north toward the Central African Republic. Buoyed by my success in getting into the Sudan, I thought that I'd see whether I couldn't cross another frontier. But when I got about fifteen miles from the border, I found the road blocked by a fallen tree. There was no way around it and no way for us to cut through it. Why had it been cut down? No one knew, but it had been there for several months. Even if the tree could be made to disappear, it wouldn't have done any good. I was told that I would still have to cross a river, and to do that I would need a ferry. The trouble was: no ferry.

So I turned back and resumed my search for Ebola elsewhere. It was raining heavily along much of the route. The roads were extremely dangerous as the red laterite grew progressively slicker. My driver seemed unconcerned—as did a lonely hitchhiker we had picked up by the side of the road. I was increasingly frightened. As the trip went on, we fishtailed with ever greater frequency. Bullets and viruses were two hazards of this part of the country I was willing to accept. But a *traffic* accident? I repeatedly urged the driver to use high four-wheel drive. On each occasion, he grunted his assent and then went on as before. Having shown no particular deference to my wishes all along, he was obviously in no mood to change his habits now. After pulling out of a small town called Bili, he took a slight angled ridge in the road at too great a speed. Suddenly, the Land Rover fishtailed and started swerving so violently that the vehicle spun like a top. The window revealed a blur of colors: the green of the trees, the brown of the mud, the red of the laterite. It was like watching a film in fast motion. The fifty-gallon drum of fuel in the back was thumping like mad against the side of the vehicle. Everything that wasn't tied down was flying in all directions. I groped for something to hold on to as the Land Rover careered and, with a shriek of metal, flopped over on one side, then flipped right over, finally landing on the other side.

When I next looked up, I was staring at the sky. I realized that I'd been thrown on top of my driver. Then my ears filled with a deafening scream. It was coming from the man we'd given a lift to.

Shaken, but unhurt, I managed to clamber out of the vehicle. I could barely stand. The driver was having an even more difficult time getting back on his feet, but it was the hitchhiker who was in the worst shape. He shrieked in agony: "My neck is broken!" But even in my dazed condition, I could see that he was probably exaggerating. People with broken necks are usually unable to shout. Nevertheless, I looked to see if he could still move all his limbs. Okay there, but his distress was real enough. After I got him to lie down, I examined him to see if there was any serious neurological injury. He seemed to be suffering from severe muscle spasm, possibly caused by whiplash. As far as he was concerned, though, he was going to die. He was completely hysterical. My words of reassurance were lost on him. In any case, there was no way to get him proper medical attention if we couldn't drive.

Some neighboring villagers came running down the road to see what all the commotion was about. With their assistance, we managed to get the Land Rover back upright. Now I was in a better position to inspect the damage. It didn't look too bad. A part of the vehicle's front had crumpled in, and a couple of windows had been knocked out—yet remained unbroken. Now I returned my attention to the hitchhiker, who wouldn't stop screaming. I decided that I had to devise some means of stabilizing his neck and then try to get him some-place where he could lie still. But I couldn't move him in his current state. While he so far hadn't complained of a loss of feeling, I couldn't entirely rule out the possibility of a severe injury. For all I knew, he might be on the verge of sev-ering his spinal cord. I dug into the medical kit I had brought with me. Well, what do you know? The kit included injectable valium. I gave the man a ten-milligram shot and waited. He began to relax and grow drowsy. This gave me the opportunity to improvise a cervical collar for him, using a shirt stuffed with other clothing. Then I tied it around the back of his neck and under his arms. Once that was done, we were able to stretch him out on the backseat of the Land Rover. When we learned that close relatives of the man lived only twenty miles or so down the road, we figured we were in luck so long as the Land Rover worked. Which, fortunately, it did. Even so, it needed repairs—a job that would take the better part of two days.

At last, I reached a town called Abumombozi, about eighty miles north of the epicenter of the epidemic in Yambuku. There was a rumor that the possible Zaire index case, a schoolteacher, had passed through Abumombozi in a car on his way north. So I was interested in seeing whether there was any evidence of an epidemic. I found a doctor who told me that, while he knew of no outbreak in the vicinity, he had seen patients with symptoms of typhoid from time to time. But was it typhoid that he'd really seen? Typhoid was the original diagno-sis physicians had made in Yambuku before they recognized the disease for what it really was. Was the doctor confusing the two diseases? I had no way of know-ing. There were no current cases and no biological evidence of previous ones.

When I arrived in Yambuku, our team was still in the midst of surveying the countryside, questioning villagers, and taking blood samples to see how many people had been exposed. My colleagues were naturally anxious to learn whether I had established a connection between the Yambuku and Nzara out-breaks. I had to disappoint them. There was no evidence at all that the two could have been related. "There are four distinct tribal areas you'd need to cross through to get from one place to another," I told them, "and there's no motive for anyone to do so." The reaction to my report was skeptical. The fact that the two epidemics had broken out almost at the same time seemed to argue in favor of such a linkage, but I remained convinced of my findings. It would take another three years, however, before I was proven correct.

I traveled next to Bumba, and spent the night at the local Catholic mission, planning to catch the flight out the next day on a military transport back to Kin-shasa. At about 10 A.M., I went to the airfield. Just because the plane was sched-uled to arrive, of course, didn't mean that it would actually be there. So it was

really quite a surprise to find it sitting on the runway. But any surprise I registered at finding the plane was nothing compared to what I felt when I saw a mob
of people surging around it. Most of them were women, their faces smeared with
white chalk, howling with grief and beating their bare breasts. Something calami-
tous seemed to be taking place at the rear of the plane, but I couldn't imagine
what. As I got closer, I saw several wooden boxes being hoisted into the craft. I
couldn't believe my eyes. They were coffins. I was going to ride back to Kinshasa
with a plane load of dead bodies.

Were they Ebola victims? I asked someone what was going on.

"They are pilots, sir," came the reply. "They were flying a helicopter, and it
crashed." The man looked at me as if he were deliberating whether to tell me
more. "The weather was bad, sir." Somehow that explanation didn't seem sufficient. So he added, "The helicopter ran out of fuel, sir."

"Thank you," I said and turned away.

But the man had one thing more to add: "They were drinking, sir. Yes, they
were very drunk, indeed."

Later I would learn that a colleague, Peter Piot, a member of the Yambuku
team, was supposed to have caught a ride back to Kinshasa on the doomed chopper. Once he took note of the pilots' condition, however, he'd sensibly declined the
opportunity to accompany them. I was later to get to know Peter well, when we
started the AIDS investigations in Zaire. (He is now the head of the United
Nations AIDS program.) I suspect he has often reflected on this episode.

It had taken several days to locate the wreckage and the bodies, which by that
time were badly decomposed. So the stench in the cabin was pretty awful.
There was no doubt that I would have preferred a different group of traveling
companions. But what choice did I have? I made my way through the wailing,
white-faced women and got on board. There was no escaping the nauseating
odor. Two hours in the air! How could I stand it? Well, I would just have to. So
I secured myself in the jump seat and took my place among the dead.

A few weeks later, another shipment followed the coffins to Kinshasa for
transfer to a plane bound for Atlanta. This particular shipment was made up of
vials containing six hundred specimens of sera and blood collected from people
who'd lived in and around Yambuku. They were to be tested for Ebola antibodies at CDC laboratories. What we had no way of knowing was that there was a
secret in those vials, one which had nothing to do with Ebola. Like some kind
of evil genie, the secret would remain locked up in a CDC freezer among thousands of other specimens, until nearly ten years had passed. Only then would
we be in possession of the knowledge needed to pick the lock. But by then it
would be too late.

CHAPTER NINE

Learning Lassa

In a poverty-stricken nation like Sierra Leone, John Kamara was one of the privileged few. In his mid-thirties, he was one of a small number who had graduated from Forah Bay College, the country's oldest and most prestigious educational institution. Upon graduation, John returned to his home in Segbwema in the remote and destitute Eastern Province and took up a position as teacher of history and French at the Holy Ghost School. He was highly regarded by his students, who looked up to him as a role model. There were few others like him in Segbwema, an area notably short on educated people. It was to John they would turn when they needed counsel. More than just a teacher, he became a friend, someone who was as likely to join in their soccer matches as he was to unravel the mysteries of French verbs.

A stout man, but in excellent health, John seldom had occasion to worry about anything happening to him, which was why he found it strange to wake in the middle of a February night in 1977 with a vague feeling of unease. His skin was warm to the touch, and he had a mild headache. Even after a strenuous game of soccer, his muscles didn't usually hurt quite as much as they did now. But, then, he'd just come back from visiting his family in a nearby village. As usual, the roads were in bad shape and, besides, it had been a particularly hot and dusty day. So, he assumed that his discomfort must be due to the rigors of the trip. He turned over and tried to go back to sleep.

The next morning, though, he was feeling worse; his head was now pounding, and the pain in his muscles had grown more intense. He was certain he had a fever. This was not an unfamiliar feeling. John had grown up in an environment where endemic malaria was an everyday event. Naturally, he decided that was what he had. Malaria is the diagnosis of choice in Segbwema—as in most

of Africa—for all fevers, headaches, and muscle pains. This is even true in the case of adults, who should have developed some immunity after spending their lives being bitten by infected mosquitoes. After sending a note to school, explaining that he was ill and wouldn't be in that day, John took some aspirin and four tablets of chloroquine, standard treatment at that time for malaria, and went back to bed. By afternoon, he felt somewhat better, though he noted that the muscle aches had settled in his lower back, and that he was now developing a bit of a sore throat.

That evening, his condition worsened. His fever climbed, and the pain in his throat intensified, along with the pains assailing his muscles. Now he was beginning to think that maybe it wasn't malaria, after all; something else might be wrong with him. The following day, with his fever even higher and the pain becoming worse, he went to the Nixon Memorial Hospital in Segbwema for a consultation. Isabelle King, the doctor who admitted him, immediately knew what he had.

It was Lassa fever.

With the end of the Zaire Ebola investigation in November 1976, I returned to the project site I had begun establishing in Kenema, Sierra Leone, to pick up where I'd left off. My primary mission was to establish a long-term program to study Lassa fever. Two essential, unfinished tasks awaited me. One was to find a place to live, the other was to get a laboratory in shape. Looking around the building the government had given me, I realized that it would take some time to turn it into a functioning lab. For a start, I would have to install the electrical system. The roof was leaking and it badly needed paint. I also had to retrieve the equipment I'd procured in Atlanta. Although it had arrived—presumably intact—in Sierra Leone, it was still sitting in storage in the port. I had to hope that it was safe. I couldn't think of anyone in Sierra Leone who would know what to do with it except me, but you never know. In Africa, particularly in countries as poor as Sierra Leone, there isn't a thing you can name that fails to get "recycled," and sometimes in amazingly imaginative ways. For instance, dead vehicles left by the side of the road—whatever their condition—are quickly picked as clean as a carcass in an African game park. So I had to get down to the port and supervise the loading process to make certain that the material was safely put on board the trucks; otherwise something might be damaged or else end up forgotten in the warehouse.

Once I reached the port, after a tortuous trip south, the shipping manager assured me that everything was in order, but that I wouldn't be allowed to inspect the goods personally. Two days later, I received an urgent message from the warehouse, informing me that the back doors of the closed-bed trucks that I'd hired weren't large enough to accommodate the crates. So now I was forced to find larger, open-bed trucks to transport my equipment to Kenema. The job was done, but at the cost of a few nights' sleep.

Now that I had the equipment I had to have electricity to run it. I didn't have time to do the work myself—although I'd gotten a good deal of experience putting in electrical lines during my days as a teacher in Zaire—so I hired a local electrician, figuring that I would be able to watch him closely enough to insure that the job was done right. When he said he was finished, I inspected the points and cabling. It looked all right to me. Now we were ready for the momentous event. Would it work? I plugged in a 110-volt microscope and was rewarded with a 110-volt shock—painful evidence that the current was running somewhere it shouldn't be. I plugged in another couple of small instruments. The result was the same. I decided it must be something fundamental. The electrician hastened to assure me that he'd done everything exactly as I had instructed. My fingers still tingling, I wondered.

Suspecting a grounding problem, I looked inside the circuit breaker boxes. Lo and behold, none of the ground wires was connected inside the box. The electrician had put in the neutral and live wires all right, but had twisted the ground wires together, a Medusa's head unconnected to anything but air. He must have been assuming I was the ground. Patiently, I demonstrated to him what ground wires were for, and what he needed to do with them. He got the message and settled down to fix the problem. I briefly wondered what went on in other buildings in Kenema. Oh, well. It was heartening to be able to sit down at my new laboratory bench and use my equipment without suffering electrocution.

With my lab up and running, more or less, I now had to find people to work in it. I needed to fill four positions. Given the conditions we would be working under, I wasn't just looking for professional qualifications. I demanded immunity—immunity to Lassa. I was planning to work with *Mastomys natalensis*, a medium-sized bush rat, about halfway between a mouse and a standard, home-grown U.S. rat in size. These rodents are the carriers of the virus. They become infected as fetuses and, though they never have the disease themselves, excrete the virus in their urine. The virus replicates merrily in the mouse for its entire life, somehow evading its immune response, possibly by tricking it into thinking the virus is mouse instead of virus. The very persistence of the infection is one of the mechanisms viruses use to keep plenty of themselves in circulation wherever they go. Viruses can be diabolically clever. Some, such as HIV, keep an army of smart people working their entire lives, and still we do not understand the virus sufficiently to cure or prevent infection.

At that time, the antibody from previous exposure was the best protection I knew of if you were to work with the disease. The antibody might not stop you from getting infected again, but at least there was a likelihood you would escape without becoming seriously ill. At all costs, I had to avoid getting my rodent squad infected. These people would be handling hot rats excreting large amounts of virus, and this is the most dangerous work. I worried about this quite a bit, but there was no reason to be unduly concerned. I had little problem finding individuals who had the antibody to Lassa virus. My real problem was finding personnel who had antibody *and* were competent enough to do the rodent work. Very few people from the Eastern Province of Sierra Leone had more than one or two years' schooling. The best-qualified applicants had only a high-school education and had mastered only basic literacy and numeracy skills. Except for government positions, mostly acquired as a result of patronage, the only other jobs available were in agriculture, and why would anyone need to go to school for that? So, right away, I was limited in my pool of applicants. I also needed people who were able to speak Mende, the local language, as well as Krio, the more commonly used Lingua Franca. Finally, some of the work that had to be done required a modicum of basic medical knowledge, but I quickly realized that I wasn't going to have much luck in finding people who met this criterion. In fact, initial efforts to find trained physicians for the project were a dismal failure. This was no surprise. Sierra Leone had no medical school, so native physicians fell into two categories: the ones who were trained in the West and the ones who were trained in the Soviet Union. The doctors who'd attended medical school in the West, if they ever came back, generally found lucrative practices in Freetown, or else they worked for the government, gaining political points, a reasonable pension plan, and a bustling private practice in the afternoons and evenings. They rarely came up-country.

Most of the Soviet-trained physicians, on the other hand, went into the government hospital system as soon as they returned home and were immediately dispatched into the rural areas. Theoretically, the Soviet-trained doctors were supposed to have first served an internship in Freetown before going into the countryside. There was little evidence, however, that this experience did anything more than weed out the grossly incompetent from the merely incompetent. The local hospital was filled with these inept doctors, many of them veterans of the Patrice Lumumba Friendship University in Moscow, where political indoctrination figured more prominently in the syllabus than anatomy lessons. To an appalling degree, a disproportionate percentage of the Soviet-trained doctors were little better than butchers. They practiced on an unsuspecting local population with little supervision, if any. It didn't take long for even the most illiterate villagers to figure out that the government hospitals were to be avoided whenever possible.

Unfortunately, there was often nowhere else to go for help. In search of a reasonable, mission-run hospital, sick people in desperate condition would travel miles over terrible roads in little poda-podas, small blue pickup trucks that function as all-purpose buses and vans, usually crammed with people, produce, and animals jammed in like proverbial sardines and pouring out of every orifice. Excess passengers balanced precariously on top or hung off the sides and the back. And anyone or anything that can't fit inside, piles on top of the vehicle. Then it's just a matter of hanging on for dear life.

Even in the absence of suitable doctors, I had a stream of job applicants. By Kenema standards, I was about to become a major employer. The Ministry of Health presented me with several candidates. The first group I interviewed was a motley bunch, all of whom—big surprise!—turned out to be relatives (thoroughly unqualified) of someone in the ministry. I quickly realized that I was expected to take whomever they sent. The successful candidates would then be expected to provide kickbacks to their government sponsor. This was a delicate situation, since I would need the cooperation of the ministry if I were to get my project off the ground.

I solved my problem by hiring more people than I needed, including some of those the ministry had sent me, and the remainder those who actually met the standards I'd set. I let them all know that I was hiring them on a trial basis and that I would then select the most qualified after seeing how well they performed. This approach, I thought, would allow everyone to save face no matter who was ultimately chosen. Blame for the failure of any individual would fall squarely on the shoulders of the individual who failed to come up to standard.

So I began my work in Kenema by running what amounted to a training course for my new employees. I put them through drills and subjected them to practical and written exams designed to assess their competence. At the end of this process of natural selection, I ended up with a group of young people I believed I could shape into decent technicians. Most of them were local high-school graduates with little work experience. That was just as well. They hadn't had a chance to develop the customary bad work habits yet. I began at the most basic level, telling them that they would have to turn up for work. Every day. And on time. And while they were in the lab, I told them, they would be expected to complete certain tasks. Unfortunately, in those days, these concepts would have been highly novel for many experienced government workers in Sierra Leone.

For all the difficulties I had recruiting and training my employees, I succeeded in finding some great people, among them dedicated individuals who would stay with us for thirteen years or more. One particularly lucky find was John Kande. John was short and plump—plumper as the years passed and served only to increase his fondness for palm wine. His round face was decorated with a mustache, his dark eyes perpetually hidden behind sunglasses. He had a jolly disposition and was known as a lady's man.

Palm wine is the local hooch of much of Africa. It consists of the self-fermenting sap of the palm tree, gathered at some personal risk by wine tappers. Secured only by a flimsy bamboo hoop, these men scamper barefoot up palm trunks, to the top, where the sap can be bled into a gourd. It is a whitish drink that is definitely an acquired taste. In sufficient volume, it can be quite intoxicating.

Although Kande was fluent in several local languages, it was his knowledge of Limba that made him especially popular. Limba is the language of the palm-wine tappers, and John's mastery of the tongue guaranteed a daily fresh supply of the libation to cheer the rodent team's evening meals. And everyone knew that, with palm wine, it was freshness that counted. How fresh is fresh? The wine you were drinking at dinner should have been tapped that morning. A single day of fermentation is all the time it needs to develop sufficient potency for a good party. By the next day the flavor deteriorates to that of kerosene.

Kande's persuasive skills extended as well to his relations with the local chiefs and other notables who needed to be convinced of the importance of our project. Some people might have been able to obtain good palm wine without much of a problem, and others might have been able to win village chiefs over to their side, but only Kande was able to do both. If there was a drawback to his temperament, it was his tendency to get into fights after putting in a full day of work and a full evening of wine. From time to time, Kande would be thrown in jail for drunk and disorderly conduct. Then his talent for making friends really counted. Undaunted, he would call upon one of the chiefs or village elders, and soon someone would materialize to bail him out. Often this task fell to the Lassa fever project director.

During his years with the Lassa project, Kande became an expert animal technician. He ran a *Mastomys* breeding colony and processed large numbers of specimens for major studies. His crowning achievement was the construction of a set for the film crew who made a documentary on the project in 1989. Rats are difficult to film in their local habitat, so he fixed a few selected Lassa virus-free animals with a minimal dose of anesthetic and then released them into his set. The rats performed brilliantly, and the crew managed to get some excellent footage. In his enthusiasm, however, Kande did OD a few of the rats, and it took quite some time for them to come around. They looked as though they had been into the palm wine as they tottered around the plates of leftover food we'd put out for dramatic effect. A couple of them actually fell off the table and had to be picked up and put back.

It wasn't enough just to hire a staff, of course. I also needed to find a place for them to live. This got to be a problem, because we had to overcome all sorts of prejudices toward outsiders. In this society, there were no houses to rent because the whole concept of putting up a stranger overnight, let alone for longer periods, simply did not exist. Anyone who wasn't born in a village was an outsider. Villagers were suspicious of strangers, fearful that they might bring some unspeakable evil into their homes. Why else would anyone have left his own home, they reasoned, if not for some malevolent purpose? Of course, all of this was in the days before migration from rural areas into the cities had become so widespread. More recent attitudes toward strangers have eased considerably, as the modern world of mobile labor and commerce sweeps into even the remotest regions of Africa.

But, back in 1976, the diamond-mining areas north of Kenema were exceptions to the rule. There you could find strangers in abundance, searching for some of the best-quality alluvial diamonds in the world. Diamonds are among the few natural resources Sierra Leone has. Mostly, the mining was government-owned and run, but the trade also brought out the fortune seekers. They came from all over, not just Sierra Leone. Villages would spring up overnight. All it took was a rumor that so-and-so was walking in the bush and found a diamond on the path....The next thing you knew, the whole bush would be stripped of trees. Where forest once stood, all you'd find were great open pits and mounds of red laterite mud. Men would stand waist-deep in muddy water, wearing only simple loincloths, their bodies slick with sweat, digging and sieving and spraying, intent on finding instant riches. Some got wealthy, but, more often, the diamonds fell into the hands of the Lebanese, who controlled most of the viable commerce in the region. They bought up as many of the diamonds as they could. Probably the majority of precious stones were smuggled out of the country.

Just as villages sprouted overnight, so they withered and died. One settlement would be drained of population if a diamond find was discovered in another village just down the road. These sudden shifts in population would later play havoc with our survey work as we sought to understand the complex spread of Lassa virus. It must have been like this when gold was discovered in California and Alaska. However, we were not dealing with a population mesmerized by the prospect of instant wealth from diamonds—just people hoping to find relief from abject poverty. The hope of even a small amount of money casts a spell; and it had a miraculous power to overcome the deeply ingrained prejudice against renting rooms to strangers. Once we made it known that we were willing to pay for rooms, any number of them were suddenly made available.

When I was assured that my staff was as ready as it was ever going to be, I decided formally to launch the Lassa surveillance program. It was early February 1977, just four months since my return from Zaire. With two months' training under their belts, my recruits were going to have a crack at the real thing, real patients—with real Lassa fever.

And they wouldn't lack for patients, either. There was lots and lots of Lassa fever about.

While we expected that we would be busy, I'm not sure that we were prepared for just how busy. In the first month of our program, we saw nearly thirty patients. What I didn't know then, but was soon to find out, was that Lassa fever peaks in January and February, in the dry season. Nine of these first thirty patients would die. For my new employees, it turned out to be a hard and all too effective education.

The system we'd put in place was designed to test both how well our newly renovated lab facilities would work and how efficient our data collection system would prove. We selected the two local hospitals that Karl Johnson and I had visited, Nixon Memorial Hospital in Segbwema and the mission hospital in Panguma, to evaluate patients for exposure to Lassa. Both were mission hospitals, considered to be the best in the region, and both were located in towns that were hotbeds of Lassa. Our hope was to better understand the variety of signs and symptoms associated with Lassa fever. Our staff, trained to take a complete history from patients, was told to record the presence of headache, muscle pains, sore throats, and so on. They made sure that the doctor measured the fever and looked for bloodshot eyes and bleeding from the gums. They checked urine for protein and blood. They took blood samples, which they brought back to my newly painted and electrified lab, so that we could test for antibodies with the reagents (substances that demonstrate the presence of viruses and bacteria by producing an observable reaction) I had brought from CDC. We went through the same processes again and again, separating the yellow serum from the blood in the centrifuge, then mixing it with the reagents. Then we looked at what we had under the microscope.

Positive, positive, positive.

It looked like we'd hit the jackpot. Everyone on the staff was on their toes. Not only had they figured out that they'd better do a good job because their future with the project was on the line, but they were getting involved in the excitement of the work itself. Beyond that, they realized that they were really beginning to do something that mattered. As we started to dig deeper into the mine of Lassa-fever stories, we came across tragedy after tragedy. Lost mothers, lost fathers, lost children. There was so much for us to do.

Most of Panguma Hospital's Lassa patients came from the diamond mines. The nuns and lay hospital staff were dedicated, but, with Lassa fever, dedication was hardly enough. They freely bedded patients with infectious diseases beside those who were noninfectious. They had no special place to discard infected needles. There was hardly any disinfectant to be found. No one bothered to wear gloves, even when handling dangerous specimens. And the wards were full of Lassa fever patients. We set about persuading the hospital staff that it would be prudent to supplement faith with a little freshly made up bleach.

With the sisters' cooperation, we set about establishing a system of barrier nursing. There is no point in going into a remote part of Africa to talk about airflow apparatus and respirators. Any airflow you are going to get will have to come from an open window. In a culture rich with ritual masks, respirators are likely to get you labeled as an evil spirit, with very nasty consequences. Even the Irish nuns had difficulty understanding the importance of our advice. They were so overwhelmed with problems of malaria, dying babies, diarrhea, and anemia, that they thought Lassa fever was largely irrelevant. Not so, we told them. We found that about 30 percent of their adult medical admissions had Lassa fever, and that it was the most common cause of adult medical death! Besides, Panguma had been the site of a hospital outbreak, which had also killed members of the staff. But memories are short, and expatriate staff turnover high. Now, there weren't nearly enough separate rooms for individual Lassa patients, so we needed to find some way to place them in the open wards yet still keep the other patients and the staff safe. The solution was quite simple: We separated the beds of Lassa fever patients from the others by mounting cloth screens on wheeled metal frames, which could be moved to wherever they were needed. Of course, the screens couldn't stop any viruses, but their presence served to remind the patients and the staff that the disease is contagious: *Beyond this point, take care.*

We went a step further, placing a table at the foot of the bed of each Lassa fever patient. The table contained a supply of surgical masks and pairs of clean latex or polyvinyl gloves, as well as a bowl of bleach solution. The bleach was there to decontaminate all used gloves and masks. Ordinary household bleach was one of the very few supplies we could buy in the local market. Diluted to a fresh ten percent solution with water, it is an excellent disinfectant. After soaking, the gloves were then washed and put out in the sun to dry for reuse. The luxury of disposable gloves was way beyond these hospitals, but labor was cheap. So you employed a glove washer. Unless the gloves were punctured, they could be used up to eight or ten times. Our project sites could be easily identified by a clothesline with many pairs of gloves hung up to dry. Over the years, under many different circumstances, we have used this same simple system safely. It was eventually incorporated into WHO recommendations for handling viral hemorrhagic fevers in Africa, and, later, for those handling HIV in similar circumstances.

In contrast to Panguma Hospital, with its commodious, well-lit wards, the atmosphere in the older Nixon Memorial Hospital in Segbwema was gloomy and oppressive. The lighting was poor, the walls uniformly drab, the toilets didn't work well, and there was often a shortage of water. The stench of feces and urine pervaded the place. As at Panguma, wards were divided by sex and by whether a case was medical or surgical. There was also an obstetric and a pediatric ward. Although there were no specialists on staff, the hospital was the most important medical facility in the area, serving all of the villages within a twenty-mile radius. Like Panguma, Nixon had a single room attached to each of the male and female medical wards where infectious patients could be isolated; however, there were always far too many infectious cases than could be contained in a single room. Once again, simple barrier nursing came to the rescue.

Our project wasn't limited to monitoring the incidence of Lassa or instituting preventive measures. We were also trying to eliminate the source of the virus itself: the rodents. For this purpose, we assembled a rodent patrol. In addition to John Kande, there was mammologist John Krebs, who, under the supervision of a man I will call Adam Sherrington, came from the States under an NIH grant. The story goes that Adam's wife had run off with a hippie a few years earlier, which might have accounted for why he appeared to dislike John so much. John was the quintessential hippie. Balding on top, he compensated by growing a luxuriant beard. He was in search of adventure, eager to discover the secrets of a rodent species about which little was known. He thrived on fieldwork, venturing off on his own for several days at a time to trap rodents and then testing them to see whether they carried the virus. He was open to whatever a given culture had to offer—especially the local food, which he consumed heavily spiced. Evening meals—heartily deserved after a long day of rodent trapping—consisted of rice with a palm oil-, peanut-, or fish-based sauce. To this concoction was added lots of pele-pele, the local red peppers. However many pele-pele went into the mixture, John added more as he ate. A free spirit, John seemed to rub Adam the wrong way. Yet his work was key to understanding the dynamics of the rodent population, especially when it came to determining how they interacted with humans. From his research, he was able to piece together a plausible scenario for the transmission of Lassa virus among the local rodent populations. He became so well known for his activities that he was later called Doctor Arrata—Krio for "Rat Doctor."

For all his accomplishments, John had a difficult time convincing Adam, who was his supervisor, of the veracity of his data. Indeed, Adam's antipathy toward John evolved into a vendetta. He wrote memos to John, accusing him of incompetence and dishonesty. Since I worked with John on a daily basis, I knew that none of these accusations could possibly hold up. In fact, I've never known anyone to be more careful and compulsive about the accuracy of his data. I couldn't figure out what exactly was Adam's problem, until I learned that he'd failed to get his own research published. This, together with his simmering resentment against John for simply being John, made his motives comprehensible—though hardly excusable. There were, after all, human lives at stake.

Internal agendas were not the only obstacles the rodent squad faced. Local customs had to be taken into consideration as well. Before going in to lay down traps, the rodent team always made sure to first consult with the residents of a village and explain the purpose of the program. Typically, the role of spokesman would fall to John, who would usually take along the other John—Kande—to assist him. Kande's powers of persuasion convinced villagers that, while anyone who wanted to relieve them of a few rats might be crazy, they were more than welcome to do so.

There were three things we wanted to learn from this survey: What species of rodent was prevalent in the village; how many rodents were present in each household; and how many of them were actually carrying the Lassa virus. The team would begin by assigning a number to each house in the village and then draw up a careful map. Once this was done, a grid could be superimposed on the map, allowing John to select houses for study on a random basis. On the scheduled day, the team would appear in the village, arriving in an instantly identifiable yellow Isuzu rat truck emblazoned with a Lassa logo consisting of a rat on which a circular image of the Lassa virus had been superimposed. Within the virus image was a miniature map of Africa. Everyone for miles around got to know this logo. In later years, as we passed through villages, people would sing, "Lassa fiva no good-o." The lyric came from the popular song our public-education team had managed to get on the Sierra Leone Top Ten.

The team would find a place at the edge of the village to establish its camp, which consisted of the truck and a couple of tents—one for personal use, and the other to process the captured rats. The members of the team then set about laying the traps in the houses that they'd selected for the survey. Traps were of two types: one was designed to kill the rodent, and the other captured it alive so that it could be tested. Usually ten traps were placed in each house, depending on the number of rooms. Then, early the next morning, the team made the rounds of the houses to pick up the traps. The rodents found in the kill traps would be duly noted in the database and the carcasses placed in a bag for incineration. The live traps were also retrieved. These were cleverly designed so that they could be opened into a plastic bag. When the collector sprung the mechanism, the rat would leap out, under the misapprehension that it had made an escape. Instead, it would find itself in a plastic bag containing a ball of cotton soaked with an anesthetic. Once knocked out, the rat would be bled using a pipette, and the blood would be analyzed for presence of antibodies to Lassa virus and for the virus itself. Once the rat's species was determined, it was killed and dissected, its organs saved for possible virus isolation. All the information we learned from these tests was then placed in a computerized data bank.

To understand just how the rats were transmitting the virus, we needed to conduct another study, which required us to trap rodents found in the homes of people who had been infected with Lassa virus. To do this, we also needed to compare rodents trapped in homes where no one had become ill. One of the things we wanted to learn was whether trapping rodents cut down on the number of new infections in the house. This particular study was conducted by Dick Keenleyside, an English physician working with CDC. A competent researcher, he was tortured by a singular handicap: paralyzing fear of contracting Lassa fever. Every evening, he would return to the lab complaining about having a sore throat. It is not that such a fear is entirely irrational; he was, after all, working closely with the rodent team. But neither rat catchers nor virus hunters can afford the luxury of phobia. In any case, he had a rather hilarious experience going through the London Airport on his way to the States.

He happened to mention to someone on the flight that he'd been working with a Lassa fever project in Sierra Leone. Apparently, he talked to the wrong person. No sooner did he walk into customs than he was apprehended by the ever-vigilant, somewhat paranoid, and usually clueless officers of the British immigration service. They decided that he should be quarantined in Coppett's Wood. This old fever hospital in northeast London had been equipped with a plastic isolator, where patients thought to have a contagious, exotic infection could be quarantined. This was an enormous, unwieldy plastic bag, which formed an airtight cage over the bed. In Dick's case, an isolator of this type was quite unnecessary, since Lassa—and, indeed, most hemorrhagic fevers—isn't spread through the air. But the British, having gone to such pains to install such an isolator, and at considerable cost, were anxious to use it. As a result, a number of unfortunate passengers alighting in London from Africa with a headache and a touch of fever found themselves imprisoned in this way.

Dick protested vigorously and pleaded with the immigration officials to get in contact with CDC. He pointed out that he wasn't even sick. Even threatening to secure a writ of habeus corpus was to no avail. Habeus corpus, which dates back to the twelfth century and the Magna Carta, is intended to prevent the king (or his agents) from detaining anyone against his will without just cause. It is one of the sacred laws of Britain—but apparently not for British quarantine officials. Only after representatives from the quarantine division at CDC were able to convince the British quarantine officials that Dick posed no risk to the British Isles, was he allowed to go on his way. He was lucky. It wasn't only Africans who exhibited superstition and "unreasonable" behavior in the face of a viral disease.

Not all of our problems came from Lassa. Politics touched our lives and work, as it does wherever we go. Most of the political tensions in Sierra Leone stemmed from a perennial feud between the two largest tribes, the Mende, located in the Eastern and Southern provinces, and the Temne, located in the Northern and Western provinces. The two dominant political parties had their power bases in these tribes, so that what appeared on the surface to be a political dispute was actually a tribal conflict. Free elections were supposed to be held in late 1976 and early 1977, but the leader of the All People's Congress (the Temne-based party), Siaka Stevens, managed to rig the elections and declare himself the winner. He then proceeded to depose the sitting president. This coup precipitated a series of skirmishes in various parts of the country, some of them turning into pitched battles. Several of these clashes occurred in Kenema and Bo, the capitals of the Eastern and Southern provinces.

In Kenema, which had at that time a population of about 25,000, one battle claimed twelve lives. Kenema was in Mende country. A curfew imposed on the Eastern Province directly affected us. We were not allowed to be outside of the town after 6 P.M. Not surprisingly, the curfew succeeded only in arousing even more animosity toward the Temne-dominated government, particularly in the provinces loyal to the opposition party, and for us the curfew was a particular hardship. It was imposed just after we'd begun our project, and it went on for four or five months. Because we were just beginning to become known in the province, we were often harassed at makeshift checkpoints by uniformed men.

Some were military, some were from paramilitary units, and some had no special affiliation at all. It wasn't unusual to encounter a well-armed and quite drunk soldier who was more interested in how much money he might extort from the driver and passengers than in making any legal or political point. With the breakdown of law and order in the area, anyone with a gun took the opportunity to exploit the situation for his own advantage.

The curfew meant that, whenever I traveled to either Nixon or Panguma hospitals, I had to make sure to be back in Kenema by 6 P.M. Since the hospitals were located twenty-five miles away and I always left at the last possible moment, this made for some hairy rides over washboard, potholed, and gullied roads. Even such relatively short distances could take one or two hours by car. We were always worried that a random soldier might have had enough afternoon palm wine to prompt him to shoot at us. While we were fortunate that no one in our project was ever assaulted, several members of our staff were taken into custody from time to time. This didn't mean that they'd done anything wrong. Arrests could occur on trivial pretexts or for no reason at all. Someone might be mistaken for someone else, or a person might vent his frustration at being stopped and provoke an angry response. When this happened, we had to find the right people to negotiate with in order to secure their release.

The political stalemate was finally resolved—violently. After a furious battle between the Temne and Mende forces in Bo, which left over three hundred dead, most of them Mende, a compromise was worked out. Since the president already came from a minority tribe allied with the Temne, it was agreed that the two vice presidents would be a Mende and a Temne. Once the settlement was reached, the curfew was lifted. But just because the political crisis had come to an end didn't mean that the roadblocks disappeared. In fact, they remained a source of harassment throughout the life of the project. You could never tell where you'd encounter a roadblock, which had become ingrained in local custom by then. They varied in place and timing. Maybe some local commandant had decided on erecting one that day, maybe it was the whim of some individual who didn't see why he should consult with anyone. Actually, setting up roadblocks became something of a national sport, open to people of all ages. You'd often come upon a ragtag group of children gathered around a makeshift roadblock that consisted of nothing more than a hole dug in the middle of the road. The dirt that they'd shoveled out would be lying in a heap nearby. When you stopped, the children would march up to your vehicle and ask for money to repair the road. And that was one way Sierra Leone educated its young.

CHAPTER TEN

Of Souls and Centrifuges

Before we could find out how much Lassa fever there really was in the area of the villages, we had to have an accurate census. You have to know how many people you started with before you can work out estimates of how many have been infected in a given population. In this way, we could then proceed to studies aimed at discovering who was most at risk for infection, and why. Normally, you could find this information by simply consulting the national census data, but in eastern Sierra Leone, no one had taken a census since the British had been there nearly twenty years before. So we had to do our own. This meant going from house to house in our chosen villages to ascertain the number of occupants, their ages and sex. Most of the houses consisted of three or four rooms with mud walls and floors, and roofs of tin, which had replaced the traditional reed or palm leaf thatch. Corrugated metal is one of white man's many dubious gifts to Africa. With worsening economic conditions, however, even corrugated tin has become a strain on the budget of rural Africans. So, nowadays, you see the traditional thatches coming back.

We *thought* we were prepared to find that each of these houses sheltered a good number of people. We were not prepared for what we actually found. In villages close to active diamond mines, houses were packed with as many as forty to fifty occupants. The only reason that there was any space at all for their bodies was the fact that miners worked in shifts: when one group was at the mines. the others were sleeping, wall-to-wall. With the miners changing shifts about every twelve hours, the houses were occupied twenty-four hours a day. This was very different from the living arrangements of most subsistence farmers, who left their houses empty during the day while they labored in the fields. Not surprisingly, these distinct patterns of behavior made a difference as far as the rates of infection with Lassa fever were concerned. Among the mining villages we saw some of the highest rates of infection we'd yet observed.

Just how high? That was difficult to say. Surveying the inhabitants of these mining villages was a sociological nightmare. Every time one miner left to return to his village, his dreams of sudden wealth shattered by the hard reality of experience, there would be someone with a new dream of his own to take his place. In some of our surveys, we found that the population of a village could double—or halve—all within a few weeks. When several large diamonds were found in one small village near Segbwema, its population of 2,500 doubled overnight. And in their desperation to be the first to discover yet more diamonds, the miners dug so fast into the loose soil that one of the pits collapsed. Between fifteen and twenty people probably died, with the exact death toll never determined because not all of the bodies were recovered. Moreover, since many of the miners were illegal immigrants, there were no families to inquire as to what had happened to them. In the aftermath of the accident, the village emptied out again. People were convinced that an evil spell had been cast over the village.

It was one thing to survey the incidence of Lassa, but what to do about the people who came down with it? That was a problem we encountered daily. As we recorded patient after patient, all beseeching us for help, we became more and more depressed by our inability to do much of anything.

This is when we came across John Kamara, the beloved school teacher from Segbwema. We watched him and gave him the usual therapy: drugs to reduce his fever and fluids to keep him from getting dehydrated. He was in intense pain; he couldn't swallow, he couldn't find a comfortable position, and he couldn't sleep. By the eighth day of his illness, he was developing swelling in his neck and face. This is an incredible sight. The face of the victim swells up until it is almost unrecognizable due to the pouring of fluid (edema) from the damaged blood vessels into the soft tissues of the face and neck. While he was still able to answer questions, his responses were often inappropriate. We knew he was becoming encephalopathic, which means his brain was being affected. We know now that this is a terrible sign, and invariably heralds convulsions and death. Back then, however, we were less certain about the inescapable gravity of the symptom.

Throughout his illness, his wife never left his side. Although we couldn't be certain that she hadn't already been infected herself, we made sure that she wore a gown and gloves just as our staff did. Naturally, she was distraught. We tried to reassure her, but we could not begin to calm her. She saw the expression on her husband's face. She understood what was happening. She saw the fear in his eyes. She appealed for us to do something, anything. As far as we knew, there was only one, possibly effective, treatment. Maybe it would work on John Kamara.

Plasma treatment was the reason for the large centrifuges I had scavenged at CDC and went to the trouble of transporting first to Sierra Leone, then up-country on the long bumpy road. Plasma therapy involved taking blood from people who had recovered from Lassa fever, separating the red cells, and giving them back to the donor. Then we would collect the plasma containing the anti-bodies and infuse it into the veins of people in the grip of acute Lassa fever. This immune serum treatment was based on the success of plasma therapy on vic-tims of another viral hemorrhagic fever, this time from South America, called Junin. Junin virus, which is found only in Argentina, is also carried by rats. Mortality is high, and the disease is very similar to Lassa fever. Most important of all, Junin is an arenavirus, which means it is a close relative of Lassa virus. If the therapy works with Junin, we thought, why not with Lassa?

We confronted a serious problem in separating the plasma, though. We had the centrifuges, but when we tried them, we discovered that they were not prop-erly wired. I kicked myself. I hadn't had the time to check the wiring before I left Atlanta—why would I?—and I hadn't gotten around to it since. Now, where was I going to find someone in Sierra Leone who would even know what a blood separation centrifuge was, much less be able to repair one? I had no idea that we were going to encounter so many patients with Lassa fever in the first month of our hospital surveillance. We were just trying to get our feet wet, and suddenly we were being swept over a waterfall.

So I found myself in a terrible predicament: no plasma and nothing to offer our patients for treatment. Did I have to stand there helplessly and watch John Kamara die?

This personal tragedy galvanized me into action. If I was going to make any impact, I not only had to do something, but I would have to do it in closer con-tact with the patients. The reason I'd established the project in Kenema was because it was equidistant from the two major hospitals and was accessible to the road from Freetown—not because I thought it had many cases of Lassa. So I would have to figure out some way for the laboratories at Panguma and Nixon hospitals also to function as Lassa fever labs. This way, I'd be sure that at least some of the basic lab work was done on the spot. But before I could do any-thing at Panguma or Nixon, I had to get those centrifuges working.

I ripped into them and soon discovered that whoever had rewired them for 220 volts had neglected to include the timer in the circuit. This was a problem in a system designed so that it wouldn't start spinning until the timer was turned on. No timer, no action. To rectify the situation, I had to overhaul the centrifuges so that the main current now went through the timer and then flowed into the motors. It took me a day to figure out the problem and then another day to get the machines repaired. There was no electrician to call. *I* was the electrician.

The next ordeal was to transfer two of these old, obsolete, but now highly functional, centrifuges to the two hospitals. Each one weighed about six hundred pounds, and we had no vehicles in Kenema capable of carrying them. Moreover, they were refrigerated centrifuges, so we had to be certain that the compressors were anchored down sufficiently to prevent the copper pipes connecting them to the cooling coils from breaking. If the pipes broke, the fluorocarbons would leak, not only contaminating the atmosphere, but rendering the refrigeration unit non-functional. So we searched until we found trucks sturdy enough to carry our precious cargo over the washboard roads. With the expenditure of a lot of muscle power, we succeeded in heaving them onto the trucks. Then we simply lashed them in place with rope, anchoring the compressors as best we could, drove them to Panguma and Nixon—and hoped.

Finally, we got the machines to their destination, but our problems still weren't over. The centrifuge we placed in Panguma worked perfectly, but the one at Nixon wouldn't cool. This was my worst nightmare. There were no reliable refrigeration people around Segbwema, and if there had been, they would have had no experiences with centrifuge refrigeration.

I had to do some hard thinking. Before I left CDC, I'd been given instruction by a refrigeration specialist on how to repair a refrigerator and refill it with coolant. Hoping that the process would be similar for the refrigerator of a centrifuge, I pulled the machine apart again, only to discover that the copper tubing from the compressor was broken at the entry point to the cooling coils. I had to cut it off, reflange it, and place it back into the proper position among the coils. Then I had to hook up the vacuum pump we used in the laboratory and press it into service to expel the air out of the compressor and create a vacuum. After several hours, I'd gotten the system at a sufficiently low pressure to connect up the can of coolant and recharge the coils. It all *seemed* to work, but the real test was to find out whether the temperature would drop low enough to cool the centrifuge. I waited. When I saw frost forming on the inside of the tub, I was elated. Now I could start. The next question was: Would the plasma work?

Even though I had the centrifuges on line, I had other hurdles to overcome. We still needed a line of supply and communication. We had no functional telephone system, and the electrical system was so erratic that we had to depend on a backup generator. If we wanted electricity, we generated it. If we wanted to communicate, we cranked up our single sideband radio. Panguma and Segbwema laboratories could only depend upon a few hours of electricity each morning from the hospital generator. The supply was limited, and the operating theater had priority, so the labs often had to do without even this minimal supply. At that, the generators had a tendency to fall apart, and spare parts were not exactly easy to come by. By scrounging, I finally managed to scrape together enough money to acquire a small locally available Honda generator for each of our labs. Independent power! There's really nothing like it.

Establishing a supply line was more difficult. Ordinarily, we relied on the assistance of the U.S. Embassy in Freetown to facilitate the shipment of supplies and equipment from CDC and to make sure that it got through customs and was transported safely up-country. Among the most critical items were the reagents for diagnosing Lassa fever. This is where we really ran into problems. The very first shipment from CDC went through Dakar, Senegal, on Pan American Airlines. Somewhere in Dakar it vanished. A bottomless pit. Though it was the only shipment that we ever lost, it taught us a lesson. We needed a more reliable route, particularly if we were to examine the virus. We contacted KLM, who had a regular Thursday-night flight into Lungi Airport, just up-coast from Freetown, out among the mangrove swamps, across the estuary from Freetown. The airline agreed to transport material by air freight via Amsterdam directly from Atlanta, where they also ran a daily service, and to ship back specimens from Freetown to Atlanta, via the same route. There was no way we could perform virus isolation in the field. The process requires tissue culture, which demands, in turn, sterile working conditions. It is one thing to handle clinical specimens with virus on the open bench, but it is quite another to grow it up to high concentration in tissue culture. After all, this virus is a Level-4 bug and can only be safely handled in a suit lab. It could only be done at CDC.

When it comes to studying a virus, the ultimate assay or test is always virus isolation. The assay doesn't just tell you whether a virus is present in a sample of blood or tissue, it tells you how much of it there is. If we were going to map and document the spread of a disease and later treat it, we had to have virus isolation, and not just on one specimen from each patient, but on several specimens from each patient. It got still more complicated. We couldn't just freeze the specimens in a regular freezer and ship them in regular ice packs. We needed dry ice, if the specimens were to be preserved so that they would make it all the way to Atlanta. There wasn't any dry ice in Kenema; there wasn't even any in Freetown. The closest source was Abidjan, Ivory Coast—two doors down the West Coast of Africa. We tried to set up a relationship with the U.S. Embassy in Abidjan, which made a stab at helping out. But the first two boxes they sent of what was supposed to be dry ice arrived empty. The dry ice had all sublimated. Nothing there but lingering carbon dioxide gas. Clearly, the Abidjan connection wasn't going to work.

So we tried our ever-efficient Dutch friends at KLM again. They told us that they could deliver the dry ice from Amsterdam on a Thursday evening if we ordered it in advance. That would allow us time to pack our specimens and have them ready for shipment back to Atlanta (via Amsterdam and Chicago) when the plane returned to Lungi the following morning. All that was left was to figure out a system, with the help of the embassy, to get the packages over the ferry to Freetown and back again before everything melted. We had to make certain that we reached Freetown on time, so that we wouldn't miss the plane, and that made for some very hectic Thursday nights. Eventually, we kept large freezers in the basement of the embassy. Here we could lodge our precious specimens, secure in the knowledge that we were hooked up to the massive embassy generator. Hosting Lassa virus in this way did not seem to faze our State Department friends. By Friday afternoon, as KLM took off with our boxes, we'd all be collapsed on the Freetown beach, one of the loveliest on the continent.

So much of our operation could never have worked without the intercession of the U.S. State Department. I have worked with U.S. embassies and consulates in at least fifteen countries on four continents, and I have always been able to count on their cooperation. Accompanied by a Belgian friend, I recall going to the American embassy in Kinshasa once to cash a check He was astounded that such a thing was possible. He said that there was no way he could obtain such a service from his own mission.

No matter what we did, no matter how many problems we were finally able to surmount, it wasn't fast enough for John Kamara. His high fevers and severe body pains continued unabated. We still had hope—hope born of our scant experience with Lassa fever. But by the eighth day of his illness, his eyes, once bright and piercing, had become dull, filled (it seemed to me) with resignation. And when I looked closely, I could see flaming red spots in the whites.

Now he was bleeding into his eyes.

John's wife couldn't take it anymore. She kept getting up and running from the bedside, only to return a few minutes later to hold his hand and try to soothe his pains.

By the ninth day of his illness, John Kamara began to alternate between periods of mere disorientation and outright delirium. While we still hadn't had much experience with this disease, we doubted that many people could survive this stage. We had to acknowledge that we were losing him. His wife seemed to share our sense of foreboding. She began to grow strangely calm, now that she'd accepted the inevitable. It was one of those times in medicine that humble all physicians. It reminds us: we make feeble gods.

John fell into a deep coma. Soon, his body was wracked by a series of seizures as the virus proclaimed its victory over his brain. Next came intractable shock, with unrecordable blood pressure. And then his heart and lungs failed.

As testimony to his legacy, all of Segbwema turned out to mourn him, their highly regarded teacher and role model. They wanted to know what kind of world it was that would rob them of such a good man. How would his wife cope alone, with two little children to raise? Who would look after his aged parents? And who would teach his students? Who would give them counsel and help them carve out a future for their country? All around me, the only response was silence.

CHAPTER ELEVEN

Magic Bullets

The ongoing struggle to get equipment and materials from Atlanta was beginning to pay off, and I was reasonably assured of a continuous supply of reagents. Now we could get nearer our goal of providing immune serum treatment for patients. The first job was to begin to survey patients who had survived Lassa fever in either Panguma or Segbwema hospital, and who might now be expected to have high antibody titers to Lassa virus. It was the individuals with the highest titers we were most interested in, because we believed their plasma would be the most effective. We also had to find people who had recovered sufficiently to be able to withstand the plasma donation without injury to their health. Of course, we had to hope that they wouldn't still have Lassa virus circulating in their blood. To be reasonably certain that the plasma was safe, we tried to choose only those who'd been over their illness for at least three or four months. Lassa wasn't the only virus we were worried about. We also screened the blood for hepatitis B the fatal hepatitis, transmitted through blood serum.

This was the late 1970s. We knew nothing about HIV or AIDS in those days, nor did we have any idea that there were other viruses in addition to hepatitis B that could be spread by blood. We knew nothing then about hepatitis C. If we had, we would have been less enthusiastic about using plasma from patients in this way. We can take some comfort from the fact that, later, when we started to look for HIV infection in Sierra Leone in the late 1980s, we couldn't find any, with the exception of a couple of prostitutes in Freetown and a few students who'd returned from the Soviet Union. It is likely that at that time AIDS was just moving into Sierra Leone from Central Africa. It is pretty reasonable to conclude that there was little or no HIV infection in the Eastern Province in the 1970s.

Things are different now. But if serum therapy was a risky business, well, so was Lassa fever. We watched John Kamara and eight other patients die in the first month of our operation. We had to find a way to do something.

Collecting the plasma was a logistical nightmare. First we needed to round up the prospective donors, which wasn't an easy job, since they'd all gone back to their villages. If they were itinerant diamond miners, we might never find them. Once we'd found the person we were seeking, and then persuaded him to become a donor, we then had to provide him with transportation and food for the day. As an added inducement, we also gave donors a two-week supply of iron supplements and a two-pound bag of rice. The whole procedure often took the field team a complete day for each donation. It didn't take us long to realize that a program like this could run into a good deal of money.

Even explaining to the potential donor what a blood donation was and why we needed it took some finesse. The whole concept was so far outside the education and experience of the people, that they just didn't know how to interpret it. Most of the patients were illiterate villagers whose comprehension of medicine was limited to the powders and potions the medicine man gave them. If that didn't work, you found a hospital with a white doctor, and instead of medicine-man magic, you received white-doctor magic. We had to develop some fairly simple explanations—another cultural hurdle: a belief in native medicine and magic lurks just beneath the surface, no matter where you go in Africa. Many people believed that blood possessed magical powers and that it could be used against them. At one stage, whispers went round that "White men drink blood." There was even talk of cannibalism, and rumors that we must be making magic potions to defeat our enemies. We were either lucky or our staff silver tongued. Against all odds, we managed to prevail upon about half of the former patients to agree to participate.

Once we separated the plasma from their blood, we transfused the packed red blood cells back to the donors. That was where the real surprise came, since they couldn't tell the difference between what we had taken out and what we put back in. Now they were often completely convinced that we were recruiting them for some kind of witchcraft. I'm not sure whether that reinforced their suspicion that we were tricking them with bad spells, or whether we'd gained their respect for superior magical powers. Probably a bit of both.

It wasn't just the cost of obtaining the plasma that made it so demanding on our collective resources. The whole procedure took enormous amounts of time. We had nothing on which to build. No blood banks—zero—existed in Sierra Leone. Blood was given as needed and usually taken from a family member or a friend, and even then it didn't come free. So how could the kind of system we were developing with plasma take root, especially when there was no obvious source of funds to pay for it? Just introducing an expensive treatment for a few patients and then leaving the rest with no way of getting hold of it wasn't our objective. Optimally, treatment had to be practical, feasible, and cheap. True, we couldn't see how plasma could be employed for widespread use, no matter in what form it was finally presented for infusion. But it was all we had. We pressed on, determined to find something that worked.

What we wanted was a simple drug—the penicillin of Lassa fever. But drugs that can be used safely against viruses are very few indeed. Most antiviral compounds are highly toxic and often have the nasty disadvantage of killing host cells as well as the virus. Moreover, the treatment needed to be cheap, safe, and easy to give to patients. In addition, it had to be very stable for a long period of time, so that it could be carried into remote areas and stored under less than ideal conditions until needed. A wonder drug—we were after nothing less.

Then a letter arrived from Karl Johnson. Karl had a way of dropping terse cables and letters on me that contained in a sentence or two something that would change my life. First Ebola. Now this.

Karl's letter mentioned a relatively new drug that had been developed at the University of Utah. It was called ribavirin, and it had been shown to have some effect against certain RNA viruses. Lassa is an RNA virus. If DNA is the blueprint of life, the genetic material that tells cells to become eyes or hands or sprout stems and leaves, RNA is the messenger, carrying information from the gene to the ribosome, a small round particle that is a kind of protein factory, producing antibodies, hormones, and enzymes that build life. However, quite a number of viruses use RNA not as a messenger but as their very genetic material. Viruses are unique organisms in this respect. What made ribavirin so interesting, so promising, was that it seemed to interfere with the way the virus makes protein from its own genetic RNA.

Arenaviruses are RNA viruses. Lassa is an arenavirus. Could ribavirin work against Lassa virus?

Karl told me that he was already testing the drug against Lassa virus in tissue cultures, adding that similar experiments were being conducted by Peter Jahrling at USAMRIID. Peter was infecting monkeys with Lassa virus and treating them with ribavirin. There was good safety data on the drug, including human use data, because it had already been used successfully as a treatment for acute viral pneumonia in infants. Best of all, it was easy to make, potentially cheap, and stable at room temperature—maybe not Sierra Leone room temperature, but we could put it in the fridge.

This could be my magic bullet. I was very excited, but knew we would need to wait for the results of the laboratory studies with Lassa virus before we could consider launching a treatment study of our own. Not that we were about to let the grass grow under our feet while we waited to hear how Peter's monkeys fared. We would need permission to use both the drug and plasma from regulatory boards in Sierra Leone and the U.S. This was going to take time, so we needed to get moving immediately.

The first step was to prepare a written protocol detailing the treatment trial. In Atlanta, Karl and his wife, Patricia Webb, a CDC Special Pathogens epidemiologist, shared the burden of this work. The protocol would then have to be approved by the National Ethical Committee of Sierra Leone, the Human Subjects Review Committee (HSRC) of the CDC, and the FDA. This is no trivial process. The protocol contained a comprehensive, detailed scientific justification for the project, as well as a clear picture of how the study would be conducted. The protocol included a precise description of the procedures for gaining the informed consent of the patients to be treated in the study. Absolutely no guinea pigs.

This was where things got sticky. Informed consent can only be given by a participant who fully understands what he is agreeing to. The majority of patients we were dealing with were illiterate, completely uneducated, and most spoke only a local tribal language. Half of them harbored suspicions, anyway, that we might be tangling with witchcraft, though they would be far too polite to say so. How do you go about explaining a complex medical procedure to such an individual? Translate into Mende? Okay. But tribal languages such as Mende have no words to describe concepts such as "clinical trial," or "adverse reactions." So how could we adequately convey what we were trying to tell them?

We simplified the concepts and used words that would somehow get the meaning across, even if an exact translation could not be made into their local tongue. In simple terms, we needed the prospective subjects to know that ribavirin was a new medicine, as yet untried for Lassa fever. At least we could assure them that it had proven safe in human beings. We would also tell them that there was no treatment for Lassa fever, and that this was only an attempt to find one. We could guarantee nothing. To explain all this, we had to depend on the skills of our interpreters. Since most of the patients were unable even to sign their name, the best we could do was to ask them to place a thumbprint and an X at the bottom of the informed consent document, signifying that it had been read to them, that they had some idea of what we were doing, and that they agreed to let us give it a try.

The National Ethical Committee of Sierra Leone had less of a problem going along with the informed consent procedures we'd devised than the HSRC. Sierra Leoneans knew their country, after all, and knew what could and could not be done. They also understood the desperate need of the Lassa fever patients. However, this high degree of cooperation might be better understood if I first mention that the National Ethical Committee of Sierra Leone was created by Dr. Marcella Davies, chief medical officer in the Sierra Leone Ministry of Health at that time. Her successor, Dr. Belmont Williams, chaired the committee when it ultimately approved our protocol. The Human Subjects Review Committee of CDC required the approval of a local ethical committee, never mind that no such committee existed in Sierra Leone. Fortunately, Dr. Belmont Williams was enlightened and well organized. She was a Krio—a descendant of slaves who had escaped from the United States and tried first to settle in Nova Scotia but found it too cold for their African blood. They eventually made their way back to Sierra Leone, settling in Freetown. They are the most educated and advanced group in the country. It is their pidgin, developed over the subsequent hundred or so years, that has become the Lingua Franca of the country. Dr. Williams received her medical education in Bristol, a very good medical school in the west of England. Not only was she competent and effective, she was also a great supporter of our project. At our request, she called together a group of respected physicians and designated them as the National Ethical Committee. She gave them our protocol to read and asked them to provide their comments. While a few had questions, they took no serious issue with the protocol.

Nevertheless, the fact that there were so few people in the country with any scientific training placed an added burden on us—and on the HSRC. We had to make certain that our procedures were not only correct, but that they did not do anything to exploit an illiterate people. There couldn't be *any* perception that we were trying to take advantage of them. This issue remains one or the more nettlesome in medical research, particularly in developing countries: How can you ever adequately inform a poorly educated population about the purpose and risks of a study? There is no easy solution.

The National Ethical Committee agreed that our procedure for obtaining informed consent was the only workable one possible under the circumstances. HSRC, however, had difficulty in coping with the idea of illiterate patients in a clinical trial. None of the members of HSRC had worked in any place remotely like Sierra Leone, and they were only vaguely aware of where it was. Their own ignorance hampered them, as they tried to judge issues on the basis of precedent established from experience in well-run American hospitals. The problem has come up time and again with HIV, especially in Africa and Asia, where, for instance, local physicians frequently withhold from patients a diagnosis of AIDS or HIV infection, arguing that there is no remedy that the patient can afford and that his family may well abandon him if the truth becomes known.

Having at length secured HSRC approval, we had to tackle the FDA. Because the use of ribavirin as a possible treatment for Lassa fever was brand new, we needed an Investigational New Drug (IND) permit from the agency. We expected little difficulty obtaining this, because oral ribavirin had already been extensively tested and found safe in a number of human trials mounted to study its effect on other diseases. We would rely on existing data compiled by the pharmaceutical company instead of trying to conduct our own prohibitively lengthy and costly tests. The FDA is not only responsible for giving the go-ahead for use of drugs in the U.S., but it also serves the same de facto function in many parts of the world. Many countries that do not have institutions or sufficient expertise of their own rely on FDA recommendations and decisions.

By November 1978, we had the necessary permissions in hand and satisfactory laboratory data (Karl's tissue cultures, Peter's monkeys, and the FDA tests) to show that ribavirin was potentially effective against Lassa virus with few or no toxic effects. We also had plasma ready for use. At this point, we had been operating for eighteen months and had learned some fundamental things about Lassa fever. We knew that Lassa accounted for about 10 to 15 percent of all hospital admissions in the region, and that, overall, mortality in hospitals hovered in the range of 16 percent. We'd also learned to better predict who would likely live and who would die. Patients admitted with certain elevated liver enzymes (AST) were at greater risk than those with lower levels.

So we began. Our first task was to establish a system to determine who would receive oral ribavirin and who would get the plasma according to the criteria laid out in our protocol. Everybody whose liver enzymes were above a certain critical level, which had been determined in our initial studies as indicative of a poor outcome, would receive one or the other treatment. We'd already determined that Lassa fever was so terrible that we couldn't leave people untreated if we had anything at all to offer. At this point, we had no idea which treatment would work better, or even if either would work at all. Usually, clinical trials required a "control group"—patients who would be given a placebo (a solution that looks like the drug, but is actually nothing more than sugar water)—in order to "blind" the researchers, so that there is no way for their enthusiasm or prejudice to influence the patient or affect the results. The control group provides a neutral index against which to measure the effects of the actual drug. The trouble is that half the patients go untreated in a controlled trial. With Lassa fever, we knew that many untreated patients would die. Because the laboratory data looked promising, our committee had determined that we should not use an untreated group in our study. We determined, therefore, to compare the results in our new patients with data from previous untreated patients.

Without a control group, we had to be imaginative. We really didn't know what plasma would do, or whether ribavirin would have any impact at all. After all, no one had ever succeeded in treating a fulminating viral infection by directly killing the virus in the patient. So we divided patients randomly into two groups. One lot got immune plasma from our precious stocks, and the other lot got ribavirin, our candidate for a wonder drug. We started with two units of immune plasma infused in an intravenous drip for the one group and ribavirin capsules for the other. The nurses had to stay with the ribavirin patients until they were sure the tablet had gone down and stayed down. Sometimes we had to open the capsules up and dissolve the contents in water for patients whose throats were so raw that it was impossible to swallow the capsule. If the patient vomited, we tried to give another dose. Children under the age of fifteen were excluded from the study. (We later did a separate study in children.) And we excluded pregnant women, because we feared we might harm the fetus.

It was a gamble for the highest possible stakes. But our patients were often touching in their expressions of relief and gratitude. At last we had something to try. Me? I held my breath. Disappointment—at least with the plasma.

The patients continued to pour in, and we had no problems getting them or their families to agree to treatment. They came to us in various stages of disease, but often so sick that they were incapable of giving their consent themselves. We turned to relatives. We sent home no one who wanted treatment, regardless of what stage of the disease they were in. Some arrived with raging fevers, racking headache, crucifying body aches, and throats aflame, with exudates of pus clearly visible on the tonsils. These were symptoms of the early stages. Others had already started vomiting uncontrollably, were becoming confused, and were developing tremors of the hand and tongue. Others were bleeding from the gums and the gut. They vomited blood, and they bled from the rectum or vagina. Some arrived with a grotesquely swollen head and neck, the characteristics of severe Lassa, together with plummeting blood pressure.

But these weren't the worst cases. The worst were those with seizures and those who had slipped into coma. Almost everyone who progressed beyond the early stages to vomiting, confusion, and bleeding died. In our experience, no one who developed seizures ever survived.

We continued our experimental plasma treatment study for nearly two years. We slipped the IV needles into arms of patient after patient, infusing our precious immune plasma. Still, the patients died. The staff became discouraged. We had worked so hard, with such high hopes, and now we began to conclude that we could have done nothing for John Kamara, even if we'd gotten the plasma prepared in time to give to him.

But this was a clinical trial. You are not supposed to prejudge a clinical trial. You just do it. When you have finished, then you can look at the results and decide if you were successful.

And there was still the ribavirin group. To me, it seemed that these patients were doing better. Yes, many were still dying, but a few patients we expected to die actually improved. Was this really because of the ribavirin, or was it just luck? The only way to find out was to analyze the data from the study. And in the days before personal computers, that could only be done at CDC. We sent all our data to them, along with the virus isolation specimens. As soon as I returned to Atlanta in 1979, having lived with the project for three years, I set about trying to analyze the results. First, I had to learn the new computer system that had been installed at CDC while I was away. After struggling with data and computer gurus, I finally understood what I was doing.

The result? More disappointment.

The first analysis suggested that neither treatment was effective. Looked at as cold, hard numbers, even the ribavirin seemed to have had little effect.

Yet I could not let go so easily. The more I thought about it, the more I began to wonder whether there might not be another way of looking at the results. I went back and reanalyzed the data. This time, I decided, we would take a different tack. I started to break down the patients into two basic categories—those who were in the early stages of illness on the day we began treatment, and those who were in the late stages. In our first analysis, we took no account of the timing of the admission: When did the patient become ill, and when did he actually go to the hospital for help? Now I took into account how much time had passed from the onset of illness to the day we'd started ribavirin.

No matter how we looked at the data for immune plasma, the result was the same. In every case, the plasma failed to work. It didn't matter how early in the disease we treated, the patients continued to die at the same rate as before. But with the ribavirin I detected a glimmer of success, the faint glow, perhaps, of at least a fraction of the miracle we sought. If a patient was admitted in the first six or seven days of his illness, ribavirin improved prospects for survival. If the patient had been sick for more than a week, the capsules had less effect. We were on to something.

But the numbers were small, and the difference was not great. We needed to try again. There were still questions. Would administering a higher dose of the drug earlier in the disease improve the outcome? By delivering it intravenously, we could get higher blood levels of the drug and get more of it to the places where the virus was hiding, such as the liver and the spleen. Would the intravenous drug prove more powerful?

Yes. I wanted to use intravenous ribavirin. There were a couple of problems, though. We would have to go through the whole approval process all over again. There was also a question of how we were to go about obtaining enough intravenous ribavirin, since it was produced nowhere on earth except Mexico. And the FDA was not about to let us to use that, because the agency insists that all drugs be manufactured according to certain criteria, which could not be assured in Mexico.

This time, we had to develop a mega-protocol to address all of the issues required to gain FDA approval. It took me eight months to draw up the new protocol and get it through all the approval stages, but we were rewarded with quick and positive action by the FDA. There were a number of factors in our favor. First, we'd already shown that the oral form of the drug worked, however modestly; and, second, we had shown it to be remarkably nontoxic. The problem of availability was still a major hassle, but eventually U.S. manufacturers were persuaded to make a preparation specifically for our study.

By 1982, we were ready to begin to treat patients with IV ribavirin. Again, we dispensed with the control group on humanitarian grounds. We had shown some effect with oral ribavirin, so we had to offer the real thing to everyone. After I returned to Atlanta, Patricia stayed on in Sierra Leone, taking over my position as director of the project—although I continued to make frequent trips to the country during the study. With her background working on investigations in Bolivia and Central America, she brought a great deal of experience to the Lassa project. She was also a consummate organizer, an asset vital in a place where chaos is the norm. She made a great impact on the field laboratory, as she had done in the laboratory at CDC. As late as 1990, we were still using her cataloging scheme in Atlanta. But Patricia's real love was getting out into the field. Patricia made the decision to move the main laboratory from Kenema to Nixon Hospital at Segbwema. With funds that I'd managed to scrounge up, she put together facilities that allowed us to do virus isolation without having to go through the trouble of sending specimens all the way back to Atlanta. It was in Segbwema where, under Patricia's direction, we would first try out our intravenous ribavirin. The study continued under the director who replaced her, Curtis Scribner.

We fought against our own wishful thinking as the patients arrived. I saw a boy, Ahmadu, brought in by his father. He was a tall, lanky kid, about seventeen or eighteen. Father and son had traveled three hours on the poda-poda from Panguma. By that time, however, we no longer treated patients at Panguma. The hospital had undergone too many changes of leadership, including a new head nun, who vanquished Lassa by declaring that it did not exist.

When the boy first became sick, his father said they had gone to the medicine man. Two or three days and many potions, incantations, and powders later, he was in significantly worse shape. His father was desperate. This was his only son.

"We deh go nah Segbwema," he told his boy.

The father deposited the youngster on the old bench outside the laboratory. A technician, James Masserly, stepped out to take some blood from him. Then Coolbra, the man who was head nurse, joined the family and began to ask them questions necessary for our study forms. When the forms were completed, his father was told to carry the boy into the ward. He wrapped his son in a hand-woven cotton country cloth with a black and white geometric pattern. It was all that clothed his nakedness.

Meanwhile in the lab, James carried out two separate tests. One relied on a spectrophotometer to test for liver enzyme level, the other required a fluorescent microscope to detect the amount of virus present in the body. The spectropho-tometer revealed an AST of 325, a dismayingly high level of enzymes. AST is aspartate transaminase, an enzyme from the liver, and normal levels are less than 40. The data from our experience showed us that a level of 150 in Lassa fever usually means that the patient has a high risk of severe disease and death. We were, therefore, treating everyone with an AST over 150. Ahmadu's AST was more that twice this level, a very bad sign. Then James stepped over to the fluo-rescent microscopes. These were kept inside a strange-looking booth built by a couple of Peace Corps volunteers out of old wooden crates. The brightly col-ored curtains—"borrowed" from one of the staff residences—that draped the front of the booth gave it the look of a fortune-teller's kiosk at a sideshow. But the curtains were functional: fluorescence is best viewed in the dark.

Perhaps the resemblance to a fortune-teller's booth was no accident. James Masserly was reading the boy's fate. Under the eye of the microscope—a gift of the German government—the boy's cells, fixed with a fluorescent dye, had become transformed into brilliant blobs of light. The virus, James saw, was still at a low titer, because the child was still in the early stage of the disease. Nonetheless, with his AST as high as it was, we would have to act immediately. Emerging from the makeshift booth, he confirmed to Coolbra what the veteran nurse must have already known just from looking at the patient.

"He needs to be treated," James said.

Without delay, Coolbra fetched a new vial of ribavirin and a new intravenous infusion set. Despite the merciless afternoon heat, he walked briskly from the lab to the ward. No relief there. The boy lay sweating on the bed. He had been vom-iting, and there was some blood in his mouth. He responded only with a groan when Coolbra turned him over. As soon as the nurse found a likely looking vein, he slipped in the needle and set up the drip. Drawing the ribavirin from the vial, he slowly fed the liquid into the connection of the drip and strapped it into place. Turning to the father, who had been observing this procedure in bewildered silence the whole time, Coolbra warned him to be careful about getting blood from the boy on him. He gave him some gloves, showed him the bleach solution, and explained how to use it. Then he left.

Coolbra couldn't linger long at his bedside. He had fourteen other Lassa patients on the ward to check on that day.

The infusion had to be repeated every six hours. The boy remained very sick. He looked as though he were about to approach the final, fatal stages of the disease.

Two days passed. On the third morning, Patricia was making the rounds and happened to be especially busy, so she was late in reaching the boy's bed. Only the boy wasn't there. The bed was empty.

There was an old man in the next bed, drinking hot tea. Patricia turned to him in consternation.

"Where is Ahmadu?"

It wasn't unheard of for the families to take patients away during the night if they believe that they were about to die. It is considered bad luck to die away from home.

The old man grinned cheerfully, slurped his tea, and gestured toward the window. Patricia looked out through the paneless opening and saw the boy sitting under a tree, with the drip stand propped beside him. Near him were three members of his family, urging him to try some of the dishes of food they had prepared.

"You deh see em. So he deh eat," the old man said.

It was amazing. The boy should have died.

We went on to treat more than 1,500 patients with laboratory-confirmed Lassa fever. From over 16 percent, mortality dropped dramatically to less than 5 percent. As time went on, the new treatment became famous throughout the district. Whenever I accompanied the Lassa fever truck, people would often appear out of their houses and stop us so that they could shake our hands. I didn't remember all their faces, but I had no trouble identifying them. They were all people whose lives had been saved by ribavirin.

In 1985, the data from our studies provided unequivocal evidence that intravenous ribavirin is capable of reducing mortality significantly, especially if it was given in the first seven days of illness. We were able to show that virus titers in blood dropped rapidly as soon as we started ribavirin. It obliterated the virus in the blood. By the end of our studies, we hardly lost anyone we treated early enough. When we published our findings in the *New England Journal of Medicine* we achieved a historic first, an effective treatment for a fulminating viral infection using a drug. Many people contributed to this landmark study. Especially important was the work of the doctors, Isabelle King, Patricia Webb, Curt Scribner, Bob Craven, and Diane Bennett, the last four of whom were all directors of the project.

But what about the failures? Most of them were patients with advanced disease at the time of admission. The virus had done so much damage at that point, ribavirin could not be effective. We would have to look for other ways of supporting them. Beyond this, there was still one group of Lassa victims we did not know how to treat. This was a population who represented a special dilemma, because not one life, but two were at stake.

CHAPTER TWELVE

Kadiatu

Kadiatu was a twenty-two-year-old mother of two, in the last three months of another pregnancy. She lived in the Tongo Field, one of the larger diamond mining areas in Sierra Leone, where her husband was a miner. Her family shared the house they lived in with more than twenty others, most of them miners, too. Like everyone else in the area, she and her husband were immigrants, who had been drawn to the Eastern Province by the lure of sudden wealth.

Although Kadiatu hadn't visited any clinic or doctor for a prenatal checkup, she wasn't especially concerned. After all, she hadn't had any problem with the delivery of her first two children, and she didn't see why she should have any difficulties now. Certainly, Kadiatu, young and strong, seemed to be in a relatively good health, given her difficult circumstances—which, compared to what many others endured, weren't even all that difficult. After all, compared to subsistence farming, there was good money to be made in the mines. True, her home was crowded, but it was relatively large, and because most of the men were successful miners, they had no trouble obtaining a plentiful store of food, most of which ended up being stacked in the rafters of the mud-brick house, under the corrugated iron roof. To be sure, there were many rats in the house, but rats were everywhere—a fact of life.

Then one day Kadiatu woke feeling somewhat weak and a little warm, with a headache. Her muscles ached a bit, too. But she couldn't afford the luxury of staying in bed. There was too much to do; she had to prepare the morning tea, feed the children, and make sure that the men ate well before going off to the mines. This morning, though, she went through her chores with less vigor than usual. When she finished with her work, she felt so drained that she had to lie down on the grass-mattress bed that she shared with her husband and her two little children. She dozed fitfully.

When she woke, she had a burning fever, and her body was wracked with pain concentrated in her lower back. Now she couldn't summon up he strength to fetch the water—the well was about a quarter of a mile away—and so she asked one of her cousins to do it for her. Dinner. What was she going to do about dinner? She didn't have the strength to pound the hulls off the rice or prepare the cassava leaf for the evening chop. Indeed, she was already late in starting her chores in preparation for the return of the men. Luckily, there were others in the house who were willing to take over her responsibilities for the rest of the day. Surely, tomorrow she would feel fine, and everything would be back to normal. She'd had plenty of fevers and chills like this in the past, and usually they would vanish in a day or two. Everyone had them. Probably she was suffering from a spell of malaria.

Kadiatu continued to run a high fever all during the night. Abdul, her husband, became sufficiently concerned to seek the dispenser. It was no problem to find one close by. Because mining made the area relatively prosperous, people had plenty of money to pay for drugs and injections. The dispenser doubled as doctor, since there was none in the village. He gave Abdul four tablets of chloroquine for malaria and instructed him to tell his wife to take them all at once.

Kadiatu was still able to swallow the pills, but she had difficulty keeping them down because she was feeling nauseated. The next day, Abdul, convinced that his wife would begin to feel better after the tablets started to have an effect, left for work as usual. When he returned in the evening, however, it was clear that all was not well. Her fever hadn't gone down; if anything, she seemed hotter. Now her throat was very sore, and she had begun to vomit. She said it hurt too much to swallow, and she couldn't even keep down a teaspoon of water. Abdul decided that he would have to take his wife to Panguma Hospital the following day.

What was God trying to punish him for? he wondered.

Like most members of the Fula tribe, Abdul was a devout Muslim. Didn't he pray at least once a day and go to the mosque on Friday afternoon? Anyway, Inshallah, he would take his wife to the doctor tomorrow. The doctor had whiteman's medicine, and he would know what to do, and she would get better.

The next morning, Abdul got his wife ready and took her out to catch the early morning poda-poda to Panguma twenty-five miles away. Today's poda-poda was a Nissan pickup truck with wooden benches in the back. Abdul was able to find his wife and himself a place on the bench next to the cab, thinking that the ride would be less jarring there than if they took one of the benches along the side.

By now, Kadiatu was in agony. The pain seemed to come from all directions, assailing her back and creeping into her abdomen. She was nauseated, and her throat was so sore that she could no longer swallow her own saliva. She still felt like she was burning up, but now there was a new sensation as well: cramps in her belly. It couldn't be, she thought. It was far too soon. She had at least two months to go.

Each time the Nissan hit another bump, more pain shot through her. There were now twenty-four passengers packed into the rear of the vehicle, not counting the four perched on top, with all their bundles of produce, as well as a goat and several chickens. Finally, it got to be too much for her, and Kadiatu began to throw up, catching the vomit in the folds of her dress. The driver wasn't aware that she was sick, but might not have stopped in any case. Life was tough for everyone; you just kept going. Kadiatu kept silent, not wanting to take the chance of doing anything to delay their arrival at Panguma.

The unhappy band of travelers continued to lurch and bounce over the rough road toward Panguma, breathing in the stench of sweat and vomit. With all the stops, the loading and unloading, the journey took three hours.

By the time they reached Panguma, Kadiatu was slumped over her husband, unable to sit up. Abdul spoke to the driver, who agreed to drop them off right in front of the hospital. Then, with no one to help them, Abdul had to carry his wife inside. Although Kadiatu was becoming less aware of her surroundings, she was still conscious enough to realize that the cramps were now contractions. The nuns received her quickly and kindly, and took her directly to the maternity ward, where she was examined by the midwife. After taking a quick history of her illness, the midwife understood at once that her patient had Lassa. It is a terrible disease in anyone, but in a woman with an advanced pregnancy, it is catastrophic. From bitter experience, the midwife knew that the new life in Kadiatu's womb was not going to survive.

The midwife was a local girl, but she was well-trained and good at her job. Her first priority was to isolate Kadiatu, using makeshift barriers consisting of a cloth draped over a frame. This setup would remind the staff and the rest of the patients that here was a woman with Lassa fever, a contagious disease. But that was as far as it went: no one had gloves or masks.

The pain was taking over Kadiatu. There was nothing but pain. Pain in her abdomen, pain in her back, and searing pain in her muscles and throat. She couldn't think of anything else. Why did she have to endure this? Her fever was now over 104 degrees, according to a reading taken by placing a thermometer in her armpit. You couldn't take temperatures by placing the thermometer under the tongue because people did not understand what you were doing, and some would bite it in two. Taking a rectal temperature was culturally unaccept-able, particularly on an open ward. The reading in the armpit—known as the axillary temperature—is usually a degree or two lower than core body tempera-ture. So, if the axillary reading registered 104°F, then Kadiatu was dangerously hot.

As the hours wore on, Kadiatu began to bleed from her vagina. Her contrac-tions lessened. Then they stopped. A check of her abdomen for the baby's heart-beat revealed that it was slowing. The midwife estimated that Kadiatu was only twenty-seven weeks pregnant. The baby would be very small, with little chance of survival. There was no neonatal unit in the whole country that could care for preterm babies. They just died. For that reason, the midwife was reluctant to induce labor.

Kadiatu continued to deteriorate. Her hands and feet were becoming cold, clammy; her blood pressure was dropping. She was slipping into shock. Breath-ing was rapid and labored. Her lungs were filling up with fluid—the Adult Res-piratory Distress Syndrome (ARDS) that so often accompanies fatal viral hem-orrhagic fevers. To her poor husband, who was standing vigil at her side, it seemed that she now had to concentrate on her breathing to the exclusion of everything else. She had lost all awareness of him or of anything else around her.

She needed blood. Abdul offered his own, of course, but he was the wrong group—B positive—and she was O positive. There was no blood bank in Panguma or anywhere in the country, for that matter. The only way to obtain blood was to find a relative or friend with the right blood group who would give it, or to locate a willing stranger. Abdul and Kadiatu had left most of their extensive family behind when they relocated to the diamond mines in search of fortune, but he did have many acquaintances in Panguma. So he immediately rushed off in search of them. He assumed that at least one must be in the right group and willing to donate blood for his dying wife. It took him three hours, but he found a friend who thought he might be group O. Together, they returned to the hospital where his blood group could be checked and—if all went well—a transfusion arranged.

By now, Kadiatu was in complete coma and deeply in shock. Shock is a con-dition in which the blood pressure falls so low that it can no longer be mea-sured. When this happens, oxygen cannot reach the brain and kidneys in suffi-cient quantity, with the result that they both start to fail. Meanwhile, Kadiatu continued to bleed. Although it was slow and not particularly profuse, it was increasing steadily. The midwife could no longer hear the heartbeat of the fetus.

The midwife decided that it was time to send for Sister Eileen, the Irish nun who was head of the hospital. It took only a few moments for Sister Eileen to make up her mind. They must induce labor immediately. The baby was probably dead already, and unless they could get it out, the mother would surely die as well. The only hope they held out was that they would be able to deliver the dead child in time. So Sister Eileen gave instructions to the nurses to start Kadiatu on fluids and give her the transfusion the moment Abdul returned with the blood.

To induce labor, a nurse administered an injection of pitocin. Then Kadiatu was rushed to the delivery room, where her condition could be monitored more closely. Additional fluids were administered to increase her blood pressure. While the introduction of the fluids caused her rate of breathing to accelerate, the midwife noted with alarm that Kadiatu's lips were beginning to turn blue.

Abdul appeared in the delivery room, breathlessly holding the fresh unit of blood, just taken from his friend. It was quickly screened to make certain that it was type compatible. There was no time to lose. The contractions were coming faster now. While no one in the delivery room spoke about the loss of the child, everyone knew it was dead. Mercifully, Kadiatu remained unaware of what was happening.

The fetus emerged awash in amniotic fluid mixed with blood. The tiny thing was gray and still. The nurses wrapped it quickly in a cloth and then returned their attention to the mother. The midwife had been so preoccupied that she'd forgotten to put on a pair of gloves as she delivered the dead baby.

Over the next hour, Kadiatu's body temperature dropped precipitously. It was now 95 degrees. Her respiratory rate was still increasing, her lips were still blue, and her hands and feet were colder than before. The midwife now noticed some twitching in Kadiatu's hands and arms. It was a dangerous sign. A few minutes later, she suffered several generalized seizures. She struggled to breathe, grunting loudly as she took desperate gulps of air.

Kadiatu could not keep up the fight for long. She had no strength. What the pain hadn't taken out of her, the fever and the strain of the delivery had. She was now in shock and heart failure. Her body gradually became still. She sighed a few times, and then the ordeal was over. The nurses slowly covered her with a sheet and left the room, exhausted.

It fell to the midwife to break the news to Abdul.

He did not take it well. How could he?

Was this really the will of Allah? He had believed that by putting his wife in the hands of the white doctors, she would be saved. He'd paid a lot of money besides. But look what happened. She died anyway. Now he would have to spend more money to take the body back to Tongo Field for burial before sundown, in accordance with Muslim custom. And what was he going to do about his young girls? His parents didn't live in Sierra Leone, so they couldn't care for them. There was no one to make their meals, no one to fetch water. And no one who would ever perform all of these chores as well or as cheerfully as Kadiatu. He was desolated and confused and very much alone. Maybe, he thought, he should go back to his family in neighboring Guinea. But what would he do once he got there? There was no work there. There was nothing at all for a man like him, who could neither read or write.

Michael Price was a general practitioner from England. He had come to work at Segbwema Hospital in 1985, partly motivated by a yen for adventure, partly inspired by his religious convictions, and partly driven by a need to serve humanity. A quiet and introspective man, Michael was equally at home in the clinic, operating theater, or in the delivery room. He was frustrated of late. No matter what he did, he kept losing patients like Kadiatu. Why, he wondered, should Lassa strike pregnant women with such special virulence? What could he do about it?

He began to study the problem.

We had established that nine out of ten of the fetuses of women infected with Lassa virus in pregnancy died in the womb. We also knew that the placenta was a growth factory for the virus. In fact, studies at CDC had shown that the highest concentrations of Lassa virus were to be found in the placentas of infected mothers. In light of these findings, Michael decided that in every case of septic abortion (miscarriage in a woman with a fever) he saw, he would test for Lassa fever. He also studied a series of previous cases of fever late in pregnancy and cases where the fetus had not been lost naturally or therapeutic abortion performed.

The data were amazing. Patient after patient diagnosed with a bacterial infection, malaria, or typhoid turned out to have Lassa fever. This was a big discovery and an even bigger problem. Michael redoubled his efforts. He gave every pregnant woman superb obstetric care, something that was unavailable elsewhere. If a woman aborted spontaneously, he would take her into the operating theater and clean out her womb to stop the bleeding. He even performed cesarean sections on women with Lassa fever. Acutely aware of the risk, he made sure always to wear two pairs of gloves.

We were astonished by the data Michael had collected. In the first six months of pregnancy, women with Lassa did as well as nonpregnant Lassa patients. Many of them did lose their baby, but they themselves usually survived. It was only late in their pregnancy, after about twenty-six to twenty-eight weeks, that the picture changed dramatically. Almost all of the babies died, and far more of the mothers succumbed as well. There were seventy-two women in Michael's study. In almost every case, the babies had been lost—that was no surprise—but what was remarkable was that, as long as the mother delivered the baby (whether by a spontaneous or an induced abortion), she had a significantly better chance of survival than if the fetus remained inside her.

There was no disputing the odds. The chances of the fetus surviving were less than 10 percent, but the chances of the woman surviving with the evacuation of the womb were better than 50 percent. So it soon became standard practice to induce labor in a woman who had Lassa fever in order to save their life, especially in light of the fact that the baby would usually be lost anyway. There were a few exceptions. In some instances, Michael would perform Cesarean sections on women with Lassa and sometimes manage to save the baby as well. The advances in treatment were so significant that if Kadiatu had taken sick three years later, and had been cared for by Michael, she would at least have had a chance of walking out of the hospital alive.

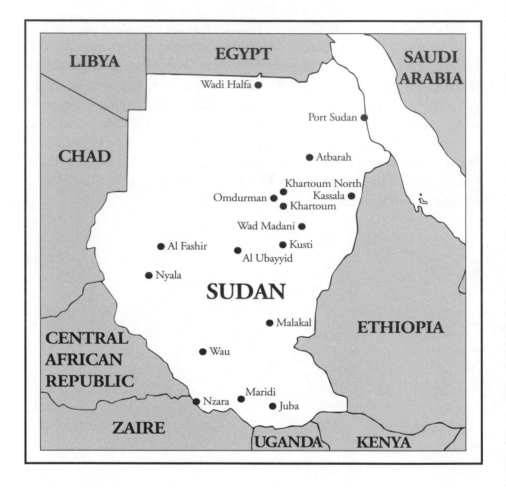

CHAPTER THIRTEEN

Nzara Revisited

In late July 1979, after three years in Sierra Leone, I returned to Atlanta, ready to settle back into CDC and family life in Atlanta. No chance. I was immediately hit by news of a fresh outbreak of suspected Ebola. We received this news of the epidemic by way of WHO headquarters in Geneva. Details were spotty. All we knew was that it involved Nzara again, the very Nzara where Ebola had previously surfaced in 1976—the Nzara I had reached on my epic journey from Zaire. A nearby town named Yambio was also involved. We heard that several victims were already dead. How many people were actually infected? Nobody had any idea. The Sudanese government in the capital of Khartoum far to the north of Nzara had once again imposed a quarantine, forbidding travel to the region. Obviously, this further limited information, in addition to imposing unnecessary hardships on the local population, since Khartoum had also instituted a complete embargo, curbing the flow of goods to and from the affected area.

From what we knew based on previous experience of this deadly, fulminating disease, we realized we would have to mount a rapid response. This meant that we would have to get a team in the air within twenty-four hours of being alerted by WHO. Fortunately, the call had come in on a Thursday night, giving us until Friday evening to prepare for the flight to Geneva. That would allow us time to meet with our colleagues at WHO on Saturday morning. But first we had several problems to solve, not least of which was to figure out exactly where we were going. We dusted off our old maps of this remote, poorly charted area and tried to determine where Yambio was in relation to Nzara. It wasn't easy.

Our problems were compounded by the fact that our most experienced personnel had just set off to Johannesburg. Ironically, they were attending a major conference on hemorrhagic fevers. This left behind only a handful of us: me and whomever else I could recruit to join the mission.

I turned to Lyle Conrad, head of the Division of Field Services in the Epidemiology Program Office. With his broad, smiling face and sprawling gray beard, Lyle reminded me of an Amish farmer. I suspect that Lyle owes his hardy nature to having been raised in the harsh winters of rural South Dakota. Any epidemiologist who had passed even briefly through CDC knew Lyle. Having participated in the original Lassa investigation in Nigeria in 1969, he was in a position to understand the urgency of the situation. I asked him if he could find anyone with the right qualifications to come along with me. This was an exacting brief: he or she would have to be adaptable, adventurous, and willing to endure primitive, possibly dangerous, conditions. The candidate would also have to be available right away.

I had known Lyle Conrad for many years. He always had a horse in his stable for whatever race needed to be run. He also possessed an uncanny ability to get people to do whatever he wanted. Lyle and I were both well aware of the consequences if we chose someone we had not properly screened. In 1976, one EIS officer on his way to Zaire decided midway over the Atlantic that hemorrhagic fevers were not for him. He turned back—and was within his rights to do so, since there was nothing in his contract that required him to contend with an outbreak of a hot virus like Ebola. It was just too bad that he waited so long to tell us about his misgivings. On another occasion, I needed someone who could take over a nationwide study of Lassa that I'd organized in Guinea. I asked CDC to send me someone who spoke French. They sent out a second-year EIS officer. The new recruit met me at the office of the Ministry of Health in Conakry, the capital of Guinea, wearing a T-shirt, a pair of leather shorts, and boots. A big knife hung down from his belt. I was wearing my usual short-sleeve shirt and maybe a tie. He turned out to be a rabid Marxist. Either he had been reading too many books, or political eccentricity ran in the family, because I later learned that his sister was working for the Trotskyite movement in London. Guinea was then under a pseudo-Communist regime, which had gone to great pains to ensure that the country's economic and social growth was the slowest in West Africa—which is really saying something. My new recruit must have figured that he'd died and gone to Marxist heaven. Taking one look at him in his getup, the minister of health, who was dapper in suit and tie, asked if he was planning to go to war or to work in the villages. Apparently, ideological purity was no substitute for competence, or maybe it was just his sartorial style. About midway through the survey, the government asked him to leave the country, and the Guineans completed the work themselves.

We couldn't afford to waste time and resources making the same kind of mistakes again. So we wanted to be certain that we found a seriously committed individual. Lyle didn't disappoint me.

"Joe," he said in his unmistakable midwestern accent, "I've got just the guy for you. He's a fearless mountain climber, and he'll do a great job. His name is Roy Baron. He's done work in immunizations."

The only mountains I had for him to climb were in the mind, but he sounded perfect to me—until Lyle added: "By the way, he's never been out of the country, so take care of him."

Nor, it turned out, had he ever been involved in a major investigation. In the grand tradition of the EIS, he was just going to have to pick up experience as he went along, learning on the hoof. In any case, he had Lyle's endorsement, and that was good enough for me.

Our first order of business was to assemble the equipment and supplies we would need. We had a secret weapon in the Special Pathogens Branch when it came to organizing an investigation in a remote location: Helen Engleman. Broad-shouldered and heavyset, she was an imposing woman who had the capacity to inspire terror in many of the technicians. Before coming to CDC, she had served in the Marine Corps. Helen presided over her domain from a swivel chair next to a telephone and a tower of ledgers filled with notes and figures that only she could understand. Her office was always shrouded in smoke, her ashtrays filled to the brim. Helen had a great, rumbling, gravely bass voice, which commanded attention from everyone, including her supervisors. You crossed her at your peril.

Helen was an incredibly meticulous organizer. She could cram more material in a footlocker than many people could fit into a pickup truck. In addition to finding us the obvious accoutrements—needles, syringes, reagents, microscope slides, and specimen vials—she also made sure to pack all the essentials of improvised life in Africa: duct tape, Magic Markers, paper, pens, and pencils. Helen succeeded in squeezing all our paraphernalia into two crates. One thing we did not take, though, was the collapsible field laboratory, which I had used in the first Zaire Ebola outbreak. We now knew enough about these viruses to realize that the major risk was from injection, not from aerosol transmission. Given that fact, the cumbersome glove-box system was a handicap, if not a downright liability. We would operate as we had in Sierra Leone during the Lassa project, working on the open bench, using basic precautions to prevent infection. Amazingly, we neglected to take a flashlight because we assumed— naively—that we could obtain one anywhere. I should have known better.

Once again, we had no time to obtain visas to the Sudan, so WHO arranged for us to pick up UN passports with visas when we arrived in Geneva. Roy and I caught a Friday afternoon flight to New York and made connections for a Geneva-bound flight out of JFK. My biggest fear was that, somewhere along the way, our baggage would get lost, leaving us in the middle of the Sudan with no provisions and nothing to wear but what we had on our backs. Luck was with us, or seemed to be, anyhow. Our baggage did not disappear en route, and, for reasons that were unclear, the airline upgraded us to first class, in which we were served an amazing meal, possibly the best I've ever had on a plane.

But we both found it difficult to enjoy, feeling that this might be the last supper.

As soon as we cleared customs the next morning in Geneva, we found a taxi to take us to WHO headquarters to receive our briefing. We had only two or three hours. The Swissair flight to Khartoum left that afternoon, so we had to be back at the airport by three to catch it. August in appearance, monumental in size, the United Nations buildings are set amid the fastidiously kept greenery of a large park. The taxi bore us past a succession of white office blocks, finally arriving at a cul-de-sac, where we got off. WHO headquarters impresses the visitor as a warren of endless hallways and escalators, a place one might easily become lost in for days at a time. We ascended to the fourth floor, where we located the Communicable Diseases Division. There we were welcomed by the director of the virus unit, Paul Bres, and his associate Fakhry Assad. The two couldn't have been more different. Paul was very French, and Fakhry equally Egyptian. Paul was quiet and measured, whereas Fakhry was enthusiastic and ebullient. Though this was my first meeting with Paul, I was well aware of his reputation. He had gone with the team to Sudan in 1976, but I don't think that he ever actually got down to Nzara. Nonetheless, he was well placed to understand just what we were letting ourselves in for. A man in his late fifties, with close-cropped gray hair, he was conservative in dress and condescending in manner, as befitted a former colonel in the foreign service of the French military. He had spent his life specializing in the investigation of arboviruses—viruses transmitted by arthropods—especially yellow fever. This trail had led him all over West Africa, and in the course of his professional travels he had actually been involved in some of the earliest hunts for Lassa virus in Nigeria. He saw himself, quite properly, as a man carrying on the legacy that the French had established in the practice of tropical medicine. However, like most traditions, it had its weaknesses. Its model was military and, as such, did not allow for much flexibility, even in the field, where flexibility was needed most. That isn't to say that it lacked certain strengths; the authoritarian model insured that investigations would be exceptionally well organized. In this respect, it was unfortunate that the French had no presence at that time in the Sudan.

Fakhry Assad, in contrast, was a round, boisterous fellow with a shock of wavy gray-white hair, who possessed a great sense of humor and a laugh that could have been measured on the Richter Scale. His secretary, June, later told us she could always locate him in the hallways by his resonating laugh He was fantastically supportive and full of boundless curiosity. It was his eagerness to learn that especially distinguished him from Paul, who behaved as if there were nothing he didn't already know. Their differences even extended to the way they dressed: compared to the somber wardrobe Paul favored, Fakhry could have been outfitted for Halloween, with his brightly colored cobalt shirt set off by an equally striking neck scarf, which he wore instead of a tie. Fakhry was one of the most delightful, kindest men I ever worked with. He would go on to become a tremendously effective promoter of research into hemorrhagic fevers at WHO.

Because none of us had actually worked together before, we carefully maneuvered around one another for a while. Paul and Fakhry began by relating to us what they had heard so far. Since I knew the Nzara area, they asked me for my assessment.

"Based on my experience," I said, "we're likely to have a number of widely scattered cases with very little or nothing in the way of support for diagnosis, quarantine, or preventing the spread of the virus. By cordoning off the area, the government has made things more complicated for us."

We all agreed that, without much information, it was difficult to grasp the extent of the outbreak, and with all trade cut off to the area, we were likely to find it tough to bring in our supplies. Paul and Fakhry were most concerned with logistics of getting us into southern Sudan with our equipment intact. They voiced concern about the local WHO representative in Khartoum. Apparently he suffered from a great sense of his own importance, and would have to be approached circuitously and with appropriate protocol and a degree of pomp. They were afraid that he would be more of a hindrance than a help.

After the 1976 epidemic, WHO had taken the precaution of storing materials against the possibility that a similar expedition would need to be mounted. So we were taken down to their storerooms and shown, among other things, a supply of protective paper suits and plastic gear. We also discovered several full-faced biohazard masks—hideous-looking things, unbelievably uncomfortable to wear. They were in the style of World War II gas masks. Nonetheless, we obligingly gathered up the masks along with the protective gear and added them to the supplies Helen had already packed for us. Then, with no time to spare, we hurried to the airport to make the flight to Khartoum.

After our experience with the 1976 outbreak, we knew that we would have to be prepared in the event that a team member became infected with Ebola in a remote part of Sudan. We made arrangements with the Belgian Embassy in Khartoum as well as the United Nations Development Program to contact the CDC if anything like this happened. At the same time, we determined that evacuation could best be accomplished by sending a plane in from Europe, which could stop to pick up a field isolator in Cairo before proceeding to the Sudan. The isolator would be necessary to keep the patient from spreading the infection to anyone who came into contact with him. So we were reasonably certain that we had a credible plan in place before setting out.

Events soon proved that we were mistaken.

Arriving in Khartoum, we braced ourselves for the pompous, stubborn official Paul and Fakhry had led us to expect. To our great relief, he turned out to be away. Instead, his assistant, an agreeable enough fellow, greeted us. He informed us that all flights (there were not many) to Nzara had been canceled because of the *cordon sanitaire* imposed by the government. With flights restricted to the south because of the epidemic, the last flight for Juba—the nearest town to Nzara—was leaving in a couple of hours. There was no telling whether we'd have another chance any time soon. We had to make a quick decision, and it was a real dilemma. We had assumed that we'd have enough time in Khartoum to gather additional information about the epidemic and sort out the local politics. I finally made up my mind to proceed to Juba alone and try to assess the situation. From there I'd do what I could to obtain transportation on to Nzara. I left Roy behind to collect the protective clothing that WHO was holding for us in Khartoum, with the expectation that he would later join me in Nzara—assuming that he could get there.

Sudan Air cannot be considered one of the premier airlines of the world, even compared with such obscure carriers as Air Senegal, Air Brousse (Zaire), and Sierra Leone Airways (now defunct); no amount of experience can prepare you for the white-knuckled, heart-pounding feeling these flights can induce in the wretched passenger. Getting aloft seems a miracle unlikely to be matched by landing in one piece on the same trip. Suffice it to say that the terror of the flight into Juba pushed any worries about the epidemic completely out of my mind.

After two unnerving hours, the plane came down on a pock-marked asphalt strip. The terminal was nothing more than a large shed with a dingy tin roof. Only a few people were in sight, and an aura of desolation hovered over the whole place. The plane immediately taxied around to fly back to Khartoum without taking on any passengers. The pilots were not about to stick around.

Although we were just north of the equator, I still wasn't prepared for heat this intense. In this inferno, it was a major challenge just to put one foot in front of the other.

Before I set out for Nzara, I needed to learn what I could about the epidemic from the UN people here in Juba. The following morning, I turned up at the UNDP (United Nations Development Program) compound. It consisted of several stucco and tin-roof houses set around a swimming pool. The stultifying heat made the pool a popular draw. Most of the UN workers in evidence were Scandinavians and French. Their job couldn't have been a happy one, and not just because of the heat. This part of the Sudan was in constant turmoil. The local populace was constantly under threat from renegade soldiers. Food, always in short supply in the best of times, was getting even harder to obtain. Bad as it was, the situation here (I was told) was far better than conditions in the west of the country.

I began asking them if they were aware of cases of an Ebola-like disease. No one I spoke to at the compound knew of any in Juba. The German and Scandinavian missionaries I interviewed were only dimly aware that something was going on in the western part of the country, but even they couldn't say for sure what it was exactly. I could see that I was going to achieve very little here. With transport cut off, and with communication with the affected area almost nonexistent, I shouldn't have been surprised. I was looking around for some means of getting to Nzara when Roy Baron turned up, borne into Juba by his own police plane, which he had somehow managed to commandeer from officials in Khartoum. I was impressed: this was style. Lyle had come up trumps. Roy was clearly a natural traveler, proving himself to be a canny and resourceful investigator. I was lucky to have him on my side. The police pilots would take us on to Nzara. This was spectacular news. Our alternatives would have been long, dusty, difficult road trips in a UN truck or, worse, hitching a ride on a commercial vehicle, along with the goats and chickens. By road, we would have had to dodge or talk our way through the roadblocks set up to prevent travel and maintain the *cordon sanitaire*.

Now that we were assured of transportation, we were ready to leave for Nzara. The sooner the better—because once they began hearing stories about the epidemic, the pilots grew decidedly less enthusiastic about the whole enterprise. Nevertheless, they agreed to continue. We reached Nzara and found lodging, and we set off for Yambio to examine the suspected Ebola patients. In the ramshackle hut that served as Yambio's hospital, the two of us got down on our knees, with only the light of a kerosene lantern to see by, and began to have a look at the patients and take blood specimens. I spent that night separating serum, so that the pilots could take the precious specimens back on their journey to Atlanta.

The following day, the pilots, happy to be on their way, returned to Khartoum with the blood samples. The first part of our job was done. Now we settled down to complete the investigation and see what we could do to try to control the outbreak.

It was two nights after our arrival in town that I had my accident. Earlier that day, the old woman had been brought in from an area where there had been a confirmed case of Ebola. She had a high fever and delirium and a seizure. It happened as I knelt down to take the blood sample from her arm. As I began to pull back the plunger of the syringe, she gave a violent lurch. That was when the needle slipped, punctured my glove, and penetrated my thumb.

When you stick yourself with a needle full of Ebola virus, the odds for your survival are just about zero. I kept working—there was nothing else to do. That evening, Roy set up an intravenous line to feed me Ebola immune convalescent plasma, which we'd brought with us. Maybe I was infected. Maybe not. We had nothing better to try.

With no choice now but to let time tell whether I'd been infected, we plunged on with the investigation. There was a special urgency now, because neither of us knew how long I had left. If I fell sick, I would have to stop, and when I stopped, so would the investigation. The same virus that might be replicating exponentially in my blood was continuing to spread around Nzara. If I couldn't do anything to stop it in myself, I resolved that nothing was going to get in the way of my trying to stop it from reaching others.

This time I was determined to be able to diagnose Ebola successfully, on the spot, using the reagents so carefully packed by Helen, but I needed some special equipment. I was lucky to find an old friend of mine from the 1976 outbreak, Simon Van Nieuwenhove. It was Simon who had taken the more southerly route in the quest for the origin of the Zaire outbreak. He was still working in the area. A somewhat portly Belgian, he looked like a prosperous sixteenth-century Flemish burgher about to sit for his portrait. Though he had a relaxed air, even laid back, he was actually a man obsessed. His goal was to stamp out sleeping sickness, a very severe disease endemic to Africa, caused by a typanosome, a single-cell parasite that gets into the blood and the brain. In some respects it is a bit like malaria, only it is spread not by a mosquito but by a particularly nasty insect called the tsetse fly. Simon craved simplicity and lived a secluded life, attended to by a devoted African staff. He generously offered us a room to use in the small building he was using for his own lab. It didn't faze him at all that we would be bringing Ebola virus into his facility. He had lived among the hazards of Africa too long to get exercised over such things. Besides, he trusted our technical expertise. His lab also had the advantage of having electricity supplied by a gasoline generator, and he had a kerosene-fueled refrigerator—an unforeseen luxury that relieved us of having to do any more hand cranking to separate blood. The refrigerator served the same purpose. Instead of mechanically separating the blood from the serum or the plasma, all I had to do was leave the specimen standing upright in the refrigerator overnight. By morning, the blood cells would have clotted and settled to the bottom of the tube. I could easily draw off the golden serum from the top. I just had to remember that it might well be laden with virus.

Every morning would begin the same way. I would walk into the lab and proceed to pipette off the serum into appropriate vials for testing. This was accomplished using the fluorescent antibody test. Helen had packed some slides with Ebola-infected cells, inactivated by gamma irradiation, fixed on them. I would then place drops of the serum on the slides, attach a fluorescent label by adding another drop of fluid, and then look at it under a microscope. If it shone brightly, it was positive. Ordinarily, I would have waited until the end of the day to read the slides. But given the circumstances, I tried to get the testing out of the way as early as possible. We needed to know as quickly as possible who was positive. I'd remove every sample from the refrigerator and then go through them one by one. I tried to concentrate on doing this as objectively as possible, but now I was looking for the name of one patient in particular. When I realized I had the slide in front of me, I put it to one side.

Finally, all the others having been read, I braced myself to look at the last slide. It was the specimen drawn from the old woman in the hospital in Yambio. If the slide shone brightly with the yellow fluorescent dye, I was infected. Unfortunately, even if it were negative, it wouldn't necessarily mean that I was free from the disease. The woman might be too early in the disease to have developed antibodies. I would have to hope she lived long enough for me to take a second specimen. My heart pounded. I saw my wife, Shannon, and three children back home. My youngest, Anne, was only three years old. I was their only support—my wife wasn't working at the time—and while it was true that I had a standard government life-insurance policy, I wouldn't want to leave anybody dependent on that. For their sake, I thought, I can't be sick.

I hesitated for a moment, then adjusted the light source and, at last, slipped the slide under the microscope. I fiddled with the knobs to get it in focus, so that I could see the cells clearly. It was all I could do to maintain my concentration.

Try to imagine that the serum belongs to someone else, I told myself. Imagine that it has nothing to do with you.

The cells began to take shape as I adjusted the microscope. I searched. The cells that passed before my eyes as I moved the slide around were all gray, green, and black. I could see their outlines and their nuclei quite clearly. Little flecks of fluorescence attached here and there. Nonspecific binding. I knew I had washed them well enough. I checked the positive control. It was brilliant yellow.

She was negative. Clearly negative.

Saved—at least until we took another serum tomorrow. I would have to check every day. Meanwhile, I had to get on with my job.

One of my major priorities was to locate the people who'd already come down with Ebola and then try to limit their contact with family members. The best way to do this was to set up a surveillance system that would identify victims as early in the disease as possible, before they had high virus titers and became more infectious. This sounded simple, but it was far from easy to accomplish. In a more developed country, you could look to the hospital for help, but not in the Sudan. In this country, the hospital was viewed as a place where you went to die, particularly of Ebola. Worse, the outbreak had caused such panic that families were being barred from taking care of Ebola victims admitted to the hospital as they died. Why condemn your relatives to die alone? Again and again, it was impressed on me that in Africa families attach great importance to the rituals that accompany death. Equally critical was the precise location of the burial site. Without the assurance that the hospital would be able to return the body for appropriate burial, families were understandably reluctant to commit anyone to its care.

It was clear that we weren't going to get anywhere unless we had the cooperation of the relatives of the patients. We would only defeat our purpose if we kept them from caring for their sick relations. We decided to encourage them to continue what they were doing, but urge them to take some simple steps to prevent further infection. By all means, we told them, go ahead and nurse your husband or your daughter. We understand how important it is. But please make sure that you wear masks and gloves and surgical gowns when you do so. We supplied them from our stocks and watched to make sure they understood what do to. To facilitate the process, we would appoint one or two designated caregivers in each patient's family. By shouldering the responsibility of caring for their relative, they were able to disrupt one of the principal routes of viral transmission by reducing the number of individuals exposed to an infected person. At the same time, family traditions could be maintained. This was the same process Alain Georges was to use successfully in Gabon in 1996, during the Ebola outbreak among young men who had handled meat from a dead chimpanzee.

Of course, this still left us with the problem of burial. We knew that in traditional funeral ceremonies, the mourners made intimate contact with the dead. The outbreaks in the Sudan and Zaire in 1976 wouldn't have been nearly so extensive if it hadn't been for traditional funeral practices. Realistically, there was no way for us to prevent the practice of purging all the urine and feces from the body before burial. Interring a body without doing this would have been profoundly offensive to the families. We figured that the best we could do was try to make certain that no one became infected while performing these rites. Why not, I reasoned, apply the same solution that we had at the hospital? After all, the level of intimate contact is much the same. Only, instead of barrier nursing, it would be barrier embalming.

We devised a series of hygienic measures to be used during preparation of the body for burial. Since we also had to supervise these ceremonies, we got to know the routine quite well. In return for their cooperation, we assured the families that, if their loved one died in the hospital, he or she would be returned to them for burial. Our compromise received widespread acceptance. This gratified but did not surprise us. After all, the people were understandably frightened of Ebola. While having to don masks and gowns might seem both inconvenient and strange, it was a small price to pay for avoiding infection. Besides, costume—whatever the style or motivation—held an important place in the culture. Best of all, it meant that they could continue their traditional practices.

We began to see benefits from our policy almost immediately. Now more individuals, possibly infected with Ebola, were willing to come forward to have themselves examined and their blood tested. Yet not all of them. No matter what we did, we still were unable to assuage everyone's anxieties. Those were the ones we needed to go out into the bush and find—and then persuade to come back with us for supervised care by us and their families.

The majority of people in the area around Nzara and Yambio, lived in family compounds, which could be reached only by following tortuous footpaths through the bush consisting mainly of elephant grass nine feet tall. There were, of course, no local maps, so we had to find someone to show us the way. Even when we found our destination, there was no telling what kind of reception we'd get, since many families didn't like the idea of surrendering a sick relative to strangers who wanted to haul him or her off to the hospital. Then we had to cope with the challenge of figuring out who was related to whom, and who was not. A man might have more than one wife. A woman might identify someone as her brother, and you would write this down. Then she would point to a second man and say, "Yes, that is my brother," and you would dutifully write that down as well; and then you would ask about a third man, and she would say, "Yes, that is my brother." Pretty soon we would have all these brothers listed—maybe nine or ten of them. It made us suspicious. Even with very large families, that was a lot of brothers. It took us some time to realize that her idea of a brother and ours were completely different. In many cultures, calling a man a brother or a woman a sister is only another way of saying that someone was important enough to be accorded all the love and respect due a blood relation. While this practice of conferring honorific status might be endearing, it certainly wasn't of much help when it came to defining a group—or "cohort"—in an epidemiological survey.

Three of us, Roy Baron, Dr. Omran Zuberi from the Sudanese Ministry of Health, and myself, spread out to search for local Ebola victims before they spread their illness to everyone else around them. It was a local hospital nurse, however, who took me into the bush and proved the most adept when it came to tracking down the sick. Because he came from one of the compounds himself, he was on friendly terms with a great many people in the area and was familiar with their habits. His instincts were especially canny when it came to determining who was trying to mislead us. He used the same techniques that any decent detective in New York or Chicago would bring to an investigation. First, he would ask people if they knew of any suspected cases. Then he would supply me with the translation from the Zande: "This man, he says we go over there, to the west, and maybe we find some woman with Ebola," he would tell me.

And I would think: Great, let's go.

But the nurse would shake his head and give me a look that suggested that I shouldn't be in such a rush.

"This man is lying, sir," he would say. "You can see the way he moves his eyes."

"All right, so what you're saying is that there is no woman with Ebola."

"No, no, no! A woman is sick, sir, no doubt. But she is living not where he says, but in this direction, to the east."

When I'd ask him how he'd acquired this piece of intelligence, he would smile and say, "So what are we doing, standing here?"

It was just such a display of intuitiveness on the part of the nurse that brought us late one afternoon to a compound that seemed to appear in the middle of the high grass like a mirage. A compound consisted of several huts built out of mud-baked bricks. They are usually arranged so that the dwellings, all with thatched roofs, are laid out in a circle around an open space, much of which would be carefully swept. The arrangement of the huts was strictly hierarchical, the chief in the biggest hut, the first wife in the next hut, the second wife in the third hut, and so on. In the central compound you usually found women pounding maize in a large wooden mortar with a pestle made of a piece of wood five or six feet long. Others prepared some other local staple, and there were children, chickens, and other baby livestock running around. A few wisps of smoke rose from a three-stone wood fire under a big pot, its belly blackened, sitting on top of the stones. There was a strict protocol as to who occupied which hut: the patriarch of the clan had the first hut, his eldest son and his family had the second, while his next oldest son and his family had the third, and so on down the line.

Our appearance in the compound caused something of a stir. They knew very well why we were there. The nurse went up to one of the men and addressed him in Zande.

"Is there anyone who is ill in the compound?" he asked.

He shook his head no. Everyone, he insisted, was well.

Even without an interpretation, I could guess at his answer.

"Lying," my guide pronounced with his customary authority. "No doubt, he is lying."

The nurse continued circling around the pen, which held a motley collection of chickens and goats or sheep (they are difficult to tell apart in Africa) tended by a small boy. My guide turned to the boy and asked him whether he knew of any woman who was very sick. The boy glanced apprehensively over his shoulder. The guide repeated the question. The boy's eyes shifted, coming to rest on a hut immediately to our right. We knew where to go.

A young woman, who I judged to be in her twenties, had been moved from her own family's compound to another where she had relatives. This was a way of hiding patients from the authorities, so that they would not be taken away. No one tried to bar our way, however. She was lying on a mat, her face and limbs glistening with sweat. She was feverish and delirious. When I asked, I was told that she had been sick for four or five days.

With Ebola, there was little that I could do to halt the progress. The only possible therapy was convalescent plasma. But would it work? We still had no idea yet, but we certainly had nothing else. Ribavirin is totally ineffective against Ebola. What better way to see if we could do anything at all than to give the plasma to someone who was already sick with the disease? And maybe, in that way, I would find out whether it might be working in me as well. The big drawback for this girl was that by day four or five the disease was well advanced and unlikely to be easy to treat.

It took us a while to persuade the family to place her in our custody. No, we reassured them, we will never try to prevent the family from seeing her once we get her to the hospital. If she died, we said that we would make absolutely certain that her body was returned to them for a proper burial.

That still left us with the problem of getting her to the hospital. Even if there had been transport available, no ordinary vehicle could get through the elephant grass. So the only thing left to do was to carry her out. Optimistic as ever, I reckoned that it would take us forty-five minutes to carry her to the nearest road—or what, at any rate, passed for a road. So we sent a messenger on ahead to arrange for our pickup, while I recruited two members of the woman's family to help us with the stretcher.

It was a struggle. The heat was merciless, the humidity was worse, and there was all that elephant grass to get through. Darkness was falling. It was vital that we reach the road before darkness overtook us. I had no desire to spend the night in elephant grass. Who knew what we would be sharing it with? Besides, we needed to see where we were going, and we needed to give her plasma as soon as possible. It took us twice as long as I had estimated. By the time we reached the road, it was dusk. Mercifully, the girl was virtually comatose; she felt no pain. The pickup was waiting for us.

Once we reached Yambio, we placed her in a separate room in one of the buildings where we were isolating the Ebola patients. The room was uncomfortable, but there wasn't much that we could do about it. There was no ventilation and no windows. Nothing but stale air and paralyzing heat.

I put on a paper protective suit, a surgical mask, and a double pair of surgical gloves. Dr. Omran, the Sudanese physician working with our team, volunteered to assist me in putting in the IV and getting fluids started, but he chose to use the face respirator rather than a mask. It was a mistake; full-face respirators have a much lower rate of air exchange and therefore become fogged very easily. If you hyperventilate, you produce a rapid build-up of carbon dioxide, and nothing will fog a respirator faster than a good panic. My colleague was very nervous. There was nothing I could say that would convince him that this wasn't a high-risk task.

He began to complain how hot it was. When I looked up at him I could no longer make out his face behind the mask. By this time, we were trying to set the IV to administer the plasma, using the feeble light of a kerosene lantern. The IV pole was nothing more than a rough upright of wood nailed on two cross sticks that acted as its base. A nail driven in the top of it doubled as a hook for us to hang the IV bottle. It looked for all the world like a crucifix.

Suddenly, my Sudanese friend muttered that he felt very faint and weak.

"I don't know whether I can continue," he said.

I stopped what I was doing and told him to go outside and take off his mask.

He never came back, so I finished setting up the IV on the young woman myself. She was still delirious. There was no way that I could communicate with her. Though she was completely soaked with sweat, she felt very cold and clammy. Her blood pressure had plummeted to such a point that it was unrecordable.

Although I was aware of the risks of giving patients too much fluid—their membranes had been rendered so permeable by the virus that you might end up flooding their lungs, drowning them—I felt that I had no choice in this instance. It was obvious that I was losing her. I had to get her blood pressure back up. An infusion of fluids might do it; from what I was told, she had been unable to take any fluids for well over twenty-four hours.

Would the plasma work? And if it did work, would it prove effective in a case this far advanced? I only wished that I'd gotten her earlier in the illness. Not that I had any way of really knowing whether the treatment would work at any stage. The only precedent that I had to go on was what had happened to Geoff Platt, who had infected himself with Ebola (Zaire) in 1976. He had been given immune plasma in England, and he'd survived. But, on the other hand, he was also given interferon and good medical care. So it was impossible to say what was chiefly responsible for his recovery. The point was simple: plasma was all we had. I gave her two units. It might not be a cure, but it wouldn't do her any further harm, either. After all, I had already done the safety test on myself.

Ironically, one of the effects of plasma is that, in the short run, it raises the antibody titers to Ebola. This means that once plasma is administered, the presence of antibody can no longer be used as a good diagnostic indicator. It is the antibody in the plasma that is supposed to kill the virus. Put another way, when you measure antibody in someone who has received immune plasma, there is no way of knowing whose antibody it is. In this case, we'd already taken blood prior to giving her the immune plasma, and we were unable to detect any antibodies to Ebola. This could mean that she was early in disease, which would be good, or it could be telling us that she did not have Ebola, which would be better. However, there was not much doubt in my mind: this poor girl definitely had Ebola.

The next day, when I tested her antibody, I was gratified to see that there was a small rise from no antibody to our being able to see it in her plasma at a dilution of only one in eight. That was the good news, at least in so far as she had something to fight the virus with. The bad news was that she had started to bleed around her gums, and there was blood in her stool. It would take a miracle to save her now.

There was no miracle. Two days after we struggled to take her out of the bush, she died. If the plasma did work, it hadn't proved itself in this case.

So, I thought, what about me? Am I incubating this terrible disease?

There was some reason for hope. The condition of the old woman, whose fate had become so entwined with my own, had changed. Now she was sitting up and chatting in a friendly manner. She didn't look like she had Ebola now. She was far too well—and far too cheerful. Even those few patients who did recover didn't start sitting up and making merry all that quickly. My bet now was that she'd had something else. So it was with enormous anticipation that I went to take a blood specimen from her. There was no question that she was convalescent. If she had no antibody, she never had Ebola to begin with. This time, I didn't put off testing her serum; in fact, I was in a hurry to see what the cells would look like with this sample.

I decided that this was something I had to do alone. As I entered Simon's lab, I tried my best to keep my composure. My palms were moist, my heart pounding. I would know at once what the results were. I'd seen plenty of good positive control sera to judge a negative without any problem.

Steeling myself, I scanned the slide, first checking the positive and negative controls. All fine. The test was working. Once again the cells in the old lady's serum were dark green. Negative. She never had Ebola, and I had never been exposed.

It is difficult to describe the wave of relief that rushed over me. I was quietly ecstatic, feeling as if I had just been given back my life and my future. As soon as I finished testing the other sera, I rushed out of the lab. I wanted to tell Roy. I would have liked to celebrate with a good bottle of scotch. But we killed all of that the night of the accident.

CHAPTER FOURTEEN

Sue's Story

It was summer 1983, a few weeks before I left for Zaire to investigate AIDS in Kinshasa. I was about to meet a researcher who would change my life in a number of unexpected ways. I'd first heard about her from David Simpson, a WHO investigator active in researching the Ebola outbreak in the Sudan seven years before. He wrote to tell me about a British woman who was interested in the pathogenesis of viral hemorrhagic fevers. Her name was Sue Fisher-Hoch, and she turned out to be a thin, supremely energetic and articulate woman with curly red hair and a face full of freckles. She called me when she was in Atlanta at the invitation of the National Legionnaires' Disease Conference to present her novel data on the spread of the bacterium *Legionella pneumophila* in hot-water systems in the U.K. Her report pointed out that once Legionnaires' disease was seen in the U.K., it was obvious to her that it couldn't be from air-conditioning systems. Air conditioning is not much used on a cold, damp island. So she set about searching elsewhere for the bacteria.

I hadn't a clue what this research had to do with the pathogenesis of viral hemorrhagic fevers. But I was soon to find out.

I learned that, in addition to her investigation of *Legionella*, she'd already done considerable research on Ebola in the Level-4 laboratory at Porton Down in the U.K. It was clear that she was brimming with enthusiasm about unraveling the mysteries of this virus. When she told me about her experiments with Ebola infections in monkeys, I was aghast. The conditions under which she and her fellow researchers had to work were appalling. Basically, the only protection that they could rely on were superfine HEPA filters to decontaminate the air, and a wardrobe that amounted to nothing more than oversized pajamas and a full-face respirator. That she had been able to carry out her research without infecting herself in such circumstances was even more astonishing.

The kinds of experiments that Sue was working on had never been done before. CDC would be especially interested because there were some striking similarities between the way in which Ebola and Lassa acted on the body. By the time she returned to the conference, we agreed that she should come to CDC at the earliest opportunity to undertake a series of experiments on the pathogenesis of Lassa fever. I assured her that she would find her work conditions a little safer than she was accustomed to. I dropped her off at the Legionnaires' Disease Conference in midafternoon.

Curiously, the same investigation that brought me to the Sudan in 1976 was indirectly responsible for involving Sue in virology. Her interest in the field was sparked by David Simpson, whom she heard lecture when she was completing her master's degree at the London School of Hygiene and Tropical Medicine in 1978. The lecture was a turning point for her, but let Sue tell her own story:

David was working on hemorrhagic fever viruses at Porton Down, Great Britain's only hot lab, continuing research on viral diseases that he'd begun in the wilderness of Uganda. He was one of the scientists who had contributed to the isolation of the Crimean Congo hemorrhagic fever virus—probably his greatest claim to fame. It wasn't as though I was entirely unacquainted with hemorrhagic fevers. I had heard of the mysterious green monkey Marburg disease, which David had also worked on; and I was aware that the related Ebola virus looked like Marburg—filamentous, coiled, and twisted—and that it could cause a fatal disease in man. It was so cryptic and diabolical that it reminded me of the fictional epidemic from outer space described in *The Andromeda Strain*. Where did it come from? Why did it behave with such virulence? I wanted to know more about it. What did the virus actually do? Why did people die so rapidly from it?

I recall David Simpson's lecture vividly. I listened to David talking about his experiences in the 1976 Ebola investigation in Zaire.

So then we'd taxied to a halt on the tarmac of Juba airport. Everything was ready to be unloaded. The trouble was that we had too much gear. Her Majesty's government was a confirmed believer in the efficacy of respirators, a tradition no doubt inspired by the heroic example of the RAF in World War II. However, what served our airmen well in combat was not exactly what you needed on your face while you performed an autopsy on someone who had died from Ebola.

You'd sweat like crazy. You could hardly see what you were doing. But we had no choice. This is the way it was done. Once we'd gotten on the ground, the pilots turned to us and said, "Here's the landing fee. You pay it for us." Then they threw us out with all the gear, and took off.

One of the principal investigators of the WHO Ebola team in the Sudan in 1976, David was a brilliant and engaging lecturer, but also something of a maverick who was able to coast through life on brains and charm. And he had lots of charm. I doubt whether he ever prepared his lectures in any detail, although they never failed to be first-rate. Perhaps it was because he was Irish and had the legendary Irish gift for words that made his talks so compelling. Yet it was neither David's charm nor his penetrating intelligence that had persuaded me to attend his lecture. What really intrigued me was the experiences he'd undergone in his hunt for viruses all over the world.

So there we were in Khartoum, champing at the bit to get going. But we can't move. We're immobilized. We beat a path to the door of the relevant officials, but they insist that there is nothing to be done. A *cordon sanitaire* had been set up around Maridi, and there's absolutely no transportation to and from the city. Everyone was so scared, they told me, there was no way I would find anyone willing to transport me. They would not, or could not, help us.

And then, lo and behold! Deus ex machina. Or if not Deus himself, then his representative. For who, of all people, should appear in Juba, but the archbishop of Canterbury! And what's more, he has his own plane. It turned out that the Church of England has a rather substantial following among the beleaguered folks in the south of the Sudan. He was on his way south to minister to the faithful. Evidently, the archbishop carried more weight with the regime than we did, because no one made any noises about forbidding him to fly wherever he pleased. So, we approached the archbishop and asked for a lift. He said, "By all means, come along."

I'm still not sure if he understood the full implications of the outbreak, or whether he was just ignoring it, protected, of course, by faith.

David then went on to expand upon his investigation in Maridi, where he found a deserted hospital, with many of the staff dead from Ebola. The remainder were cowering in their houses, but cases were still occurring. He performed autopsies in the middle of an open field. He finally reached Nzara and ended up clambering along the roof of a cotton factory, catching bats in the hope that one of them might harbor the virus that causes Ebola. He then packed six dead bats in a crate and sent it back to Porton, where it was opened by a researcher named Ernie Bowen. Ernie hadn't been told what was in the crate, so he was rather surprised to discover six dead bats. Regrettably, researchers at Porton could find no evidence of Ebola in the bats. But, as far as I was concerned, that was beside the point. As David talked on, my only thought was: *This is what I want to do.*

It had taken me quite some time to reach this decision. And it would take longer still actually to become involved in an investigation similar to the one David described. But, then, I was always a late starter, and I had a lot of obstacles in my path—more than I would ever have imagined.

I was born in August of 1940 in Denby, England, in the middle of the only bombing raid the town experienced during that long hot summer. My mother tells of gathering blackberries before I was born, which is very unusual because blackberries ordinarily don't ripen in the north of England until the end of August. My earliest memory is of lying on my back on the stairs leading to the air raid shelter, as enemy planes flew overhead.

When I turned eleven, I was sent to boarding school in the north of Wales. It was like being in exile—the isolation was virtually complete—and the six years I spent there were the unhappiest of my life. To escape the loneliness, I read voraciously, devouring all the literature and poetry I could lay my hands on. I took up music, playing the piano and organ.

When I graduated, with A levels in English, French, and history, I found myself at a watershed. My parents were moving close to Paris because my father was to work with NATO in Fountainebleau. I now had a chance to study at the Sorbonne and do some real traveling.

After spending two years in France and Italy, I had developed a facility for languages and a passion for foreign lands. I wasn't looking forward to going back to the cold and wet grayness of England, but I needed to make a living. By this time I considered myself practically a Mediterranean at heart. I just couldn't readjust to life in my native country. I married a man older than myself, who had traveled widely, particularly in Africa.

The restlessness that my own travels had stirred in me did not go away with marriage. On the contrary, it only became stronger as time went on. At the age of twenty-seven, I knew I had to take some kind of action. I decided to do something that I'd been told was quite impossible for a married woman, especially one with a child. I would become a doctor.

In spite of the bias prevailing in Britain during the late 1960s and early 1970s against women entering medicine, I attended physics, chemistry, and zoology classes at a local technical college in order to satisfy pre-med course requirements. Perhaps those in charge thought that the work would be sufficient to discourage me from following my aberrant road. Indeed, I endured a difficult first term, sitting at the end of a row of fifteen-year old boys. In spite of their immaturity, they had two years of physics under their belts, which meant that they knew all the answers, while I didn't.

I was, however, determined. I wrote down everything I could, and when school was over, went to pick up Hannah, then got back home to prepare dinner and take care of my family. Only after all this was done could I settle down to do my homework. And that didn't come easily, either. There were no pocket calculators in 1967, and I was too ashamed to tell anyone that I had forgotten how to do logs. So, for a whole semester, I did my calculations by long division and multiplication, right across the page, like the mouse's tail in *Alice in Wonderland*, until my brother took pity on me and gave me a slide rule.

For all the handicaps I was operating under, I still managed to earn straight A's in A-level physics and chemistry—good enough, I thought, to expect easy entry into medical school. There were quotas for women; their admission was limited to 15 percent of incoming classes, and most of the successful girls were chemistry applicants straight out of the exclusive private schools. But, even so, I believed that I could get in. I wrote to the dean of every medical school I could think of. All the responses were the same: I was deemed "unsuitable" and told, in so many words by one school, to go back to the kitchen sink. In the meantime, I watched medical schools accept my younger male classmates, with poor grades, on the basis of their rugby skills.

Thankfully, there was one exception: Dame Frances Gardner, Dean of the Royal Free Hospital. She was intrigued enough to call me in for an interview. After we had a chance to talk, she agreed to admit me, even though she was clearly doing so against the wishes of the rest of the admissions committee. At that, she posed one inviolable condition: I had to submit a letter from my husband guaranteeing that he wouldn't divorce me before I had completed the four-year course. He was kind enough to comply.

Dame Gardner was married to a man named George Quist, who was one of the most eccentric people I'd ever met. His name alone seemed to recall a character from Dickens. In fact, he was rumored to have been the model for Sir

Lancelot Spratt, the eccentric surgeon in the comic novel by Richard Gordon, *Doctor in the House.* He often declined to scrub in the operating theater, although (quite reasonably) he never performed an operation himself unscrubbed. He would walk into the theater and put on a large plastic apron. To demonstrate a surgical technique to his students, he would take a huge set of forceps with a bloody swab and proceed to illustrate his point by using it to draw on the front of his apron. He had a large girth, so there was plenty of space for the illustration. According to one (alas, probably apocryphal) tale, George Quist once told me as, driving erratically, he gave me a lift from one of our hospitals to another, that he had given beer money to laborers working on the new hospital and encouraged them to stay on strike. He said he didn't "hold with" new buildings.

It was with a considerable degree of anxiety that I entered medical school. I was sure that I was going to find myself among a group of highly intelligent kids, who would wipe the floor with me. Then there was the matter of the hour and a half commute each way—we were living fifty miles outside of London—which took a big chunk of time out of my day. But I felt so privileged to be able to attend medical school that I didn't mind juggling the demands of family and work. I was very fortunate to have my family's support and something even more important: I had the capacity to surprise myself. And, as it turned out, I was able to do much of my reading on the commuter train. The only problem was the curiosity exhibited by my fellow commuters, in their pinstriped business suits, when I pulled out my anatomy books—particularly when I got to the intimate bits. One memorable day I even placed the box with my real skeleton on the luggage rack above my head, hoping that it would not fall down and spill its contents. Such an accident, I had been told, would have me arrested under the Anatomy Act, a law dating from who knows when designed to stop anything medical from unduly shocking the general public.

My career was also influenced by another indomitable woman, Dame Sheila Sherlock. It was she who guided me through my internship. Majestic and demanding, she was known as the Yellow Queen, because of her international standing in the field of jaundice-inducing hepatitis. Under her dominion, it was simply out of the question to make a mistake. As eccentric as she was brilliant, Dame Sherlock was the sort of character Lewis Carroll might have invented. She once came to a Christmas party thrown by her residents and immediately stepped up to the punch bowl, a large chromatography tank filled with dry ice to make it smoke invitingly. The residents grew silent.

After tasting the concoction, she puckered her lips and frowned: "Not enough alcohol!" Turning to one of the residents, she said, "Boy, go and look under my desk. You'll find a bottle of whisky there. It's a gift from a grateful patient. Fetch it." Once the bottle was in her hands, she poured it in. All of it. It turned out to be a very good party.

I was fortunate to have a friend in Teresa Tate. A lovely girl with long ash-blond hair, she had come from a wealthy family and had even gone to school with Princess Anne, so it wasn't the prospect of future earnings that accounted for her presence in medical school. Rather, she was driven by her own ambition to become a surgeon. She was spectacular—nothing could deter her—and when the two of us teamed up together, we were a force to be reckoned with, which greatly pleased Dame Sherlock. This was the one time in my life when my boss favored women over men.

On one occasion, Teresa and I had the bright idea of using colored crayons to mark patient charts: red for hemoglobin, yellow for bilirubin. Bilirubin is what causes jaundice. We taped the results to the end of the patient beds. Dame Sherlock thought these were great and held them up to the senior residents. "There you are, boys, this is the way to do it!" she proclaimed. (She always called the males "boy.") "Now you can see what's happening." Our colleagues just loved us. We could see them grinding their teeth.

After medical internship, I had to serve six months as a surgical resident. But surgery was not for me. Surgeons didn't really seem to me to do much in the way of thinking. They just cut. Besides, my interest lay elsewhere. Ever since I'd worked under Dame Sherlock, caring for patients with hepatitis, I had developed a fascination for viruses. So, as soon as I finished my ordeal as a surgical resident, I was naturally drawn to the Public Health Laboratory Service (PHLS), a good place in Britain for training in virology.

The problem was that viruses were the Cinderella of science at that time. The medical profession didn't take them seriously because they were not thought to be a very significant factor in causing disease, and, in any case, "you couldn't do anything about them." The focus of public health on disease prevention was a foreign concept in those days. You waited for the patient to get sick, and then tried to cure him. It was also widely believed that modern medicine had conquered diseases such as polio. One colleague even reproved me for my interest, insisting that it was quite "inappropriate" to specialize in viral diseases. What I needed to do, he said, was bacteriology—as he had done—which would provide me with all the knowledge I needed to become a virologist. Somehow, though, I was convinced that he was wrong, and that, in fact, most everyone was wrong. I had a gut feeling that virology was about to become a growth industry.

CHAPTER FIFTEEN

Virology, Plumbing, and Ebola

After completing my Master's degree at the London School of Hygiene and Tropical Medicine in 1978, I went to Oxford, where I was employed by the Public Health Laboratory Service at the Radcliffe Hospital. As soon as I could, I requested a transfer into virology. I was informed that there was no room. Disappointed, I turned my sights in another direction, and found a position working with Bob Mitchell, a very good bacteriologist and teacher, who directed the bacteriology laboratory at neighboring Churchill Hospital. My overall boss was John Tobin, director of the Public Health Laboratory Services and chief of virology. John was in his late fifties or early sixties. Fast-moving, enthusiastic, and witty, he was something of a maverick. Anything new, and he was in there, slightly wild-eyed. He enjoyed sitting at the lab bench and doing technical stuff himself. He would be happy to draw in anyone else who wanted to get involved and work with him. I was one of those who volunteered. John was the first person to teach me virology.

"In virology," he told me, "it doesn't matter what you do as long as you do it fast."

At that moment we were preparing tissue cultures—cells grown in tubes—and we had to be very careful not to contaminate them with bacteria and fungi. It was good advice: speed works. We were using the fluorescent antibody test—a technique called immunofluorescence (IFA)—to detect a newly described organism, Legionella pneumophila, which causes Legionnaires' disease. John wasn't fussy; it didn't make any difference to him that Legionella wasn't a virus. All that mattered was that it was new. Anyway, the immunofluorescence technique is used more by virologists than bacteriologists. It identifies or quantifies the antibody or antigen of infecting organisms. The fluorescent dye allows you to visualize the organism, because it shines yellow under a certain wavelength of ultraviolet light. Antigens are substances that the body perceives as "foreign" invaders. These can include viruses, toxins, incompatible blood, bacteria, or organ transplants. Their presence in the body stimulates the immune system, which in turn manufactures protective antibodies against them. Without this mechanism, human beings would long ago have vanished from the planet. One of the most notorious antigens at the time was, of course, Legionella, which was responsible for a mysterious outbreak of fatal pneumonia in Philadelphia among attendees of an American Legion meeting there in 1976. Even though two years had passed since the outbreak, John was the first person to work with the bacterium in England. Bob Mitchell was also very involved, and the three of us worked together to find out more about the disease.

Our research was conducted mostly after hours—and for good reason. Every day, around 5 P.M., when the lab technicians were about to head home, John would come up to me and say, "Sue, let's fluoresce." Others might find this an odd invitation, but I responded eagerly. This was fun. I would put the routine of the day behind me and get out the cultures and work with them until late at night. The main reason we did this after hours was because the technicians were getting upset and anxious. What did we think we were doing, putting their lives at risk by handling highly pathogenic new organisms? And few organisms, they thought, were more pathogenic than Legionella. This was the beginning of my "Level 4" life—handling things that other people thought were dangerous!

I didn't worry about the danger, though. I figured I'd be safe so long as I had a healthy respect for the organism I was working with and followed reasonable precautions. Besides, I had confidence in John. He had been around organisms for a long time and knew how to deal with them.

Our work with Legionella began to pay off. John and I were invited to address the clinical respiratory diseases team. We explained what we knew about the disease, emphasizing its symptoms. Physicians would often mistake Legionella for more common pneumonia and never realize what was really the matter with their patients. We were sure it had to be much more widespread than people thought, and we were right. The disease we were studying under the microscope was about to make its presence felt right inside of the wards of Churchill Hospital itself. And the person who was going to alert us to that fact was sitting in our audience. He was a young resident named Martin Muers.

A few days after our talk, Martin called me. He said he was phoning from the Radcliffe Hospital, which was just down the hill from us and located in the center of Oxford University in the old city of Oxford.

"I'm about to do a bronchoscopy on a patient who has had a renal transplant," he said, then added: "I think she might have Legionnaires' disease. What do you need to do a test?"

"I'd like you to get a specimen from as far down as possible in the lungs, whatever you can extract without contaminating it on the way up."

There was silence on the other end. Undoubtedly, Martin was trying to figure out how he could meet my specifications. I made some suggestions. Finally he said, "Okay, let me see what I can do."

An hour went by. It was late, and the lab technicians had already gone home. Then I looked out the window to see Martin come bicycling up the hill. He had with him a 10-milliliter sterile centrifuge tube. Inside were three strips of bronchoscopy tubing. Great. Just what I had asked for.

"I hope you cut them with sterile scissors," I said.

I put the flask in the safety cabinet until I could figure out what to do with it. What I needed was located somewhere inside the pieces of tubing, but how to get it out? I couldn't just cut it open because that would contaminate it with all sorts of bugs from the respiratory tract, which would be scattered along the outside of the tube, and they would overgrow my Legionella, and I would lose it. The tube had a very fine bore, and I had nothing thin enough to reach inside. The material inside was also too viscous to be sucked up with a sterile hypodermic needle, and, anyway, the piece of tubing was longer than any needle I had available. I wanted something long and thin and sterile, which I could use to slip inside and suck out the contents.

So I made what I needed. I heated a glass pipette over a Bunsen burner, then, when it became red in the flame, I drew out the glass until it was as fine as I could get it. I had to practice a few times and throw away some broken glass before I got it right. I now had a long, thin, hollow glass tube, fine enough to insert into the bronchoscopy tubing. I went to the cabinet, turned on the fans so that I could be assured of getting proper ventilation, and put on a pair of gloves. I opened the flask, carefully removing one of the pieces of tubing with sterile forceps. Then I cut the end of with the scissors, and inserted my fine glass tube. I slowly pushed it in.

It worked.

I managed to extract some blobs of what looked like goo. Then I placed it on the special *Legionella* medium, which Bob Mitchell and I had prepared that same day. The medium was a reddish mixture made up of a sort of semi-solid agar jelly, with added nutrients such as iron, vitamins, and some blood. If the goo grew something with a dark halo round it on this medium, then we'd know that we were on the track of *Legionella*.

Three days later, it looked disappointing. Not much was there, but at least I was gratified to see that no contamination had occurred. But Bob Mitchell, who had the experience to see what I could not, noticed that something in the medium had changed. To the naked eye, it didn't look like much—just a little browning of the medium underneath the blob of mucus. He picked up the plate and showed it to me by shining a light through the agar. There it was! You could just make out a tiny halo around a tinier pinpoint of something beneath the mucus blob. That was what we were looking for. This was the first *Legionella* ever isolated outside the U.S. from a live patient. We called Martin in delight.

"Congratulations," I cried. "You've got it! It's *Legionella*. Start her on erythromycin." It was the drug of choice for Legionnaires' disease.

We were then faced with the job of finding out how the patient had become infected. After an intensive investigation, John and Bob were able to demonstrate that the patient had likely been infected from a shower unit in the renal transplant suite where we isolated the same strain of *Legionella*. The patient herself, I am happy to report, was successfully treated with erythromycin and recovered.

A few months later, I moved to St. George's Hospital in South London—a district known as Tooting—to take a real virology job. But like a relentless suitor, Legionnaires' disease followed me, or maybe I took it with me. I certainly was accused of that. I was setting up a virology service at the Kingston-upon-Thames District Hospital when the chief bacteriology technician, Malcolm Smith, came to me and said that he was bored with having nothing but routine work.

"With all this virology starting up, I want something new to do, too," he complained.

So I brought him my own *Legionella* strain and reagents that John Tobin had given me, together with the medium that I'd smuggled out of Churchill. I had to be discreet about this because people were terrified to be anywhere near *Legionella*. But not Malcolm. Holding out the tube with its deadly contents, I said, "There you are, Malcolm. Make up the reagents, and see whether you can find a case of Legionnaires' disease." With an offer like that, who could refuse?

Two weeks later, I was attending a meeting in Kingston Hospital. The door opened. We stopped talking and looked up to see Malcolm's face. He was grinning broadly.

"I've got one!" he declared triumphantly.

"One what?" I asked.

"A case of Legionnaires' disease," he replied.

He was waving a slip of paper, which he held between his finger and thumb.

It was a lab report. He had grown up the organism I had given him, made slides, and examined serum from a patient with pneumonia. It revealed high-titer antibodies to *Legionella*. He had discovered the patient after going through several request slips for lab tests until he found a description of a case that sounded like Legionnaires' disease. Then, he told me, he'd called the resident who was responsible for the patient to tell him the news. At which point the resident said "What's Legionnaires' disease?"

"Never mind," Malcolm had replied. "Just send me some serum." And he did.

I immediately got in touch with Chris Bartlett at the Communicable Diseases Surveillance Center at the Public Health Laboratory Service headquarters in North London. Chris was one of the very few epidemiologists in England at that time. I had learned a great deal from him the previous year while I was in Oxford. When he learned the news he was as excited as I was, but at the same time he was inclined to be conservative: one case, after all, did not make an outbreak. We decided to get as much information as we could about this case, and look further.

A week later, Malcolm reappeared in my office in Kingston. There was a familiar grin on his face.

"I've got two more!" he said, clearly very pleased with himself.

He sounded like a hunter who'd bagged another pheasant. All three cases, it turned out, had probably become infected in our new hospital building. That was it.

"Okay, Malcolm," I said, "I think we really do have an outbreak."

I put in another call to Chris Bartlett. This time there was no hesitation. Chris said, "I'm on my way."

Suddenly, I became a detective. What was the source of the outbreak? What was the means of transmission? I soon began to realize that an epidemiological investigation involved more than tissue cultures and pathogens. It could also involve engineering. I was now about to learn how a hospital plumbing system operated. First thing: no obvious air-conditioning system. The Americans said *Legionella* came from air-conditioning systems. But we didn't have any. Back to the drawing board—literally. My guide to blueprints and the technology of operating a hospital was an engineer named David Harper. There wasn't anything he wouldn't do for Chris, Malcolm, and myself. If it was a piece of plumbing I was interested in testing, he would simply say, "Okay, Doc," and then heave his large frame under the pipes to reach the particular set from which I wanted a water sample. One of the things I learned was that the design of a large institutional building is more random than you would expect. Although the building was new, not everything was installed properly. When we went back and examined the blueprints, we found that much of the piping did not go where it was supposed to. This proved to be as much of a surprise to the engineers as it was to me. So David explored the system with us in tow, gathering specimens here and there, wherever we thought there might be stagnant water.

Twice a week, I would attend the engineers' meetings. I became so immersed in the nuts and bolts of their work—literally—that once I had to stop myself from giving them advice as to where to locate a one-way valve.

To detect evidence of the bacterium, we would collect samples of water in twenty-five-liter containers. Once filled, containers of this size are quite heavy. So, in order to transport them, we needed some kind of conveyance—and no conveyance is more easily acquired in a hospital than a wheelchair. We hijacked them wherever we could find them. Although we did our best to be discreet— we didn't want to alert either the staff or the patients as to what we were up to—the theft of the wheelchairs, however temporary, did not always escape unnoticed.

"Hey, bring that back!" we'd hear people call after us.

We simply pretended not to hear them, and moved fast.

Isolating the organism from water was difficult. Nonetheless, Malcolm performed magnificently. First he had to pass the contents of each container, all twenty-five liters of it, through a single filter. Then he had to inject the deposit that remained on the filter membrane into guinea pigs to see whether they would become infected with *Legionella*. At that time, that was the only way to do it.

But why this intense interest in plumbing systems? It was simple. American investigators had nailed air-conditioning systems as the source of Legionnaires' disease. Although at Oxford we had been able to show that *Legionella* could also be found in a shower, the evidence was still circumstantial. In any case, no matter how exhaustively we searched, we could not find an air-conditioning system at Kingston Hospital. Not that we really expected to find one: After all, it is damp and cold in England. True, we did discover an air-cooling system, which was used for the operating theaters, but neither of the three patients who fell ill had undergone surgery.

The first patient who'd become infected was a student who had traveled down from Edinburgh by British Rail and taken up a vacation job cleaning floors at the hospital. He'd worked on the top floor, where the windows were left open, so he might have been exposed to the exhaust of the ventilation system, which was located on the roof. Another possibility was that the infection had occurred on the train. It was an interesting idea, but not one supported by the available evidence. When other patients came down with the disease, none of whom had had any contact with the ventilation system, we were forced to rule out that hypothesis, too. Eventually, we found twelve cases, most of whom had been infected in the hospital. Four of them died. One was a baby, the first baby ever reported to have been stricken with Legionnaires'. Actually, it was the baby's mother who made the diagnosis. The child had been admitted to the hospital and then released. Shortly afterward, the family went on vacation. It was then that the child fell ill with pneumonia. When the baby was admitted again, this time in a hospital in a resort on the southern coast of England, the doctors couldn't figure out what was wrong. Finally the mother asked, "Could it be that Legionary thing?" There are some advantages in publicity.

Not only did I become familiar with the mysteries of architectural engineering, I also had an opportunity to acquaint myself with the arena of aerobiology. Aerobiology is something of an archaic art, a relic of World War II, when biological warfare and gas masks were all the rage. To determine whether our elusive pathogen was airborne we recruited World War II experts from Porton to come and measure air patterns. Although they'd made few changes in their methodology in thirty years, they possessed a nice little machine that blew bubbles from the top of the hospital roof. They wanted to see whether the bubbles would float into upper floor windows where their appearance would be recorded by an elaborate detection system. Apparently the idea was to register air flow. Whatever its purpose, the experiment had little chance of working, not on an island as naturally windy as England. Their data, unsurprisingly, proved inconclusive.

A relative of one of the patients who'd died worked for a local television station, so it wasn't long before the news of the outbreak became known to the public. A television interview was set up with one of the clinicians. Under the guise of a friendly chat, the reporter began to probe him, asking him hard questions for which he was obviously unprepared. His statements were taken out of context, so that he seemed to be endorsing opinions that were the exact opposite of what he meant. For example, the clinician wiped his brow in sheer fatigue at one point. But the way it came across on TV made it look as though he were responding guiltily to a question about clinical negligence.

The reports in the media sparked a rally by incensed relatives outside the hospital. Actually, the demonstration didn't go too well; it was raining, so only about a dozen people turned up. Nonetheless, Chris and I found ourselves under siege in coroner's courts, trying to explain to the group why the patients had died. In fact, we assured them, the hospital had behaved in a highly responsible way. Our doctors had diagnosed and treated the patients appropriately, and we had gone to great lengths to deal with the source of the infection. Of course, the irony was that if we had never made the diagnosis, there would have been no accusations, because no one would have known anything about the existence of Legionnaires' disease to begin with. The cases would have been labeled simply as "pneumonia, origin unknown." This is one of the penalties of investigating new diseases.

There was some panic among staff as well. Chris and I had to cope with their apprehensions with constant reassurance. We did our best to persuade them that it was perfectly safe for them to go into the hospital building and look after the patients. They gradually got over their fears. In many respects, the recognition among both the staff and the patients that we were actually trying to do something succeeded in restoring confidence. However, we couldn't please everyone. There were still some people who held us responsible for the disease just because we told them it was there!

Meanwhile, Malcolm was continuing to grow *Legionella* out of the water samples we were bringing him, but the results of his tests weren't conclusive enough for us to nail down the source of infection. Then, one evening, David Harper, who was on call for engineering, responded to a complaint from the nurses that they weren't getting enough hot water on the wards. He went into the plant room to see what he could do; the plant room fed the entire hospital with hot water. There were three large cylinders in this room, known as calorifiers. Only one of the calorifiers was on line at the time and producing hot water. A second was empty—down for maintenance. The third was stagnant, but still ready for use. He turned on the steam supply to the stagnant calorifier to increase the flow of hot water. As it heated up the water, layers started to flux and mix as currents developed within the tank before the water could be drawn into the hot-water supply system to the ward.

A few days later, David Harper became ill with pneumonia and was admitted to the hospital. We suspected that he had Legionnaires' disease, though we were never able to confirm this. After being treated with erythromycin, he began to recover. With little to do but lie in bed and think, he searched his mind for the key to the mystery that had eluded him throughout the investigation. Suddenly he had an inspiration. He sat up in bed, shouting out, "I've got it!"

It was a Saturday morning, and there was no one around. That was fine with him; he didn't want to say anything until he'd had a chance to confirm his hunch. As soon as he felt well enough, he got dressed and went to the plant room. He located the calorifier that had been left open for routine maintenance. He peered in. Seeing the thick, scaly, waterlogged deposit at the bottom, he scooped out some of it, placed it inside a sterile jar, and took it to Malcolm.

David's inspiration had been right on the money. The deposit yielded an astonishingly high titer of *Legionella*. In fact, it was almost pure *Legionella*—and it had been sitting there the whole time in the hospital's hot-water system. What David had figured out was that the organism must be flourishing on the bottom of the cylinders—*Legionella* was an organism that lives in water, after all—and because it grew so far down in the cylinder, the water never became hot enough to kill it. It was only when David remembered that the same night he'd turned on the stagnant calorifier two new cases of Legionnaires' disease had occurred that we had our real breakthrough. By stirring up the crud at the bottom of the tank, David had inadvertently allowed *Legionella* to get into the hot-water system. We knew that at least one of the infected patients had taken a shower that night. He would have had no way of knowing that he was showering in high-titer *Legionella*.

We ran several experiments to see how we could stop this sort of occurrence from happening again. Could we sterilize the cylinders and kill off all the organisms they might contain? We decided to test one cylinder by raising the temperature of the water until it was boiling and then see what would happen. The whole time that the water heated up, we continued to measure the temperature of the cylinder's outer surface. As the gauge began to creep towards 212°F, I had to fight back a growing sense of panic. These cylinders weren't designed to withstand temperatures so high. I was terrified that David was going to blow himself up. He was crawling around under the tank with temperature probes, trying to see how hot he could get it. The needle continued to creep up the gauge, and I was sure the whole thing would burst and deluge him with boiling water. I held my breath, but nothing happened. We all survived. The insidious organisms inside the cylinders, however, did not.

In the course of our investigation we came to rely on a scientist from the Thames Water Authority named Jenny Colborne. She had a lovely apparatus that she used to test the ability of plumbing materials to support bacterial growth. Sounds somewhat dull until you start putting *Legionella* into her system. It grew beautifully on plumbing materials like washers and caulking. She discovered that when a faucet or shower head hadn't been used for some time, the water that first emerged was stuffed with *Legionella*. Now we had a good idea how people were becoming infected; it was from the potable-water systems. To prevent future infections, we put together a series of guidelines that called for chlorinating cold water and heating hot water so that no organisms could possibly flourish.

My epidemiological experience with *Legionella* became the basis for my doctoral thesis, which I presented in 1981 at London University. By this time, too, I had met my virology membership qualifications at the Royal College of Pathologists—roughly equivalent to being board certified in the States. My life was beginning to change. But I had no idea just how dramatically.

The same year, 1981, I went to see David Simpson at Porton Down to ask him for a job. He was more than happy to oblige, but he needed to secure the necessary funding. So, together, we applied to the Wellcome Trust for a fellowship that would allow me to become a full-time research virologist. The fellowship also allowed me to work abroad for the first time. A rabies investigation was in progress in Thailand, and I was invited to join it for three months. A team of scientists, led by David and Mary Warrell, had set up a research program in Bangkok's Mahidol University. Mary is a virologist expert in rabies. She and I had worked together in Oxford before she went to Bangkok, and I left for St. George's, so we knew each other well. The Warrells were trying to find out whether it was possible to treat rabies with interferon, a naturally occurring protective protein that the immune system produces to fight viruses. No one had ever survived rabies, but David thought that, with the right kind of clinical support supplemented by antiviral agents, it just might be possible.

It turned out to be a riveting experience.

One of the things that made it so difficult was having to work with rabies patients. One in particular made a lasting impression. He was a man who was already in the advanced stages of the disease. What especially got to me was the look of terror in his eyes. He knew full well what was happening to him. Just trying to keep him sedated was a major problem. The nurses kept having to moisten his dry, cracked mouth. Even after he became unconscious, dabbing his lips with water would trigger terrifying convulsions. I thought he was going to shoot out through the window. The disease had made him hypersensitive to the slightest touch. There was nothing that we could do for him. The interferon therapy failed, he died, and I met my first real virus.

Mary established a virology laboratory, where we isolated rabies viruses from brain biopsies from our patients and then grew them in mice. I had no problem working with mice. I did have a problem with snakes, though; in fact, I am pathologically afraid of them. Unfortunately for me, David was fascinated by problems associated with snake bite and was working with snake venom. The lab only had two animal rooms—one for David's collection of snakes and one for the rabbits. David declared that snakes didn't get rabies, and that rabbits did, so we put the rabies-infected mice in the same room with the snakes. Because I had to work with the mice, it was necessary for me to walk past the snakes twice a day. David had some very nasty snakes. I would greet every morning with terror as I contemplated having to enter the hut on the roof that served as our animal house. These weren't ordinary run-of-the-mill poisonous snakes. Among them were large cobras. They took as dim a view of me as I did of them. As soon as they saw me, they would begin to hiss and lunge in my direction. David assured me that there was no way that they could get out. The only concession that he made to my antipathy toward the creatures was to exempt me from participating in the venom-milking sessions held once a week. I later found out that David's safeguards weren't all that they were cracked up to be: one snake actually did succeed in escaping. They took care that no one told me at the time.

Putting snakes and rabies behind me, I returned to England to take up the study of yet another virus. It was one of the most intriguing of all, because so little was known about it. It was called Ebola. My interest in it was primarily to figure out how the virus wreaked such catastrophic effects on the human body. But as I was really getting into Ebola in the spring of 1983, another opportunity arose that would have a lasting impact on my life. Both Jenny Colborne and I received an invitation—accompanied by round-trip tickets—to attend the CDC's first Legionella conference. The outbreak of Legionella at Kingston Hospital had attracted considerable interest because of our work with plumbing systems, and as two of the researchers who were most intimately involved in the investigation, we were being asked to present our findings.

A couple of months in advance of my departure, David wrote to two people he thought I'd like to meet while I was in the States. One was Karl Johnson, who was then at USAMRIID. The other was the branch chief of Special Pathogens at CDC. His name was Joe McCormick.

When I returned to England from Bangkok to begin my Ebola research, I knew that I would have a great deal to learn. I had never worked in a Level-4 situation before. Not being able to allow anyone else to help me because of the danger of handling the virus, I knew that I would have to develop the necessary skills myself. I was interested in how the virus worked, and therefore needed to be able to set up assays that looked at the pathophysiology of the infection. Pathophysiology is the study of how a virus causes disease. One aspect of great interest was the effect the virus had on platelets and endothelial cells in the tiny blood vessels of the capillary system. Platelets are involved in stopping bleeding, and there was evidence that platelets were affected in Ebola infection. Endothelial cells line every blood vessel in the body and keep the fluid and blood cells inside the blood vessel. In Ebola patients, the blood vessels become leaky and stop holding in the fluid and blood cells. Basically, in Ebola, both platelets and endothelial cells cease to function. I needed to figure out why.

To help me get up to speed, I was referred to Guy Nield, who worked in the renal unit at Guy's Hospital. Guy had done some interesting work on platelets and endothelial cells in renal disease and was willing to teach me techniques that I believed would be applicable to Ebola research as well. I also had to learn how to work in a Level-4 lab, using the simplest, most reliable techniques and apparatus, and I had to learn how to do everything myself. I was preparing to work with live viruses, not killed viruses. This was because the techniques I would be using relied on bioassays—tests that measure the function of living cells. If we were to kill the virus, we would have to kill the cells. So, if I wanted to work with live cells, I would have to work with them with the virus still alive inside them.

Once I got to the Level-4 lab at Porton Down, I was fortunate to have the assistance of the best technician they had. David Simpson told me that he was assigning Geoff Platt to help me with my Ebola studies. Geoff was a marvelous gift. An experienced and skillful arbovirologist, he was also level-headed and delightful to be with. I couldn't have had better support. Without Geoff, I could never have completed the monkey experiments that were at the heart of my Ebola research. The work was grueling. We had to work with monkeys, because these were the only animals that we could infect and that provided a good model for what happened in humans. We could control the disease process in monkeys, and we could set up the delicate tests we had to perform. Human patients were, fortunately, not available at that time, and had they been, they would have been in places like Nzara or Yambuku. It would have been difficult to set up the experiments in such places. We had to have controlled conditions. The objective was to gain insight into the disease, which would allow us to figure out how we should treat it.

We'd spend long hours laboriously performing clotting tests, endothelial tests, and platelet function tests, all carried out before an intent audience consisting of several caged monkeys. They were a rather voluble audience, too. As soon as they would recover from the anesthetic we gave them, we couldn't get them to shut up. From time to time, they would throw things at us.

In addition to the experiments, we also had to do autopsies on the monkeys. Which was how we came to rely on Arthur Baskerville, a veterinarian and a histopathologist. It was Arthur who did most of the cutting. What astonished me when I witnessed these dissections was the size of the masseter, the monkey jaw muscle. It is huge, many times larger than any human jaw muscle. It was a graphic reminder why you should never be in a position to let a monkey bite you. The most perilous part of the autopsy came when Arthur began to dissect the brain, using a saw. As the saw began chewing up the bone, I would grow increasingly anxious: any flying debris might contain Ebola. I would always count heads daily for several days after each autopsy. When five had passed without anyone showing signs of fever, I knew I was home free. At least until the next dissection.

Improbably, one monkey actually survived my first experiment. As expected, it became extremely ill from the Ebola we had given it, but while the other monkeys died, this one made a remarkable and complete recovery. No one could explain it. With only two days remaining before I left for Atlanta to attend the *Legionella* conference, I couldn't delay any longer. I had to determine what to do with the monkey. It was a difficult decision. I felt that, having made such a miraculous recovery, the animal deserved a break. Only the rules at Porton were rigid. If they could have, they would have autoclaved Geoff and me whenever we left the laboratory, so they were hardly about to allow a monkey to go free. Reluctantly, Geoff and I came to the sad conclusion that we would have to sacrifice it. But I simply couldn't do it myself and decided this was David's job. He understood and came down to help me with this miserable duty.

The following day, Jenny and I were on a plane bound for Atlanta, ready to describe our findings in the Kingston Hospital plumbing system. Upon my arrival at U.S. customs, an immigration official asked me if I had been in any contact with any infectious disease. Luckily, Jenny answered for me. She was concerned I might start telling him about my Ebola monkey, and I would find myself dispatched back to England as a health hazard.

The two of us were surprised to make quite a stir at the conference. For one thing, we were bringing news to the gathering. Up until this point, the commonly accepted wisdom was that *Legionella* was spread exclusively by air-conditioning units. We told them that *Legionella* could also thrive in hot-water cylinders and in shower heads. For another, it was unusual to have the expertise of someone like Jenny, who was so conversant with plumbing materials. Her impact was enhanced by being young, blonde, and good-looking.

That evening I called Joe. He said that he'd come by and pick me up, but then admitted that he didn't know where I was staying.

"It's been years since I've been downtown," he said.

He managed to find me, rolling up in a tiny beat-up Honda, later to be dubbed the "Joemobile." He came into the lobby of the hotel, introduced himself, and then took me to see the CDC. I was very excited. CDC was Mecca for me. I was first amazed at how big it was, and then at how relaxed and friendly people were. In Joe's office we fell into an intense discussion about the pathophysiology of viral hemorrhage fevers. It was a subject that we would return to, separately and together, many times afterwards. I was impressed with his grasp of any subject that we covered, and found him both a good listener and a good talker. This was the scientific feedback I needed.

Naturally, I was eager to tell Joe about the experiments I'd just completed at Porton.

"We have data indicating that platelets and endothelial cells failed to function long before they looked damaged microscopically," I explained. "Essentially, I've shown that our monkeys died because their circulatory systems failed to maintain their own integrity."

In simple terms, the blood didn't clot because the platelets, which normally stop bleeding, didn't work. At the same time, the lining of the blood vessels, the endothelial cells, also failed to function. So what resulted was a one-two punch. First the blood wouldn't clot, and then it leaked through the compromised lining of the blood vessels. The patient starts to ooze blood from multiple sites, and the lungs and tissues become boggy with the fluid leaking from the blood vessels. Curiously, though, for all the devastation the disease can wreak on the circulatory system, it doesn't destroy the organs themselves. Contrary to a common misconception—spread by certain movies and best-selling books—the vital organs to not liquefy or turn to gumbo, as one author described it. Actually, the great mystery of these viral hemorrhagic fevers is that the organs appear relatively intact, both to the eye and under an electron microscope. There is plenty of healthy-looking tissue left. In fact, sometimes the only way you know that the patient is dead is because there is a dead body in front of you.

What I was proposing was that the disastrous failure in dying patients was functional, not biochemical—that is, there was no destruction of cells. This was why the disease was so fast, and also why, if the patient recovers, the recovery is also fast. If we could get the patient over the critical phase, then recovery is complete. This gave us a chance to work out a way to treat Ebola.

Joe was intrigued by my findings. He'd seen many patients make a rapid and complete recovery, but had never quite figured out why. Now I was offering an explanation to better explain the histology and pathology of viral hemorrhagic fevers. Joe had done numerous autopsies in West Africa, and been repeatedly stymied in his efforts to understand how the virus acted.

Such an intense discussion about deadly disease had naturally whetted our appetites. Joe took me to lunch. I was honored.

I left Atlanta the next day to visit USAMRIID, in Frederick, Maryland, at Fort Detrick. I was met at the gate by a soldier who insisted on keeping my passport for the duration of my stay on the base. I met Karl Johnson and C. J. Peters, chief of the Disease Assessment Division, who seemed very pleasant. While I couldn't tell what he thought of my work, he did talk to me about it. The third person I met during my visit was Gene Johnson, the army's Ebola man and, as such, the scientist responsible for the monkey experiments they were conducting. He told me that he was going to "crack this pathophysiology thing, all nine yards." I wondered about this. His approach struck me as problematic, since he relied mainly on autopsies. I had figured out that once the patient is dead, the critical events are over, and they leave few footprints.

I wouldn't see Joe again until November 1983. He was coming through London on his way back from Zaire, where he'd gone to conduct the first investigation of AIDS in Zaire. When he got to London, he could not find me. Apparently, David Simpson had given him the wrong telephone number. After checking into the Charing Cross Hotel, he tried to track me down. When Joe is determined to do something, he won't let anything get in his way—certainly not the British phone system. Finally, he succeeded in locating me. In light of all the trouble he'd gone to, and because I was also flattered by his interest in my work, I was delighted to take him out to a good dinner. We spent the latter part of the evening, sitting on the floor of my living room in Wimbledon, animatedly talking about science until the small hours. The next day I gave him a lift to the airport for his flight back to Atlanta. Just before it was time to board the plane, he said, "Sue, you must come to Atlanta and repeat your Ebola experiment with Lassa."

That was when we made our bet. The secret of Lassa, I told him, was to be found in the platelets.

"No," he said, "the platelets are normal in Lassa fever."

The winner, we agreed, would get a good bottle of wine.

It wasn't until early in 1984 that I had a chance to claim the bottle. With technicians Sheila Mitchell and Donna Sasso, I set up an experiment on Lassa in the CDC suit laboratory. We found the same results that I had predicted. Joe very gracefully conceded the point and took me out to dinner. A gracious winner, I didn't mind sharing the bottle of wine one bit.

Level 4—British Style

The monkey grabbed my hand and tore my outer glove. Now, for the first time, I was scared. The monkey had been infected with Ebola virus from Zaire four days earlier.

I swore. But my voice was muffled by my gas mask. I doubted whether Geoff had heard me, but even so, he realized at once what had happened. He could identify. He'd accidentally infected himself under similar circumstances in 1976—and had barely lived to tell the tale.

That incident had occurred one Friday afternoon late in 1976 while Geoff was working in the same laboratory in which I was standing. He was injecting mice with the original Ebola isolates from Zaire. To do this, he had to hold a very tiny mouse between his finger and thumb, while simultaneously placing a very fine needle full of live virus into the creature. Geoff had been doing this for many years without any problem; I know of no other virologist who has developed as fine a dexterity for these procedures as he has. But the outbreak in Zaire was still going on, and they were all under pressure, the needle slipped, and it stuck his thumb. The needle was full of high-titer Ebola. It is one of the most lethal substances in the world.

He quickly tore off his gloves and checked for leaks. Finding none, he checked his thumb and squeezed it to see if he could spy any blood. There was none. He washed his hand in chlorine disinfectant. There being nothing more he could do, he discussed the likelihood of infection with Ernie Bowen, a short, very fat Welsh virologist who was working with him. (Ernie was the one who'd opened David Simpson's crate and found the six dead bats.) They decided it was a "non-event." Geoff put on a fresh pair of gloves and finished his work.

As soon as he was done for the day, Geoff reported the accident to the administration. This was at a time when Ebola had first been identified, so there was a barely contained sense of panic and fear about the disease. This was mixed with the usual aura of excitement, occasionally bordering on hysteria, that surrounds these epidemics, and that sometimes clouds the judgment of the less experienced. The scientists kept cool, but the excitement infected the committee that had been formed at the lab to deal with any secondary cases. The committee members didn't do anything dangerous, such as actually handle the virus, nor did they have any virological experience to speak of. Nonetheless, they made it their business to tell the scientists, who did work with Ebola, what they should be doing. Upon learning of the accident, the committee convened and deliberated, concluding that since there was no rip in the glove or any break in the skin, nothing needed to be done at all.

So Geoff went home to his family in Salisbury, where he spent the rest of the weekend. On Monday he was back at work. He felt fine. On Tuesday evening, he took his young son to an archery contest that was being held near Salisbury Cathedral. That night he couldn't sleep. He had a terrible headache, muscle pains, and a fever.

On Wednesday morning, he had to drag himself into work. He admitted to everyone that he felt sick—an announcement that created chaos. For most of the day he sat around while everyone tried to decide what to do with him. Later that afternoon, he was taken to Coppett's Wood, the fever hospital in North London, and placed in the isolation bubble. From that point on, Geoff has no further memory of his illness. He told me that those two weeks are missing from his life.

He was given immune plasma, just as Mayinga had been given when she took ill with Ebola in Kinshasa. He was also given the entire U.K. supply of interferon, a naturally occurring drug that has antiviral properties, but which has yet to be shown effective in human hemorrhagic fevers—though theoretically it should be expected to have some antiviral effect. For several days he hovered near death, but survived. Perhaps it was because of the nursing care he received, the interferon, or the plasma, or just because of his own innate strength. When we later used his platelets as controls in some of our experiments, we would joke about how vigorously they functioned. Perhaps that was why he survived. Of course, he also had the advantage of receiving much better medical care than he could have got in a grass hut in a remote corner of Africa.

Sometimes, over a beer, we would tease him, pretending to offer him various inducements for a "rechallenge" experiment. This way, we said, we could find out whether anyone could develop immunity to Ebola after a first attack. Curiously, he always declined this opportunity to contribute to science.

When Geoff reported for work again at Porton Down, the committee decided to hold an inquiry into the accident. Geoff and Ernie were summoned to appear. After Geoff got through recounting the incident, the designated safety officer turned to him and said, "If you were aware of the danger, why didn't you cut off your thumb?"

And he was serious.

Even years later, the attitude of the safety officer hadn't changed. He seemed to be of the mind that you could never be too careful: in his view, it would be better to autoclave the scientists along with everything else that came out of the laboratory rather than risk the spread of Ebola.

So now I was in the situation that Geoff had been years before. The first thing I thought of as I wrested my hand from the monkey's grip was what had happened to him. Was I going to have to go through his torment as well? I tore off my outer gloves and inspected the pair of inner gloves. The rip didn't seem to have penetrated the second layer. To be certain, I filled the inner gloves with water. There was no leak. There was no injury on my skin. But did that mean I was safe? Geoff had noticed no leak, either. Of course, the monkey claw was highly unlikely to have remotely as much virus on it as the needle Geoff stuck into his thumb.

All this took place in the Level-4 laboratory at Porton Down, located in the south of England. It is the only facility of its type in the U.K. The lab was a small part of a major civilian research complex that occupied several vast concrete and brick structures set on a hill in the middle of the great plain of Salisbury. It made for a commanding sight. The Level-4 lab itself owes its origins to the British military, which carried out top-secret experiments in chemical and biological warfare during World War II. The principal focus of their research concentrated on aerobiology—the potential for transmitting infectious biological material by some form of airborne vehicle.

Geoff and I worked in a large monkey room. David and I obtained grant money to purchase the hematology (blood analysis) equipment I needed for my experiments. Indeed, our operation was highly improvisational: Geoff and I learned to do everything ourselves.

I'd begun working with Geoff on Ebola soon after I arrived at Porton Down in 1982. Since the experiments sometimes called for us to work for six hours at a stretch, I was lucky to have him at my side. The day the monkey grabbed my glove, the two of us were conducting a series of complicated tests intended to find some way to treat Ebola, based on our earlier findings that the platelets and the endothelial cells were damaged by the virus. We reasoned that if we could discover how to treat the disease in monkeys, we could apply the results to human victims of the disease as well. I was again using the Zaire strain of Ebola virus because it was this virus that had caused fatal illness in virtually all monkeys infected with it. This was the same virus that had struck Yambuku in 1976 and had carried off nearly three hundred people. The mortality rate was almost 90 percent. This was the virus Geoff had survived.

Although we didn't like to use monkeys in this way, it was the only means by which we could gain the information we needed about the disease. In fact, most of what we know about Ebola disease processes today is in large part a result of the monkeys that Geoff and I worked with at Porton Down.

We decided to take a sample of blood from the monkey that had torn my glove; it was due for a blood test, anyway. When we tested the serum for virus, we found it contained four logs of Ebola virus. That comes to one thousand virions per milliliter of blood. A lot of virus.

I would now have to wait five days, the length of time that the virus takes to incubate. That meant five days of examining myself in the mirror for telltale rashes, five days of swallowing continuously to see whether I had a sore throat, five days of wondering whether a dull throb in my temples meant the onset of a headache, one of the first symptoms of Ebola. Mostly, though, I was more angry than scared, and neither Geoff nor I thought that I had actually been exposed. Nevertheless, it was hard to forgive myself for my negligence. I should never have allowed my hand to be in a position where the monkey could seize it. At the same time, I knew that the fault wasn't entirely my own. Much of the problem lay with the design of the laboratory itself. Because the lab had been built expressly for the purpose of testing airborne agents that might be useful in biological warfare, we were obliged to wear biological masks. Trapped inside one of these things, your face becomes drenched in sweat and breathing becomes an ordeal. Even worse, it makes it impossible to talk. You cannot, for instance, communicate with a fellow worker to tell him, "Look, I'm now going to place a needle in this monkey's arm. Make sure you keep your hand out of the way." Moreover, the mask is so heavy that by the end of the day you have a terrible neck ache. I could understand why Joe had abandoned his mask when he worked in the Sudan in 1979. Cosmetically, too, the mask is a disaster for the face, since the skin is constantly rubbing up against an airtight seal. And because of the placement of the two eye pieces, you are robbed of all peripheral vision. Unless you turn around, you have to sense what is happening beside or behind you. That includes someone with an infected needle, or a monkey with intentions to grab hold of you.

But aside from our gloves, the mask was the only real protection we had. Not that the gloves were very useful, either. The gloves extended to where the sleeves ended, but there was no seal to connect them to the sleeves. That left skin exposed. The gloves were also cumbersome, being far too heavy and awkward to use in delicate operations involving animals and specimens. They were actually regular yellow washing-up gloves. Nor was there protection of any kind for the rest of the body. Before you entered the lab, you had to strip naked and then put on what were essentially theater scrubs. Because it was assumed that only men would be working in the lab, all the scrubs were about four sizes too big for me. They all had to fit Ernie Bowen. I had to wind them at least twice around my waist. More than once they threatened to fall off while I was working. But losing my clothes because they didn't fit was the least of my concerns.

What it came down to was that the whole system was inherently unsafe. The only reason we didn't have more accidents was because of the high level of skill displayed by the researchers. Without Geoff's patience and skill and my own driving desire to know how these viruses worked, I'm not sure I could have made it.

The countdown began. Despite my objective assessment of the incident, I was watching the days and watching myself. Every day, I was driving seventy miles each way from my home in Wimbledon to Porton Down, on the M3 motorway. I had plenty of time to think. Five days . . . four . . . Was that the beginning of a rash on my arm I was seeing or some blemish that had always been there? Was it dryness that was making my throat feel a little sore, or was I really coming down with something? Did I need a few hours more sleep, or was that tingling in the back of my neck signaling the start of a headache? My imagination was becoming my greatest enemy. I had to be careful about distinguishing what was actually happening to me from what my lingering fear was conjuring up in my mind. Three days . . . two . . .

I concentrated on my work and the tasks that I had to get through at home. I reasoned. I felt I would be okay. Throughout this whole time, Geoff maintained his customary cheerful attitude, and never mentioned it again. We continued to work with our monkeys as though nothing had happened.

One day left.

I went to bed that night, determined to go immediately to sleep. I still felt fine, but I couldn't be certain. It took some hours before I dozed off. The next morning broke, and I rose from bed and walked into the bathroom. I hesitated for a few moments, then turned on the light and looked at myself in the mirror. Nothing, no rashes. No headache or fever, either. And no sore throat. I was home free.

But I was still angry, not so much at myself any longer, but at the whole system at Porton Down that put everyone working there at risk. Yet it wasn't until I had a chance to work in its counterpart at CDC for three months, later in 1984, that I realized just how deplorable conditions actually were at Porton. When I saw what a suit laboratory could be like, how good the protection was, and how much easier it was to work in, I was even more appalled by what I'd gone through. In 1984, therefore, after returning from CDC, I asked to see the director of Porton Down. He was a middle-aged scientist, very stiff in manner. He had no experience of Level-4 organisms, and he wasn't about to start. Certainly he wasn't interested in hearing about them from me. I proceeded to tell him that his system was unsafe.

He was livid. How could I dare to confront him? He'd never seen anyone so impertinent.

What made the director's indignant reaction even more astonishing was that neither he nor the safety officer had ever been inside the laboratory the whole time we were working with the virus. Neither had any idea of what we actually did, and they cared little about the problems we faced. I never saw either of them in a full-face respirator. He went on to say that it wasn't "proper" for me to challenge the experts, even though these experts had done most of their work during World War II. He added that the Americans had nothing to teach the British when it came to running a lab. Later, when one of his "experts" visited CDC, he told Joe that I was a troublemaker.

"Great," Joe replied. "Those are the people who get things done. I'll keep her here if I can."

I continued to campaign for a proper Level-4 suit facility even after my experiments with Geoff had come to an end and David Simpson had abandoned the field of hemorrhagic fevers to become chair of the Department of Microbiology at the Queen's University of Belfast. I held out some hope that a new lab being built at the Central Public Health Laboratory at Colindale in North London would meet some of our needs, and that, in due time, space and resources would be made available for a suit laboratory. But I was disappointed on both counts. The new lab was arranged in the form of a cabinet line, which is a "double shell" system. The laboratory walls compose an airtight outer shell, while the inner shell contains a sealed system of interlinked cabinets. It resembled the system that CDC had long ago abandoned with a sigh of relief. To work inside these cabinets, you need to insert your hands into a series of ports. While this layout freed us from the necessity of using full-face respirators, we still had to wear gloves, which were so big and clumsy that it was virtually impossible for us to do any real science. For all its cost and elaborate engineering, the new lab at Colindale turned out to be nothing more than a series of very expensive connecting boxes. It was out of date before it was even functioning. The people who were in charge of the design had little experience working in Level-4 facilities. Obsessed by the need for safety, they lost sight of the real risks, while simultaneously making it difficult for any significant research to be achieved.

Since Colindale was the best I could expect so long as I stayed in the U.K., I resigned myself and started work there in 1985. I did manage to make some improvements: I prevailed on the Public Health Laboratory Service to purchase a gamma irradiator for use in inactivating the virus. But there was only so much I could do. Toward the end of 1985, Joe made me an offer: by now I had spent three months working at CDC, and a further three months in Sierra Leone, and we were preparing papers for publication. Would I like to take a position at CDC? He recognized that I was unlikely to make much progress in England, given the limited facilities I had to work with, especially in light of the restrictive atmosphere. He asked me to set up a pathophysiology program to study mechanisms of disease in viral hemorrhagic fevers in his lab in Atlanta and also held out the possibility of working in a field program, particularly Sierra Leone. We had nothing of this kind in England.

It was a difficult choice. For one thing, in spite of the problems I was having, I still had the support of my colleagues who appreciated what I was doing and were trying to help me get set up. For another, with David Simpson gone, there was no one else who had the medical and scientific experience to lead the hemorrhagic fever initiative in the U.K. I consulted with people who had helped me get established in England. To my surprise, they were unanimous in their opinion. Every single one of them urged me to accept the offer. They told me that it would be a grave mistake on my part to turn down the opportunity, reminding me that if I stayed, I would continue to face more frustrations and endless obstructions. I realized that they were right. I made up my mind. On January 4, 1986, I left for Atlanta.

CHAPTER SEVENTEEN

The Outbreak That Escaped

While Sue was struggling with Ebola at Porton Down, I first became interested in the issue of HIV/AIDS. It was early 1983 when a colleague, Jan Desmyter from Antwerp, told me about a group of patients from Zaire he had seen with a disease similar to AIDS. By March 1983, he and other colleagues in Belgium had treated more than thirty such patients. This was an impressive number. Although Belgium had relinquished control over Zaire two decades before, there were still close ties between the two nations. Any Zairian who was really sick would find his way to Belgium for treatment—provided he could afford it. Having been in Zaire, I knew that far less than one percent of the population could afford this trip for any kind of medical treatment, so it was clear that there must be many more people in Zaire with AIDS. Sensing that there was the possibility of an epidemic in the making, I discussed the situation with Jim Curran, then head of the AIDS task force at CDC. He, too, realized that there might be a great many more cases as yet undiscovered in Zaire. Agreeing that Jan's report ought to be followed up, he assured me of the agency's support. Accordingly, in July 1983, I sent a cable to the U.S. Embassy in Kinshasa, which found its way to Seth Winnick, the science attaché posted in the Commerce Section. I asked him to contact Kalisa Ruti, *premier conseiller* to the minister of health. Using the embassy as a go-between, I was able to bring Ruti up to date on what was happening in Belgium. Then I asked him to see if he could arrange for government permission to come to Zaire to lead an investigation. In September we received his agreement.

The AIDS virus had not yet been isolated in 1983, so there were no specific tests we could use to prove a patient really had AIDS. All we had was the arduous and highly technical T4/T8 ratio test, which measures the loss of T cells in the immune system—a telltale sign of AIDS infection. In fact, at the time, we didn't even know for sure that AIDS was a viral infection, and without a virus, there could be no antibody test. For the T4/T8 testing, I had to find a technician who would not only be able to do this tricky test, but who could also transport the materials for it to Africa and make it all work there as well as it worked in a CDC lab. I chose Sheila Mitchell, who had already put in two years working with me in the lab. Although this would be her first trip to Africa, I felt she was most capable of tackling this difficult task. I'm happy to say I made the right call. She not only did a superb job for me, she later went on to develop a unique career helping to establish laboratory facilities for HIV testing in developing countries.

About ten days before we were ready to leave for Zaire, I received a call from John Bennett, my former chief in Bacterial Special Pathogens, who was now deputy director of the Center for Infectious Diseases. John informed me that, in addition to my team, another AIDS investigation was being mounted in Zaire led by Tom Quinn from NIH. He would have the assistance of a young entomologist/epidemiologist named Fred Feinsod, who was working on Rift Valley fever virus in Egypt, and Peter Piot from the Prince Leopold Institute of Tropical Medicine in Antwerp. Although I'd never met Tom, I knew Peter well from the time we'd worked together in the Zaire Ebola investigation of 1976. He was the epidemiologist who had wisely declined the ride in that ill-fated helicopter manned by a pair of drunken pilots. I had flown back to Kinshasa with their coffins and their grieving families. These are things you don't forget.

John made a good case for all of us to work together. When I called Tom and let him know of my plans, he agreed that it made sense to combine forces, since the strengths of CDC and NIH could complement each other. Possibly our greatest advantage at CDC was that we had an official invitation from the Zairian Ministry of Health, which Tom did not have. On the other hand, Tom could get hold of a supply of reagents for the T4/T8 testing that were superior to the CDC's, while Peter had established personal contacts with hospitals in Zaire that none of the rest of us had. We decided to all meet in Antwerp at the Prince Leopold Institute for Tropical Medicine before heading on to Zaire. I was expecting to have a serious discussion with Peter and Tom concerning strategy, but that wasn't what happened.

Several more people found their way into the act. In addition to Peter and Tom, I found Dick Krause, the director of NIAID (National Institute of Allergy and Infectious Diseases), Luc van Eyckmens, director of the Prince Leopold Institute, and an epidemiologist from Johns Hopkins, who was part of the NIH team. None of these additional people would be coming with us to Zaire, but this was all getting a bit heavy nevertheless. I could understand what Krause and Eyckmans were doing here; it was their organizations that were providing funds for Tom and Peter. But the others? I guessed that the size of the gathering was just another sign of the intense interest that the scientific community was beginning to show in AIDS—an interest that was both a blessing and a curse. A blessing because a great deal of good science got done. And a curse because it provoked some turf wars, with all the attendant strife and jealousies that such conflicts always entail. The role of the epidemiologist from Johns Hopkins was unclear, and became less so when he launched into a totally useless and pedantic lecture about how we should go about choosing control groups. I thought I was back in the EIS introductory course in epidemiology. The rest of the meeting, however, proved more valuable. At least we came away with the feeling that we would be able to work together. That we remain friends to this day demonstrates the truth of that assessment.

The following day we left together on board a Sabena flight to Kinshasa. We had two new recruits to our party: a lab technician from Antwerp and Henri Thaelman, a clinician from the Prince Leopold Institute. The only member of the team who didn't travel with us was Sheila Mitchell, who flew directly from the U.S. to meet us in Kinshasa. In her luggage was the gear for the lab we planned to set up to do the T4/T8 ratio tests. The Belgians had arranged for us to be put up at the Fometro, the same institution where we were housed during the 1976 Ebola outbreak. Nothing about it had changed in seven years. For Peter and myself it was a little disquieting; the place conjured up ghosts of the past. There was clearly something special about Central Africa and its relationship to viruses. We would see what this one had in store for us.

We got to bed early in hope of recovering from the journey. The following morning we were scheduled to meet with Seth Winnick, the science attaché at the American Embassy. He was surprisingly young, in his late twenties or early thirties, with a trim mustache and a full shock of red-brown hair. His primary function had nothing to do with science at all—his real job was to track information about commercial activities in Zaire—and "science attaché" was a title added to a long list of others he had been given. Seth admitted that he never believed he'd actually have to do anything related to science. Now he realized that this was about to change. He had arranged for us to see Kalisa Ruti that afternoon, paving the way for an interview with the minister of health himself the following day. For a beginner, he was doing very well.

Kalisa and I had met before at meetings of WHO in Geneva and Nairobi. He greeted us in the safari suit that most Zairian functionaries had adopted for workday wear. While Western custom dominated fashion among the elite, there were some intriguing departures from it. For one thing, ties and jackets were eliminated by presidential decree from formal wardrobes, a move that made a great deal of sense in a tropical country. Christian names, too, had been abandoned, also by decree, several years before in favor of traditional African ones.

Kalisa was clearly unaware of the implications of our visit. He was highly cooperative, and he assured us that, no matter what the minister might say to us at our meeting tomorrow morning, the government would allow us to carry out at least the initial phase of our AIDS study. In Zaire, it was never possible to guess just how the government would respond. There was always some sociopolitical crisis brewing. The latest one was economic, and our arrival coincided with a dramatic devaluation of the zaire, the nation's standard unit of currency, which had plummeted from five to the dollar to thirty to the dollar the day before we arrived. The first problem brought about by the devaluation was that the largest zaire note available was a *one* zaire note. Suddenly, people were unable to go to a shop without taking along satchels and suitcases full of cash. Our second night in Kinshasa we needed to lug briefcases stuffed with half-zaire and one zaire notes so that we could all eat at a Greek restaurant. We felt like mafiosi on our way to make a payoff. But then it turned out that no matter how much money we had, there was only so much we could order. Many items on the menu were unavailable due to inflation and a lack of foreign exchange. In addition, certain commodities couldn't be found because of the devastating increase in the cost of fuel needed to transport them. We were inconvenienced, but, as usual in such cases, it was the poor who really suffered. The majority of the people in Zaire subsist on cassava, a root consisting primarily of starch. A tuber that grows easily, it looks like a large yam covered with bark, but African cassava also contains a toxic alkaloid. To dissolve out this toxin, the root is soaked in running water for two days. In the process, it absorbs a great deal of water, making the cassava much heavier—and, therefore, that much more expensive to transport. With devaluation and the soaring coast of fuel, shipping costs for cassava became prohibitive. And so the city poor went hungry.

Over dinner at the Greek restaurant, I began to question Tom and Peter further about the nature of the risk factors for AIDS in the U.S. We were still early in the epidemic and not a great deal was known about it. They were more than happy to fill me in. As they began to tell me stories about the bath houses in San Francisco and the propensity on the part of large numbers of homosexuals to engage in anonymous sex with hundreds of partners, I didn't know what to think. I'd never heard of anything like it. And I wasn't alone. Out of the corner of my eye, I could see that everyone seated nearby seemed to be hanging on every word, not least because Tom and Peter were unsparingly graphic in their accounts. When you specialized in the study of sexually transmitted diseases (STDs), you were bound to pick up some surprising information.

There was nothing in my experience that allowed me to relate to the gay life of the 1970s. Even my experiences in Zaire—where sexual customs like polygamy, abhorrent in the West, were perfectly acceptable—hadn't prepared me for the revelations that Peter and Tom were offering over dishes of moussaka and stuffed grape leaves. Because sexual behavior in Zaire was so different from what Tom and Peter were describing in San Francisco, I found it difficult to believe that the disease could be spread the same way in both places. And while it was true that the majority of patients from Zaire who'd sought treatment for AIDS in Belgium were men, they seemed to be heterosexual. Moreover, I doubted whether the disease affected men exclusively. Rather, I assumed that it was simply a matter of economics: since men controlled most of the money, they were much more likely to be able to travel all the way to Belgium and pay for medical care.

The next day we collected in the office of the health minister, Dr. Tshibasu. He was a tall, stout man with graying hair, who gave off an air of studied sophistication with his fluent French and his affable manner. I knew enough about the political situation in Zaire to realize that he must have been hand-picked by Mobutu. He'd already been in office for more than six months—unusually long for officials in the Mobutu regime. Generally, they served less than a year, which was enough time for them to line their pockets before they were sacked and had to go back to parliament, where it was more difficult (but by no means impossible) to get rich. So I had to figure that Dr. Tshibasu was reaching the end of his ministerial career. And if he were in danger of losing his position, so was my friend Kalisa Ruti. Which meant that we would have only a small window of time in which to conduct our initial study and develop an agenda for future action based on whatever we found.

The meeting with Dr. Tshibasu started on a very interesting note. He greeted us somewhat stiffly, but he was cordial just the same. Yet he made little effort to conceal his skepticism. He told us that he already had more problems to deal with than he could handle: malaria, malnutrition, diarrhea, tuberculosis, sleeping sickness, measles. "Don't count on finding much interest or support from us for the problem you are interested in," he said in his polished French. "We can't even cope with the ordinary problems I just told you about." AIDS, I soon realized, was an unknown quantity to him; he had no idea what kind of a threat it posed to the inhabitants of his country.

So I began to tell him about the disease. I told him that several of his wealthy countrymen had AIDS and were currently languishing in Belgian hospitals;

they included an army colonel, a banker, and a vice president of a large local brewery. The minister began to show more of an interest, yet we were unable to persuade him to reconsider his position. That would come later, after we had gathered some startling information.

I would hear the same kind of argument from officials in other countries over the next few years. It was very difficult for people, especially those in authority, to understand the impact of AIDS until it was already sweeping through their country. Several years later, in Pakistan, I urged health officials to travel to an AIDS-infested African city and see for themselves what could happen if they failed to take measures in advance of an epidemic. At least Dr. Tshibasu was sensible enough to allow us to go ahead with the initial phase of our investigation. He'd done exactly as Kalisa said he would.

Our first task was to find a location in Kinshasa where we could set up a laboratory. It had to offer us convenient access to clean water and a sufficient supply of electricity. We also had to decide on which hospitals we would survey for our study. The two major hospitals serving the city were Mama Yemo Hospital (named for Mobutu's mother) and University Hospital in the suburbs. While University catered to a relatively wealthier clientele, Mama Yemo served the indigent, who constituted the majority of Kinshasa's population. This is the same institution my friend Bill Close had directed at the time of the Ebola epidemic. Bill had left Zaire in 1977 to settle in Big Piney, Wyoming, with his wonderful wife, Tine, where he is now a family doctor and a community icon.

Mama Yemo is a sprawling institution. Its design is typical of hospitals put up in the colonial era, with vast wards and high tin-roofed ceilings that had begun to grow weak with rust. Its cement floors were dark with the stains of countless miseries. Air circulation was supplied by ancient overhead fans and paneless windows. Each ward contained about thirty metal beds, which were rarely empty. Mattresses were stuffed with cotton or grass. Sheets were hard to come by. Meals for the patients were provided by their families, and, with staff in short supply, the families also assumed much of the responsibility for nursing care. Bathrooms were scarce and rarely worked anyway, creating a rank odor that greeted you when you walked in and dogged your every step until you left. The wards were generally choked with people afflicted with the most desperate illnesses. They were jaundiced, bloated, cachectic, comatose, and vomiting. Many had diarrhea. They arrived at Mama Yemo with unhealed wounds, covered with crusts and pus, and emitting a terrible stench. They came in endless droves. Screams and moans echoed through the dank hallways. This is the face of disease and death for the world's poor.

It was in this infernal setting that we went in search of patients with AIDS.

We would assess whether they had the disease both by the symptoms they exhibited and by the their low T4/T8 ratios. HIV selectively kills T4 lymphocytes but spares the T8. The T4 lymphocytes are the cells critical to the body's mechanisms for killing invading organisms. If the T4 cells are reduced relative to the T8 cells, we know that the patient is progressing to AIDS, or has AIDS outright. Since AIDS is the only disease we know that does this, the T4/T8 ratio provided us with something of a diagnostic test. To carry out the T4/T8 tests, Sheila began setting up her lab at the University Hospital. There were insufficient lab facilities at Mama Yemo Hospital.

Our strategy was quite straightforward: for three weeks running, we planned to survey the male and female medical wards in Mama Yemo and in University. Whenever a new patient was admitted, we would examine him or her and take a history, and then draw a sample of blood in order to perform a T4/T8 test. At the same time we would be checking on the patients we'd already seen to find out how they were doing. We were struck at once by what we witnessed. It wasn't just that there were so many AIDS-like illnesses. That would have been bad enough. But the evidence before us suggested that we were always seeing the late stage of the disease. This was a major difference between what we knew about cases of AIDS in the U.S. and Europe, and what we were observing in Kinshasa in 1983. These patients were all suffering from untreated opportunistic infections that had been allowed to become terminal. It was very sad, for example, to see someone whose entire foot was swollen three times its normal size, because it was invaded by an oozing yellow fungal infection. In the West, an infection like this would have been cleaned up quickly. Here, the patient waited until an advanced stage of disease. In any event, the antifungal drugs that might have helped are unaffordable in a country such as Zaire. One daily tablet of antifungal drug used by AIDS patients now costs $15. This could be two weeks' pay in Kinshasa, and that would be a man with a steady job. Someone with AIDS would have no hope of affording treatment.

The day after the meeting with the minister of health, we went on our first tour of Mama Yemo Hospital. I had been told to look for Dr. Bela Kapita, chief of the medical wards. I approached one of the nurses, saying, *"Je suis le Docteur McCormick du CDC qui est venue avec l'equipe pour l'enquete sur le problem du SIDA,"*— I'm Dr. McCormick of the CDC, and I've come with the team to help investigate the AIDS problem. *"Pourai je parlez avec le chef du service?"*— Can I speak to the head of the service? "Dr. Kapita isn't here right now," came the response. I took little consolation from the nurse's assurance that he would return the following day. Uh oh, I thought. This is going to be one of those nightmares where the guy in charge is always off doing his own thing. That meant we were going to have a hell of a time getting anything done. I knew from experience that in such cases the rest of the staff was unlikely to act without the express consent of their superior. And if the superior was absent, then you were out of luck.

But, in fact, Dr. Kapita did appear the next day. He'd gone to see his ailing father who lived in a village not too far away. A cardiologist trained in Belgium, Kapita would often visit his father and treat him for congestive heart failure. There was no one else available to do this. In fact, as soon as I got to know him, I realized that I'd been profoundly mistaken in what I had assumed about him. Dr. Kapita turned out to be a saint. Once, I accompanied him to the local post office. A little girl came up to us, begging for money. Kapita was clearly embarrassed. Tears began flowing from his eyes.

"It was not always this way," he said. "Things are different now. They're much more difficult, and now our children are forced to beg. I don't know what to do."

I felt terrible for him, for his sense of helplessness.

Dr. Kapita was a small, thin man, with a large round head and penetrating eyes. Highly intelligent, he was anxious to cooperate with us, especially since he was astute enough to realize early on, well before we arrived, that he had cases of AIDS on his ward. He welcomed any assistance we could give him, but he had a great deal to offer us as well, which was why I recommended that he become part of our team. We needed a Zairian colleague like Kapita with a sense of humanity and a deep understanding of the plight of the people we'd come to Zaire to help.

By the time we actually began our work on the wards of Mama Yemo and University hospitals, our number had diminished by one. Tom Quinn was called away to a meeting in Denmark and was unable to return to Zaire before we'd completed our study. (Nonetheless, he remained an important part of the team over the next several years.) So the team consisted of myself, Peter, Henri Thaelman and Fred Feinsod, with Sheila running the lab. Our investigation was a grueling one. We had to examine patients and collect specimens each day and then try and get the material to the laboratory by early afternoon. The more time we could give Sheila the better; her lab procedures took several hours. We were using a jury-rigged apparatus to freeze and preserve cells and specimens in reasonable condition for future use. It consisted of a tank with liquid nitrogen, which is normally about $-200°C$ ($-328°F$). However, if cells are put directly into the liquid nitrogen at that temperature, they freeze too fast, expand, and burst—something that defeats the purpose of preservation by freezing. So we put the vials of cells, immersed in a special medium, into thick paper envelopes, which we then lowered, not into the liquid nitrogen itself but into its vapor. At −80°C (−112°F), the vapor is significantly warmer. It's the equivalent of getting your feet wet before diving into a pool. But this was much, much colder; your feet would not even survive immersion in the vapor. We'd leave the cells exposed to the vapor for several hours until they had a chance to freeze slowly and evenly. The process was time consuming, and we seldom finished before eight or nine at night. The lab work was taxing as well as tedious, and Sheila bore the brunt of it. Although she was gaining considerable expertise in reading the T4/T8 cells, the need to sit at the microscope for two to three hours every night took its toll on her eyes and back. So much of the work we do is hard labor— sensitive, delicate labor but hard labor nevertheless.

On the fifth day of our investigation, a twenty-one-year-old woman was brought unconscious into Kapita's medical ward. Her family told us that she had been ill for some months, and that she'd been experiencing fever, significant weight loss, and a racking cough. Over the last two weeks, they went on, she had also begun to develop severe headaches and suffer from drowsiness. When they could no longer arouse her at all, they decided they should come to Mama Yemo. And so it was that we met Yema.

CHAPTER EIGHTEEN

A Free Woman from La Cité

When Yema's family had moved from Kananga to Kinshasa about nine years earlier, they naturally gravitated to La Cité. (With a population of more than a million people, Kananga has the dubious distinction of being the largest non-electrified city in the world.) La Cité was a sprawling slum in the middle of Kinshasa, made up of blocks of houses constructed out of wood, cement, mud, tin, and cardboard. Indeed, any material would do, so long as it kept out the rain, vermin, intruders, and evil spirits. This part of Kinshasa was filled with small shops where you could buy cheap toys made in China or Thailand, or have your car or bicycle repaired, or get your shoes resoled with tire rubber. La Cité was also where people without money settled in the belief that, as soon as they found work, they could move on to something better. It seldom turned out that way.

One of the few means that women had to make money in La Cité was to sell themselves. The so-called *femmes libres* (free women) weren't necessarily unmarried women; they were more often widowed, divorced, separated, or simply abandoned. They were women with no other sources of support and were desperately in need of money to feed themselves and their children. They were "free" in the sense that they provided casual sex for money or gifts but were not, in a commercial context, considered prostitutes. The distinction was important. "Professionals" were those women who worked constantly, while the *femmes libres* practiced their trade only occasionally. The phenomenon of the *femmes libres* became more common as conditions in La Cité deteriorated because of increasing population and a declining economy. It was just one more instance of how a woman's low status in society can create increased health risks.

Yema was the second oldest of seven children. Although her father was illiterate, he'd managed to find a job carrying crates of beer for a brewery. He had to travel by foot or bus for several hours a day to go to and from his job. Sometimes he wouldn't return home for days at a time. As a result, Yema's mother was left to fend for herself and her family. Try as the mother might, she often couldn't make ends meet. Her meager wages barely made a difference, and she was so overwhelmed by her responsibilities that she was unable to keep track of all of the children. Certainly, she couldn't expect them to be in school. Anyway, there weren't enough schools available for the number of children who needed them. So it was inevitable that they would begin to learn at an early age how to forage for themselves. Following the example of other young adolescent girls in La Cité, Yema began to make money by having occasional sexual relationships with local men. Although this behavior would not have been permitted in the rural villages of Zaire, it was tolerated, perhaps by default, here. By the time she was twenty, Yema had had two abortions. Then in late 1982 and early 1983, she started to lose weight while her menstrual periods became irregular, finally ceasing altogether by May 1983. As her weight declined, she became increasingly fatigued. She frequently experienced chills. Her appetite waned, and she developed a dry cough. Having no strength to work, she was completely dependent on her family for support. Without the money for a doctor, Yema received no medical treatment until, late in September, her mother could no longer wake her up. In desperation, her family took her to Mama Yemo Hospital.

Yema was the first patient we'd seen with what was then thought to be an unusual manifestation of AIDS. But one of our colleagues, Dr. Nyst, a Belgian physician working in Kinshasa, had observed similar cases. He told us that the number of patients who had been admitted with cryptococcal meningitis in the last year was continuing to increase in frequency. Possibly Yema, too, was infected with this type of meningitis. In normal people it occurs only rarely. It is a disease of the brain caused by a yeastlike organism and is sometimes seen in patients with advanced cancer whose immune systems are suppressed from their radiation and chemotherapy treatments.

Dr. Nyst conducted a spinal tap on Yema and took the specimen of fluid from her spinal cord to the lab. Under the lens of the microscope, hundreds of round, translucent organisms began to show up against a background of black India ink—a clear sign that Yema did indeed have cryptococcal meningitis. This organism is literally a sugar-coated fungus, which grows in the spinal cords of patients. The big problem for us was that we didn't have the necessary drugs to treat the disease. In any event, no one could afford these drugs in Zaire. The injection is very toxic, and the pills cost over fifteen dollars a tablet! Helplessly, we watched as Yema began to die.

To outward appearances, hers seemed a peaceful death now that she'd lapsed into a persistent coma. This impression was deceptive, though. In fact, a raging battle was under way inside her. It was a struggle between the cryptococcal yeast, protected by its sugar coat, and her wasted immune cells—those few remaining T4 cells that Sheila was measuring. This yeast is nothing like normal baker's yeast or even candida yeast, a common fungal infection of the skin and mucus membranes. The natural habitat of the cryptococcus is in the soil, where it thrives amid bird droppings, particularly from pigeons. Given how little we could do for Yema, our role in her care was even less significant than that of the hospital janitors. At least they could clean up afterwards and possibly derive some satisfaction from having accomplished their job. We, on the other hand, could only stand by and watch all of our sophisticated training rendered useless.

Yema's family seemed to accept her death with more stoicism than I'd expected. I had a feeling it was because they had long ago understood that she was destined to die, and that they could do little to prevent it. They had come to Mama Yemo Hospital probably in the remote hope of some miracle. We had none to offer.

Yema was only one of several similar cases we saw in the three weeks of our investigation, and meningitis was only one manifestation of the disease. The range of symptoms we saw in patients was staggering. In a country as poor as Zaire, the sort of medical support a patient with AIDS might by rights expect is quite unaffordable. These poor people suffered until their symptoms were so extreme that they had to come to the public hospital. These were the human tragedies we found before us. Some developed such exquisitely sore mouths and tongues that they were unable to eat. Those who could manage a few bites of food were suddenly stricken by cramps and disgorged a copious amount of diarrhea. Their skin would break out in massive, generalized eruptions. Infected fungating masses would appear inside and outside their bodies. When the infection didn't consist of voracious yeast cells, there were many other parasites ready to eat the brain alive. None of the victims could comprehend in any way what was happening to them or why. And we? All we could do was watch in horror, our roles as physicians reduced to scrupulous observers and accurate recorders of documentation. Our one hope was that if we could understand the processes we were observing, someone, somewhere, might find some solution.

The conversations we had after work were our only source of emotional release from our daily distress. Our friends, who were experts in STDs, continued to regale us with tales of the excessive and often bizarre sexual practices associated with HIV in the West. We, the more widely traveled if less worldly wise, would entertain them, in turn, with stories about culinary delicacies peculiar to

Africa, describing in vivid detail hors d'oeuvres like raw termites, large juicy fly-ing ants, and deep fried caterpillars (but only of certain species, of course). We even went so far as to organize a feast that included these regional dishes. But once they'd come face to face with the real thing, the neophytes balked. I sus-pect that this type of African cuisine still has a way to go before it turns up on menus of finer restaurants in the West.

Two very important discoveries began to emerge from our investigation. For one thing, we were recording far more cases of AIDS in women than we were used to seeing in the United States or Europe, where the disease was still striking gay men in disproportionate numbers. For another, we were also beginning to see a direct correlation between the number of sexual partners and the rate of infection. This was a finding similar to what physicians had observed among the gay men in San Francisco early in the AIDS epidemic. While we recognized that this was a disease spread by sexual transmission, the shock for us was that, in Zaire, it was almost entirely due to "normal" heterosexual intercourse. But that didn't mean that there was no homosexual transmission. It was just that our investigation revealed that it was relatively rare among men living in Kinshasa. The same situation appears to be true of most of Africa. Certainly there was no organized or visible gay community in Zaire, as there is in the West. On the other hand, compared to the West, heterosexual contacts in Africa are frequent, and relatively free of social constraints—at least for the men.

Our findings in Kinshasa were supported by similar investigations through-out Africa, notably in Rwanda. The world now had to face an uncomfortable, frightening reality. We thought about the ramifications of our discovery and dis-cussed them at length. There was every reason to believe that, having found het-erosexually transmitted AIDS in Kinshasa, we were likely to find it everywhere else in the world. Until this moment, especially in the United States, AIDS was linked almost exclusively to gays and drug addicts and other marginalized groups such as Haitians. It wasn't something that was supposed to affect "main-stream" people.

In 1984, our team and a Belgian team in Rwanda, led by Van der Peer, pub-lished our conclusions in *The Lancet*, the venerable British medical journal. These articles would change the way that people looked at the disease. After reviewing the situation in Africa, we then posed a question: "Will this be the face of AIDS in the West within the next decade?"

To some at the time, the very question was itself blasphemous. But we know today, for many countries in the West, the answer is yes. In 1996, AIDS is now the primary cause of death in women between the ages of 25 and 40.

In my report to the director of CDC, I suggested that AIDS was endemic in Zaire, and that it may have been present since the mid 1970s. I based my conclusions on accounts by physicians who had seen a number of undiagnosed cases of weight loss and diarrhea, invariably resulting in death over at least ten years. Although they'd attributed the cause to TB, it seemed in retrospect that the cause of death was probably AIDS related. In what was the most controversial part of the report, I went on to characterize the disease as one that was spread by heterosexual contact in Zaire, adding that there was no evidence that homosexuality or drug abuse had played a significant role in its transmission. I recommended that CDC undertake a long term collaboration with the Ministry of Health of Zaire to establish a system of surveillance for the disease in that country. Finally, I called for WHO to convene a workshop on the problem to be held either in Kinshasa or Brazzaville, in the neighboring Congo. These recommendations were subsequently accepted.

I returned to Atlanta on November 8 and immediately reported to my chief, Gary Noble, and to the director of the Center for Infectious Diseases, Walter Dowdle. After listening to what I had to say, they both agreed that I should meet with Bill Foege, director of CDC. This was the year that Dr. Foege had announced his departure as head of the agency, and because his successor, James Mason, happened to be visiting CDC that day, he was invited to sit in on the meeting. In addition, we were joined by Jim Curran, director of the HIV/AIDS Division, and Fred Murphy, director of the Division of Viral Diseases. Serendipity had brought together in one place and time many of the major players in AIDS who were associated with CDC .

Bill Foege had lived and worked in Africa, so right away he recognized just how grave a situation we were confronting. He decided that we should put in a call to Dr. Edward Brandt, U.S. assistant secretary for health. I was put on the speaker phone with him. I did not know who he was—other than that he was a Ronald Reagan appointee—and I had no idea how he might respond. I began by describing our data and went on to outline our major conclusions. I tried to spell out everything as simply and as clearly as possible.

There followed a long silence on the other end.

Brandt began by saying that I must have got it all wrong.

"There must be another explanation for your findings. Have you considered other vectors, like mosquitoes?"

Mosquitoes were obviously easier for him to talk about than sex.

"I don't think that the evidence supports that, sir," I said. "So far, we've found very little disease in children. And children get just as many mosquito bites as adults—probably more. That's why they suffer so much more from malaria.

And if AIDS were transmitted by mosquitoes, we wouldn't have seen the sort of random pattern of distribution of the disease in the population that we did. When you look at malaria, you can see a random pattern. We all know anyone can get malaria; it just depends on who gets bitten. But what we saw with this disease were definite chains of infection as well as clustering around sexual contacts. There were hardly any cases with children or with old people."

My explanation, as well-reasoned as I thought it was, failed to sway Brandt. He seemed bent on coming up with another theory, just so long as it would let heterosexual intercourse off the hook. Our discussion went on like this for about twenty minutes. But nothing I could say seemed to make an impression on him. I was stunned by the depth of disbelief—or, rather, denial—on the Washington end of the line. Certainly, everyone sitting in the room with me understood the compelling nature of the evidence and realized that it was imperative that we take action.

Evidently the conclusion the administration had reached was very different. This was the Reagan era. If AIDS was going to have an explanation, it seemed then it would have to be politically and socially more acceptable than what we had to offer. Voters were not going to like our message. They rested easier with the notion of a "gay plague," as the disease was called when it first became known to the general public. There was a self-satisfied, ugly moralism about that notion. What we proposed to tell them is that AIDS was a plague all right, but that no one was immune.

By steadfastly refusing to acknowledge the true dimensions of the AIDS crisis, the Reagan administration made itself an ally of the virus. It would take another year before Washington's policy would begin to change, with the appointment of C. Everett Koop as surgeon general. Koop, a political conservative with a strong sense of right and wrong, was a great physician and an objective scientist. He refused to contaminate public health with ideology.

Two years later, in 1987, I attended the now-famous "Meeting on the Potomac," which was held in a magnificent tent alongside the river. I was among the several hundred guests fortunate—or gullible—enough to have accepted an invitation to attend. Not that the sponsors of the gathering didn't have good intentions. To her great credit, Elizabeth Taylor had put the whole thing together, and when Dr. Koop stepped into the tent, he was greeted by thunderous applause in recognition of what he was trying to do. But when President Reagan got up to address us and for the first time acknowledge in public that AIDS was a major U.S. public-health problem, he was greeted by perfunctory applause as well as a few heartfelt boos and hisses. Admittedly, the audience, which included a number of academics, might have been partisan, but there were a great many physicians present who were not exactly noted for their support of the Democratic party. They were appalled by the inexcusable neglect that the Reagan administration had shown in handling the AIDS crisis. Surely, there have been few more blatant examples of the triumph of politics over truth—with the possible exception of Health Secretary Margaret Heckler's bold, if foolhardy, declaration in 1985 that an AIDS vaccine would be ready in two years!

For me, the Meeting on the Potomac was a turning point. I had established myself in viral hemorrhagic fevers; I was developing a program I loved and was excited by the research I was doing. I wasn't interested in abandoning my field of interest to work on AIDS. Yet I had been the one to recommend a long-term study of the disease in Zaire. Not unreasonably, I decided that perhaps I ought to be the person to set it up. I was torn. What should I do?

CHAPTER NINETEEN

Projet SIDA *Begins*

Assigned the task of finding a director for our new AIDS project in Zaire, I thought of a man whose path I had crossed from time to time. His name was Jonathan Mann. Although at that time we didn't know each other well, I was aware of his work and held it and him in high esteem. You wouldn't necessarily expect a slightly built man like Jonathan, with his dark curly hair, wire-rim glasses, and small, neatly trimmed mustache, to be so intense and energetic. I had heard him at several CDC conferences and was always impressed by his capacity for clarity, logic, and thoughtfulness. He was working as the State Epidemiologist in New Mexico and living in Santa Fe, but I had heard that he was looking for a challenge. And I couldn't think of a better one than what I had to offer.

At the end of January 1984, I gave Jonathan a call and asked him if he was interested in going to Africa.

"I've never been to Africa," he said. He had many reservations. He pointed out that he had three young children. How would he educate them? I assured him that there was a good American school in Zaire—there really was—and that that shouldn't be a factor in his decision. I must have been persistent enough, because, in the end, he agreed at least to give my proposition some thought.

A week later he called back. He'd made up his mind. He was in.

In March, Jonathan and I made our first trip together to Zaire. It was the beginning of the AIDS project. In Zaire, though, AIDS has another name, derived from the French acronym for the disease, SIDA.

When we arrived, we met with several Zairian health officials who were involved with the epidemic. Although most of them were supportive of Projet

SIDA, there were some notable exceptions. One was a senior Zairian physician named Dr. Lurhuma, who made it clear that he was less than enthusiastic about what we were doing, notwithstanding the fact that his government lacked both the resources and manpower to mount a program on any scale without foreign assistance. A thickset man, probably in his forties, he was ostensibly an immunologist, but I'm not sure where he received his training. In any case, he was more of a politician than a practicing physician or scientist. He appeared to have very important connections in the government, which may explain why he was accorded such respect. National pride apart, however, Dr. Lurhuma had other fish to fry. He was one of the first of many Africans who claimed to have a "cure" for AIDS. He never actually presented any data supporting this claim, but the press picked it up, and in a country like Zaire that was sufficient. His claim became something of a cause célèbre, and, as a reward for his alleged contribution to the health of Zaire, President Mobutu was reported to have given Lurhuma a large amount of money and a house in a fashionable district of Kinshasa. His cure, as far as we could figure out (which was not very far at all), was some sort of undefined fluid, origin unknown.

In Kenya, also a few years later, David Coech, director of KEMRI (Kenyan Medical Research Institute), contended that *he* had created an effective remedy for AIDS. At least he seemed to have some data to back up his claim—though he never revealed it all. Supposedly he had taken some material from cell culture and used it for therapy. Apparently this material contained a modest amount of interferon, a naturally occurring human substance known to inhibit virus growth under some circumstances. Others who tested this material found no evidence of its efficacy. Even so, his assertion that he'd found a cure caused a number of scientists, who should have known better, to voice their support. There were even calls for WHO to conduct a major trial of his reputed therapy, though nothing ever came of it.

Such is the desperation to find an effective treatment or prevention for AIDS. These purported cures were a bitter pill for the many Africans laboring daily to stem the tide of AIDS in their countries. By the time of Coech's assertions, the full impact of AIDS in Africa was just beginning to be felt by its people and medical practitioners. They wanted to show the West that they could produce someone who could make a major contribution to medical science. They didn't want constantly to be put in the position of dependent children, always relying on outsiders for help and new discoveries. What they got instead was another quack. The phenomenon of AIDS has created other charlatans, probably more in developed than in developing countries. For the most part, they are wealthy opportunists who exploit the despair of ailing people. They drive a hard bargain: what they give in exchange for money, and perhaps fame, is false hope.

In spite of sporadic opposition, Jonathan and I still managed to make headway in setting up a long-term AIDS program in Zaire. In this, we were fortunate to have the support of a doctor named Muyembe, a friend who worked with us in 1976, and who would later earn a large measure of recognition as the chief spokesman for the Zairian authorities when Ebola broke out in Kikwit in 1995. Although he was accustomed to coping with human misery, he somehow always managed to maintain a sunny disposition. Infectious disease wasn't the only thing that he had to fight against, either. He was also forced to contend with a deteriorating infrastructure, violence, factional uprisings, and political intrigues. He has served as an infectious disease expert and the dean of Kinshasa's only medical school. In fact, there have been few important positions in medicine in Zaire that Muyembe hasn't held. When he was dean, at the beginning of Projet SIDA, his budget was so strapped that he didn't even have enough paper for his faculty, let alone for the students. Whatever the obstacles, though, he remained undaunted; he was there when Ebola broke out in 1976; he was there when AIDS ravaged Zaire seven years later; and he was there again during the outbreak in Kikwit. His efforts were monumental, his persistence nothing short of superhuman. Muyembe was so anxious to contribute what he could to the health of his people that he spent several months in my laboratory at CDC, learning techniques for serological diagnosis of hemorrhagic fevers. It was his hope that he could perform these tests when he returned to his laboratory in Kinshasa. Tragically, there was never enough money to keep his lab operational. For that matter, he couldn't even depend on a steady supply of electricity—and without electricity, his reagents quickly became useless. Maybe if the money had been available, the June 1995 Ebola outbreak that killed 300 people in Kikwit, Zaire, might have been a very different story.

Jonathan and I had several problems to address immediately. One of the most important was to determine the primary location of Projet SIDA. Everyone had their own ideas. People were falling over one other to get involved in AIDS research. They realized that AIDS represented an opportunity for grant money, training, and the possibility of professional advancement. Not surprisingly, these opportunities created a fertile environment for intense competition. The stakes were substantial. Already, scientists in any number of countries— Belgium, France, England, Canada, and the United States—were mobilizing to study the disease. A certain bandwagon mentality took hold. Careers and reputations were riding on the outcome. In some way, the scramble for acclaim and funding resulted in progress being made. We have acquired an astonishing amount of knowledge about how the virus is constituted and why it has so far been able to trick the human immune system—and the immunologists—by mutating with chameleon ease to avoid detection and destruction. But the same competitiveness that has driven so many researchers around the world to find a cure for AIDS has also often destroyed professional friendships and generated destructive rancor and controversy. The bitter dispute between the Institut Pasteur in France and the NIH in the U.S. over credit for having discovered the AIDS virus is only the most notorious example of the divisiveness among scientists created by the public attention this disease commands.

Settling in Kinshasa with his family, Jonathan found that the excitement of putting together the SIDA project more than made up for the rigors of life in Zaire. After helping him get set up, I visited him a couple of times each year to continue our collaboration. The project grew in scope, and Bob Colebunders from Antwerp and Skip Frances from NIH were later collaborators in a tripartite project among CDC, NIH, and PLITM (Prince Leopold Institute of Tropical Medicine). It was because of Jonathan's pioneering research that we now better understand the role of the urban environment in the transmission of AIDS. He also conducted some of the earliest studies on disease transmission from the pregnant mother to the fetus. In addition, working with Alan Greenberg from Atlanta, he was able to discover a link between malaria and AIDS in children. The link was not direct—malaria, of course, does not cause AIDS—but malaria does make children anemic, and that may necessitate a blood transfusion. In places that lack the ability to test for AIDS in donor blood, it is very possible that a child will receive a contaminated transfusion. Jonathan's research also underscored some important differences between how AIDS behaves in Africa and how it acts in the West. The incubation period in Africa for the infection appears considerably shorter. The reason, we believed, was because the African population was more susceptible to disease to begin with. They were likely to have experienced far more infections, particularly by parasites; as a result, by the time they were exposed to HIV, their bodies had become ripe for the virus. Finally, Jonathan's research was among the earliest to point up the emergence of TB in association with HIV.

Of equal importance with his research, Jonathan's experience in Zaire gave him the opportunity to understand, at ground level, the impact of AIDS. It was this firsthand knowledge that made him such an effective leader of the Global Program on AIDS, which would later be established by WHO. Currently, his new organization at Harvard, The Institute of Health and Human Rights, is at the cutting edge of our understanding of the social basis of individual risk of disease caused by agents such as HIV.

Although no one in 1984 or 1985 had much of an idea about the incubation period for AIDS in Africa, evidence suggested that in the United States it was between two and five years, or even longer. (It is now estimated to be ten or more years.) This gave the virus its terrible advantage. By the time the virus has sabotaged the immune system sufficiently to express itself, the patient's fate is not only sealed, but he or she may have passed the virus on to many others. With Ebola and Lassa viruses, in contrast, the incubation period is frighteningly—but mercifully—brief. If you get infected, at least you know what's happened: toes up or toes down in two weeks. If you survive, you are cured, and you won't get it again.

AIDS is something else. Sue maintains if she were threatened at gun point with two syringes, one with Ebola and one with AIDS, she would take Ebola. Recalling how I'd stabbed myself with a needle in the Sudan in 1979, I thought myself lucky to have escaped infection by Ebola. Now I had another worry. I was acutely aware of how Roy Baron and I had infused plasma into my arm after my needle-stick accident. Who had contributed that plasma? I now wondered. I had no idea, but there was one thing I knew for certain: it had *not* been screened for HIV. Quite possibly, I had become infected with a disease even more terrifying than Ebola. As soon as the test was available, I sent in some blood. Then I had to wait. It was like the Sudan all over again. And when the result came in—negative—I felt like I'd won the lottery, twice.

The ability of AIDS to confound and mystify us even after all these years is uncanny. Why Africa? we wondered. And why now? To try to determine how long the disease had been around, a serological test for HIV was run on serum specimens stored for many years in Kinshasa. The sera had been drawn by Belgian colleagues some years previously—and for other reasons—from women attending prenatal care clinics. The tests on the sera showed that the infection had been present in the city at least since the late 1970s. However, this fact did not give a complete picture of the history of the disease. In spite of the risks of heterosexual transmission, pregnant women are not among the highest risk groups for AIDS. We now know that we would have had to run similar tests on prostitutes, drug addicts, and truck drivers to derive a clearer understanding of it.

If pregnant women were being infected as far back as the late 1970s, AIDS couldn't be that new. The epidemic, however, was. So where had it come from and how did it become so devastating?

CHAPTER TWENTY

HIV Reveals Its Past

Determining where AIDS came from was important in figuring out how the epidemic might have evolved. It soon became apparent that the infection was concentrated in Central and East Africa, and that the majority of infected people lived in the cities. When a mysterious monkey virus, SIV (Simian Immunodeficiency Virus), turned up that seemed to bear great similarity to HIV (though monkeys themselves did not seem to develop any AIDS-like symptoms), many people immediately seized upon the notion that the disease had jumped from monkey to man, and that was how humans had become infected. But in epidemiological terms, this hypothesis seemed weak. Monkeys live mainly in the jungles, whereas AIDS flourished in the cities. We could look for a "smoking monkey" all we wanted, but I doubted that such an approach was going to get us very far. So what *was* the explanation?

Studies continued. All of those involved in this research were beginning to understand that transmission of the virus was not highly efficient. Sexual contact or blood contact (a contaminated transfusion, a drug addict's shared needle, an accidental stick with a syringe) was required. What did this tell us?

Again, I returned to the question that had been gnawing at me ever since we embarked upon *Projet SIDA*. But I turned the question around a little to see whether it might lead my thoughts in a new direction. Now the question was not how long had the virus been around, but how long could it have been around without having been observed before the early 1980s? Although I'd been shown many different types of infections in the 1960s, it seemed to me that I had never seen never seen anything like AIDS, other than tuberculosis, which, in retrospect, would be the likely diagnosis for someone with AIDS in rural Africa. But Zaire, like the rest of Africa, had changed profoundly and rapidly since that time. Most of the changes were in the form of massive urbanization.

Millions of people, in countries throughout the continent, had flocked from the villages to the cities, all looking for work. In African villages, for example (at least in those areas where I had worked), sexual promiscuity was not an everyday occurrence, as it had become for young girls like Yema in cities like Kinshasa. Such behavior is not readily tolerated in remote rural Africa, and social pressures to conform are very strong. Failure to adhere to the rules could have severe consequences for the individual. The practice of polygamy, which is widespread in Africa, ensured that women would marry at an early age. In this respect, the same social system that constrained the freedom of women also had the effect of curbing promiscuity. That, in turn, likely minimized the spread of AIDS. Under these circumstances, I reasoned, a virus with a poor efficiency of transmission, and with a relatively long incubation period, could have been circulating at low levels for years without causing more than a few cases of overt disease. And those cases of AIDS that did occur might easily be missed, especially in primitive conditions, where physicians were rare and diagnostic facilities practically nonexistent. Most of rural Zaire, as I knew too well, scarcely had any medical care. So it seemed reasonable to suppose that AIDS might go unnoticed in the countryside.

It occurred to me that it might be possible to test this hypothesis with what is known as a longitudinal study. Such a study would allow us to compare rates of infection in rural Africa over a significant period of time. It was then that I recalled the Ebola investigation in 1976. We'd collected about six hundred sera from the area around Yambuku at the time, which we then stored for safekeeping in a freezer at CDC. I knew what we had to do.

I contacted people working in the HIV lab at CDC and asked them to perform an HIV antibody test on our 1976 sera. This test was harder to do then than it is now. No commercial test kits were available, and all the reagents had to be made in the lab. Moreover, the confirmatory test was radioimmunoprecipitation, a test as difficult and lengthy as its name, and much more tedious than the simpler Western Blot, which was later to replace it. (The Western Blot detects antibodies in serum against a specific viral protein.) With six hundred specimens to test, we wouldn't know the results for several weeks.

The radioimmunoprecipitation (known as RIP for short) test measured the ability of the patients' serum to precipitate certain proteins of the virus. The virus had to be tagged with a radiolabel, a radioactive tracer; the strength of the radioactive signal is proportionate to the amount of antibody in the serum being tested. This test had the advantage of being very sensitive and at the same time very specific

We waited with great anticipation. My bet was that we would find HIV, but that the numbers would be fairly low. But just how low? I hoped that it would be more than one in six hundred, so that we would have a chance to find it at all. I needed just enough evidence to be able to make a reasonable estimate of the prevalence of HIV infection in 1976 in remote northern Zaire.

Finally, I received a call from the HIV lab. The tests had been completed.

The Elisa (Enzyme-Linked Immunoabsorbent Assay) tests—which search for virus antibodies—confirmed with radioimmunoprecipitation, showed that five of the six hundred sera had antibody to HIV. That suggested that approximately 0.8 percent of the inhabitants residing in the Yambuku area in 1976 were infected with the AIDS virus. (In a study, conducted with Françoise Brun-Vezinet at the Claude Bernard Hospital in Paris in 1986, we determined that sera from Sudan in 1976 had a 0.9 percent antibody prevalence, meaning that about nine out of every thousand people were infected. This result was remarkably similar to what was found in Zaire.) But we didn't stop there. At CDC, we wanted to see whether we could actually isolate the HIV virus from any of the five sera. Most viruses, including HIV, do not survive well outside the human body. They are usually dead in samples within minutes to hours after they have been removed. In order to keep viruses alive, the specimens have to be collected and preserved with great care. This we had done in the first study in Yambuku. In the case of HIV, it is preferable to preserve cells from the patient's whole blood, but the specimens we'd taken from the Yambuku area did not hold preserved cells, only serum. However, these specimens had been carefully transported in dry ice for their long journey to Atlanta, and though they had been maintained at $-80°$ C, the temperature of dry ice, there was no escaping the fact that ten years had elapsed. That is a very long time to keep a virus alive, particularly in conditions that were possibly far from ideal, and at relatively low titers. We also knew that the five people with antibody who had given the serum were well at the time it had been drawn. They were the *survivors* of the Ebola outbreak. None of them had exhibited signs of a disease like AIDS—at least, not then.

Jane Getchell was the young woman working in the HIV lab who took on the slippery job of trying to extract the virus from the five sera. A tall, thin woman in her midthirties, she had worked her way up from a medical technologist to a doctoral degree at the University of North Carolina. She had done the work for her thesis in our lab. After getting her degree, she joined the new CDC

AIDS laboratory. Now we offered her another opportunity. She seized on it with the same energy and enthusiasm she brought to every project she undertook.

Jane had the assistance of a short, stout lab technician I knew only as Donald, and, together, they nursed the tiny aliquots of precious serum through their tissue culture systems. To discover whether any of the five sera contained a virus, they first needed to harvest fresh human donor lymphocytes: the cells found in lymph tissue—lymph nodes, spleen, tonsils—which constitute a crucial part of the human immune system. If a virus was present, the lymphocytes would show a definitive response. But there had to be enough lymphocytes to grow the virus, so Jane and Donald had to stimulate them into multiplying to a sufficient number of cells in order to seed them with the specimens containing the suspected virus. The procedure was arduous, requiring prodigious dedication, organization, experience, and not a little good luck. Jane and Donald checked the cultures every day, changing the medium (a fluid containing the nutrients in which the lymphocyte cells are grown) and assaying them regularly for the presence of reverse transcriptase enzyme. This enzyme is the calling card of HIV virus: once you find it, you know you're on to something. It is an enzyme that makes a DNA copy from the RNA, which is the normal genetic material of the HIV. This copying of DNA from RNA is not something the cells of a body ever do, so the enzyme is not normally found in body tissues or fluids. When you do find it, you know that there must be a retrovirus (such as HIV) in the mixture. Of course, we were not sure that we were onto anything at all. It was a long shot. And because the amounts of remaining sera were so small, it was likely a single shot as well. We had isolated several viruses the year before (in another laboratory) from our specimens taken from patients at Mama Yemo Hospital in 1983, so we knew it was possible to isolate virus from Zaire. But again, no one had ever succeeded in isolating a virus that was this old.

Finally, after several blind passes of the material into fresh cell cultures, there was a hint of an increase in reverse transcriptase activity in one specimen. Trying to contain their excitement, Jane and Donald passed the specimen again through fresh human lymphocyte cultures, until at last it blossomed into a full-fledged growth, pumping out reverse transcriptase as it multiplied in each fresh batch of cells. We now were in possession of the oldest HIV virus in existence.

The virus that Jane and Donald extracted has been used as the prototype strain for studies of the evolution of HIV viruses. Because HIV has appeared in the form of many different strains, this virus has proven a boon to those scientists who can use it to help sort out relationships among HIV isolates from different places and different times. Several other scientists later established the genetic code sequence of this virus. With this information, still others have been able to determine just how much its descendants have changed over time. In simple terms, this tells us how old a particular strain is relative to other strains, and it provides an accurate means of determining which strains are closely related. Information obtained in this way is then translated into a sort of family tree, called a dendrogram. By charting the family tree of the virus, scientists can clearly see where HIV diverged from the monkey viruses it resembles. Although SIV and the human virus (HIV) may once have come from the same source, their genetic history reveals that the two went their separate ways a long time ago. Whether the virus first appeared in monkeys, which then transmitted it to humans, or whether it was the other way around, we will probably never know. All we can be certain of is that this did not occur any time recently.

We also hoped that our isolate could be of some benefit to the living, which was why we were happy to send it to the Salk Institute for use in their effort to develop a vaccine. Sadly, no vaccine—produced by the Salk Institute or anyone else—has so far proven effective.

Although we had isolated the virus in the lab, we still did not know to what degree it had proliferated in the African population. Because AIDS had been masquerading as tuberculosis and so many other illnesses, there was no clear way that we could measure the extent of its spread. To begin to cope with the disease, we needed to know how many people had it. That was the dilemma that now confronted us.

Things now started to happen fast. The investigation we had conducted in Zaire in 1983, together with one carried out at the same time in Rwanda, established that HIV was widely spread throughout Central Africa. Fakhry Assad, the ebullient Egyptian I had first met on my way to Sudan in 1979, now headed the Division of Communicable Diseases at WHO. This was his bailiwick. A vigorous correspondence, conducted by phone and letter, ensued between Fakhry and myself. Fakhry had already become convinced that he needed to get a program started under the auspices of WHO, but his chief, Dr. Hafdan Mahler, was more difficult to persuade and was slower to grasp the significance of what was happening. Once he did, however, he, too, became very supportive.

In early 1985, I began to work on HIV in the Central African Republic (CAR), a country of about 2,400,000 people, bordered by the Sudan and Zaire. This was the same country I'd attempted unsuccessfully to enter, over the northern border of Zaire, in my search for Ebola in 1976. This time I came by a different route—via the capital city of Bangui. I can only assume they had cleared away the tree that had blocked the path in 1976 and had also replaced the defunct ferry.

My chief collaborator was Alain Georges, director of the Institut Pasteur in Bangui. This laboratory is part of the worldwide network of Institut Pasteur labs, most, if not all, in former French colonies. Alain is a russet-haired, energetic and enthusiastic Frenchman, who likes nothing more than getting things done. He is also a gourmet cook, an expert on fine wines, and a consummate scuba diver. He is still working in Africa, currently in the Gabon, on HIV and viral hemorrhagic fevers, which we both share an interest in. It was Alain who, in early 1996, was responsible for isolating Ebola virus from patients infected from a dead chimpanzee. Alain recognized and contained the outbreak without any secondary cases. In the course of our investigation, we quickly realized that HIV had also arrived in CAR. In the course of our 1985 investigation, we quickly realized that HIV had also arrived in CAR. In April 1985, while in Bangui, I asked Alain if he would like to host an African AIDS conference in CAR, which would be sponsored by WHO. He agreed in principle, and I then faxed Fakhry Assad in Geneva, proposing a workshop on AIDS to be held in October 1985, one month before another conference already scheduled to take place in Europe. Fakhry had no objection. To maximize the participation in the workshop, we decided to invite representatives from sixteen African countries, as well as from Europe and the United States. We intended the meeting as a forum where we could disseminate information about AIDS and educate non-African scientists about what life with AIDS in Africa was really like. There are many people, including scientists, whose knowledge of geography is abysmal. They could only benefit from an understanding of the context in which the African epidemic was taking place. Whatever illusions visiting Western scientists might harbor about Africa, it was likely that CAR would open their eyes. It is one of the least-developed and most severely poverty-stricken countries on the continent. Until 1979, it was dominated by Emperor Jean-Bédel Bokassa, who squandered millions of dollars on his coronation. In imitation of Napoleon, he crowned himself and then proclaimed the country an empire. He was ousted, but not before he had plundered the nation's meager wealth and murdered his opponents, some of whom, it is said, he later ate. Yes, even in a continent infamous for its tyrants, Bokassa stood out.

We still had an urgent need to begin to estimate the size of the AIDS problem in Africa. Only then could we figure out what needed to be done—and where. This is what is known as surveillance. It involves counting the number of cases of AIDS. But we had a peculiar problem with AIDS. Few AIDS cases in Africa receive any medical care at all. No diagnostic tests, suited to widespread use, yet existed, and even the wealthier hospitals wouldn't be capable of performing sophisticated tests like T4/T8 white-cell counts, which we used in the U.S. In the absence of any of these markers, we needed a clinical case definition—that is to say, a set of guidelines a clinician could follow in order to decide whether a certain person had AIDS or not. This was my major goal: if I could get everyone at the WHO meeting in Bangui to agree on a single, simple definition of what an AIDS case was in Africa, then, imperfect as the definition might be, we could actually start to count the cases, and we would all be counting roughly the same thing.

The U.S. ambassador to CAR, a Bobby Kennedy look-alike, threw his weight behind the meeting and offered us the use of the embassy's facilities and communications system. We needed them. In 1985, Bangui was not a modern city—not that it is one now—and it didn't figure prominently on the itineraries of many travelers. The ambassador had been a Peace Corps volunteer in Africa and had come to know the continent well. He was even instrumental in starting a basic education program on AIDS for Peace Corps volunteers as well as for embassy staff.

As preparations for the October conference got into high gear, Alain, his wife Marie Claude, known as Claudie—a lovely, petite woman with even more energy than Alain (and infinitely more patience)—and I, together with Sheila Mitchell, continued our surveys of various populations in Bangui. Set on the Ubangi River, Bangui is a major port and commercial center with 340,000 people. One of the most interesting groups on whom we focused our surveillance efforts were the *femmes libres*. Like Yema and her counterparts in Zaire, these were women who had no family to support them. In Bangui, the *femmes libres* could be found in a circumscribed area of town and were readily identifiable. When we interviewed them, they became quite cooperative after we assured them that their names weren't going to be used in our survey.

The places where they practiced their trade were pitiable. For example, one of them was an old Volkswagen van body, furnished with sheets of cardboard. Rags doubled as curtains to preserve a modicum of privacy. Fortunately, all of the women we surveyed were seen regularly by a local physician. That made it easier for us to educate them about the risks they were running. It also made it easier to trace the level of infection. Already in 1985, 4 percent of the *femme libres* had HIV. And that, we feared, was only the tip of the iceberg.

That summer, Fakhry had called several people to solicit ideas about who might set up a major AIDS program in Geneva. It would concentrate its efforts on the developing world. I suggested Jonathan Mann, who was already making a major impact in Zaire with *Projet SIDA*. I couldn't imagine a better candidate for the position. Fakhry would have a chance to meet him when we all convened in Bangui in October and sound him out. As I anticipated, Fakhry and Jonathan hit it off well. But when Fakhry asked him whether he'd have any interest in setting up a WHO program on AIDS, Jonathan didn't immediately agree, although he did tell Fakhry that he would help him put together some ideas for the new organization. Eventually, Jonathan did go on to found and direct WHO's Global Program on AIDS. Over the next half-dozen years, this program had probably the greatest single impact of any program on the AIDS problem in the developing world. Since Fakhry died prematurely and suddenly just a couple of years later, this was probably his greatest legacy to WHO.

In the meantime, the debate over what would become known as the "Bangui definition," used by WHO to identify AIDS cases, was a heated one. The definition was reached by consensus based mostly on the delegates' experience in treating AIDS patients. It has proven a useful tool in determining the extent of the AIDS pandemic in Africa, especially in areas where no testing is available. Its major components were prolonged fevers (for a month or more), weight loss of 10 percent or greater, and prolonged diarrhea. There were also several minor symptoms, including tuberculosis and repeated herpes virus infections.

Several scientists who attended this groundbreaking conference were leading AIDS researchers in Africa and Europe. Among them were the two Françoises from Paris: Françoise Barre-Sinoussi and Françoise Brun-Vezinet. Barre-Sinoussi of the Institut Pasteur was a pivotal figure in the history of AIDS research. She should by rights be much more widely known, because working in the laboratory of Professor Luc Montagnier, she was, in 1983, the first scientist in the world to have ever isolated the HIV virus. Her contribution became obscured in all the media hype about the "discovery" of the virus. From what I was later told, her achievement was due to her obsessive temperament and remarkable persistence. She was later recognized with a King Faisal Award for Medical Sciences, a prestigious prize. Françoise remains actively involved in HIV research, particularly in developing countries.

The second Françoise—Brun-Vezinet—was a co-author of the famous paper, written by the first Françoise, which originally reported the isolation of the HIV virus. A leading researcher on HIV in her own right, Brun-Vezinet was also the scientist who tested the sera from the 1979 Ebola investigation in the Sudan and determined that the infection rate there in 1979 was almost exactly the same as it had been in Zaire in 1976. She inadvertently was responsible for a near tragedy during the Bangui conference, which might have set AIDS research back a decade. During a break from the conference, the sponsors arranged for a helicopter tour of the countryside. Both Françoises were among the group who signed up for it. About twenty minutes into the flight, Brun-Vezinet, who was seated near the pilot's seat, accidentally bumped against a lever that disengaged the rotor. Suddenly, the helicopter began to plummet. The pilot reacted quickly, though, managing to stabilize the craft before it lost too much altitude. The scientists were still shaking when they stepped out onto solid ground. An accidental infection in the lab was a risk they were prepared to face, but not a joy-ride helicopter crash in the wilds of the Central African Republic.

By the time I returned to Atlanta from the CAR, I was still obsessed by the question of the origin of the AIDS epidemic. At least now we had one fundamental and unique piece of information: we knew that the prevalence of HIV infection among the village population of northern Zaire in 1976 was less than one percent. In addition, we now were in possession of a virus from one of the villagers. What we didn't know was what had happened to the same population in the intervening years. If our hypothesis were correct, then the infection would have remained the same in the countryside, even as it rose markedly in the urban population. We reasoned that transmission of the virus was just effective enough in such a remote rural society to maintain a presence in the community without exploding into an epidemic. We were anxious to know, though, whether our hypothesis could be proved, because we felt that it might indicate where HIV had been lurking all of these years. We also wanted to test our other contention and find out whether, in fact, the outbreak of HIV was linked to such factors as increasing urbanization, migration from the countryside, and changes in lifestyle, particularly involving sexual practices. Someone would have to go to Zaire and investigate. Enter Kevin DeCock.

CHAPTER TWENTY-ONE

HIV Goes Downriver

Kevin's challenge was to make his way back to one of the most isolated areas of the world to ask a decade-old question: *what happened to HIV in a remote rural society in Africa over ten years?* We also wanted to know about our original HIV-positive people, especially the young woman in her twenties, from whom we had isolated the oldest virus. What had happened to her?

At the time, Kevin DeCock was an Epidemiology Intelligence (EIS) Officer serving in my branch. He'd worked in Kenya and was eager to return to Africa. In his late thirties, he had the sinewy build of a long-distance runner, which, in fact, he was. Kevin was truly a citizen of the world. On the last occasion I saw him, he was carrying a Belgian passport, a U.S. green card, spoke with an impeccable British accent, and was married to a lovely Kenyan wife named Sopiatu. He was born in Belgium of a Belgian father and an American mother, who had come together in a wartime romance. After obtaining his M.D. from Bristol Medical School in the U.K., he moved to the United States to train as a specialist in liver diseases. He'd been drawn to CDC—specifically my branch at CDC—because he wanted to work with viruses in Africa.

Before Kevin set off for Zaire, I provided him with the complete list of the six hundred individuals from whom we'd taken the sera. The most important ones on the list, of course, were the five we'd found to be positive for HIV. It was one of his objectives to discover what had become of them.

Things hadn't changed for the better in the area over ten years. We were still going to need help with logistical support for Kevin. I couldn't ask Jonathan Mann any longer; he'd already left to assume his new position in Geneva with WHO. His replacement was Robin Ryder, who had been at CDC for a number of years, and then had left for a university position. He is an energetic, wiry fellow so obsessed with running that he runs long distance morning *and* evening. He had developed some familiarity with Africa, working on hepatitis infections in the Gambia. Although his unfamiliarity with the French language was something of a handicap, he threw himself into his new job with the same energy he brought to his daily runs through the park. We were initiating our HIV study just after Robin had taken over the helm in Kinshasa, so we needed to convince him that this study should have high priority in his long list of important things to do. We were able to do just that.

Kevin was fortunate, because flights were erratic and infrequent, to procure a plane seat for himself from Kinshasa up-country. The flight gave Kevin a chance to observe a part of Zaire that travelers seldom reach. On the way to Lisala, the plane touched down briefly in the northern village of Ghadolite, birthplace of President Mobutu, located on the Ubangi River in northern Zaire. Kevin said he had never seen anything like it, not in Zaire, not anywhere in the world. How many airports can boast of a gold-domed reception area? I had also seen it many years before, on my way to work with monkeypox in the north. The village is located in the middle of nowhere. There are no major highways that connect to it. It produces nothing of importance. Yet it has street lights, twenty-four-hour electricity, and shops with shelves groaning with goods, all flown in on Air Zaire planes. In short, it has every convenience and amenity that the rest of Zaire lacks. The whole place was a monument to the extravagance, corruption, and waste that has long marked Mobutu's rule.

Back in Atlanta, I was on tenterhooks. Could Kevin do it? Could he find out whether AIDS had spread or not? I was envious; I would have preferred to be out there myself, rather than sitting in an office waiting for news. Just before he left, Kevin had signed a contract on a house in Atlanta. I distracted myself by acting on his behalf and helping him insure that the closing proceeded on schedule—a pathetic vicarious involvement in Kevin's adventure.

Kevin's task was a painstaking and often daunting struggle to pick up the spoor of a ten-year-old trail. In the end, he succeeded in finding out what had happened to all five HIV positive individuals. This achievement was a testament not only to Kevin's skill and ingenuity, but also to the stability of the rural African society. Three of the five were dead. To determine if their deaths were attributable to AIDS, Kevin interviewed people who had known them. The friends and relations of the deceased described an illness marked by severe weight loss and other ailments that left little doubt in Kevin's mind that they had succumbed to AIDS. From the two positives who were still alive, Kevin took specimens of lymphocytes to bring back to CDC. Once the tests were conducted, any remaining shred of doubt vanished. They all still had antibody to HIV.

This was definitely AIDS.

Kevin then turned to his main task, which was to undertake a random cluster survey of the same villages from which the original specimens had been collected. The idea was to conduct a mirror image of the 1976 survey, and then compare the results. After collecting and preserving over 300 sera, he brought them back to CDC. The whole study took Kevin about six weeks. We turned the specimens over to the HIV laboratory, where they were tested by the same means that had been employed on the specimens from 1976. Once again we waited in suspense.

When the results were in, we examined the data. Right on the money. The prevalence of infection was exactly the same in the Yambuku area in 1986 as it was in 1976: 0.8 percent of the population had HIV. Now we had strong evidence to indicate where HIV was hiding, as well as a means of understanding the origin of the modern AIDS epidemic in Africa. This isn't to say that the origin of the disease was necessarily located in the vicinity of Yambuku, but we believed that it had come from *somewhere* in rural Central Africa and that it managed to coexist with the population, infecting a few individuals from time to time without ever seriously threatening the vast majority of inhabitants. AIDS wasn't new. It hadn't suddenly emerged out of the jungle. Our hypothesis—that it had been around for a long while—was confirmed. It was later further strengthened when Françoise Brun-Vezinet found almost identical (0.9 percent) prevalence of HIV in the remote area of southern Sudan, from several hundred sera collected in 1979.

But while the disease had stayed at a relatively stable level in the rural societies, it was running at epidemic proportions in the cities. That was certainly something new. Was our other hypothesis correct? Did rapid urbanization account for why AIDS had suddenly become such a crisis in such a short period of time? *Femmes libres* and other urban phenomena did not exist in the villages. But where did urbanization actually start? To examine this issue, Kevin wondered what would be the prevalence rate in a river town like Lisala, which was about one hundred miles from the villages. He collected more sera in Lisala. Our tests at CDC would reveal that, here, the HIV prevalence jumped to 11 percent among single women. So what was special here? This was a town where the river traffic from Kinshasa docked. Its population had exploded in recent years because of migration from the villages toward the large cities. It was no coincidence that there were also many *femmes libres* in this town, nor did it come as a surprise that they had a much higher rate of infection than women in the villages, where there were no *femmes libres*. This supported our second hypothesis. A combination of migration, dramatic social change, and sexual promiscuity, all related to urbanization, were the driving forces of the African epidemic. In Kinshasa, HIV prevalence was 8 to 10 percent, but in femmes libres it was now a staggering 30 to 40 percent. Thus we had an HIV gradient from very low over a ten-year period in remote villages, higher in at-risk populations (such as *femmes libres*) in larger towns, and even higher in the big city, Kinshasa. In those ten years, HIV had spread silently down the river.

One thing I still needed to know. I asked Kevin what had happened to the young woman who contributed the oldest HIV virus ever recovered from a human being.

I didn't have to wait for him to answer. I could see it in his face. She was one of the three who hadn't made it.

Her tragedy is, of course, part of one that is much larger, and even *Projet SIDA* has all but died, a casualty not of viral disease but Zaire's political squabbles. In the early nineties, the situation in the country turned explosive. As tensions between Mobutu and the opposition increased, elements of the army went on a rampage because they hadn't been paid for months. The crisis atmosphere made it almost impossible to continue the AIDS program. A veteran of CDC named Bill Heyward had taken over as project director from Robin Ryder. Like Robin, he attacked the French language with great energy in order to be able to do business in the country. However, he soon recognized that no matter how much fluency he attained, he still faced nearly insurmountable obstacles. Eventually, the political unrest made it too dangerous to remain any longer. After only a year, Bill was forced to leave. *Projet SIDA* still survives on paper—but it is little more than a ghost program. Nevertheless, as one of the earliest programs in Africa, it had already contributed greatly to efforts against AIDS in Zaire and the rest of the world.

CHAPTER TWENTY-TWO

The Lassa Project Revisited

In 1986, eight years after Joe had established the Lassa fever Project in Sierra Leone, it was my turn. With the support of WHO, he had arranged for me to go out there to try to replicate the study we had performed in monkeys at CDC, examining the impact of Lassa virus infection on cells and platelets. This time we would observe the same things in patients coming to the hospital with Lassa fever. We were hoping that the monkey experiments we had completed would give us some ideas about what it was that made Lassa victims bleed and go into shock. That, in turn, might lead to a more effective treatment.

On the advice of Guy Neild in London, I was taking with me a drug called prostacyclin, which had been used in patients with septic shock in London and North America. We thought it might offer some hope in patients with severe Lassa fever, too. By protecting platelet and possibly endothelial cell function, prostacyclin might prevent the leak from the blood vessels and, therefore, prevent shock. One of the reasons Lassa victims die is because of respiratory failure resulting from the lungs filling up with fluid. This type of pulmonary edema is called Adult Respiratory Distress Syndrome (ARDS). When liquids leak out of the bloodstream, they swamp the lungs, and the patient ends up literally drowning in his own fluids. Patients who had been given prostacyclin for septic shock responded well in the London studies and suffered no bad side effects. It seemed reasonable to try the drug with Lassa. Joe had helped me get through the difficult process of ethical approval by all the various bodies, and the company making the drug had been very helpful in providing us with appropriate safety data and some free drugs.

My journey began in London. I had just made my way through immigration at Gatwick when I ran into a delightful, handsome blond Californian. He seemed to be looking for me.

"Would you be Sue by any chance?" he asked.

So, I thought, this was Don Forthal. Don was the EIS officer currently training in Joe's branch at CDC. He, too, was on his way to Sierra Leone, where he planned to study children with Lassa fever. Although he was obviously tired after his trip from the States, he still made quite an impact on me. I hadn't quite expected such a stunningly good-looking guy as a traveling companion. Don was quite a lady killer in those days, so his trip turned out to be something of an event for the girls of Segbwema. As for Don, he was highly receptive to the Mende girls, who are tall and slim with long necks, fine features, and wonderful eyes. They dress stunningly in bright colors with turbans piled carelessly but elegantly on their heads. They thought Don was the most exotic thing they had ever seen. He broke more than a few hearts—though not all of them. Ultimately, he ended up marrying a dazzlingly beautiful Ethiopian air hostess he met while working for WHO after leaving CDC.

For both of us, it was our first trip to Africa. It did not start well. In fact, it almost did not start at all. No sooner had we greeted each other than we learned that our British Caledonian flight was delayed for twelve hours because of "mechanical problems." This didn't come as reassuring news, not the least because West African airports had only limited mechanical expertise available. One wrench and a single screwdriver do not exactly inspire confidence.

It was a cold winter morning at Gatwick, and an inch of snow covered the ground. We had no choice but to make do in clothing suitable for tropical Africa until we were able to take off late that evening. It was our introduction to the vagaries of flying in Africa. It would only get worse.

Actually, getting into the air proved to be the easy part. Getting to where we were going was a real problem. When we reached Banjul in the Gambia in the early morning hours, the crew informed us that, because of the delay in departure, they had "run out of flying hours," which meant that we would now have to wait for a new crew to take over for the rest of the journey to Freetown. But once the new crew had settled in, they announced that there had been a slight change of plans: we would now be flying to Monrovia, Liberia, and then they would drop us off in Freetown on the way back!

It was still dark as we took off from Monrovia, but as we reached Freetown the sun was just rising—so we assumed we were about to land. The pilot came on the intercom to tell us cheerfully that we were cruising at 32,000 feet above Freetown, and he hoped we were enjoying our breakfast. He said that it was a bit misty down there, so he was taking us back to the Gambia.

We had now flown over our destination twice without landing. When we landed at Banjul for the second time, the pilot announced that he now had to go back to London.

A riot ensued.

There were about sixty of us bound for Sierra Leone. We had been traveling now for twenty-four hours and were determined not to end up back where we started in the cold and snow of Gatwick! We threatened a strike. We told the pilots that we had no intention of taking our seats or buckling up for takeoff. The crew hastened to pacify us, offering to serve us a second breakfast, and this time they suggested a calming dose of free whisky. We declined.

Negotiations resumed on the steps of the aircraft as the sun rose higher in the sky, gradually burning off the mist in Freetown. Finally, the captain agreed to see whether he could establish radio contact with Freetown (not easy) and find out whether conditions had improved sufficiently to attempt another landing. It took him some time, but in the end he was satisfied that the mist had dissipated sufficiently to try again.

As buoyed as we were by this news, it did nothing to cheer the passengers who had just boarded at Banjul. They were Brits who'd come to the Gambia on a winter sunshine holiday. They'd assumed that they were headed directly back to London. That's what they had paid for. Freetown had appeared nowhere on their itinerary. The idea of a detour put them into such a state of alarm that they threatened to start a riot of their own.

Fortunately, they were sufficiently confused about the geography of the region that they failed to understand just how far out of their way they would actually be going or what the place was like. They'd come to the Gambia for the sun, the beach, and the beer; for all they knew, they could have been on the Costa del Sol. The captain convinced them that they would only have to put up with a slight inconvenience, nothing more. We took off once again.

Now that we were in a better mood, we consented to another breakfast. By the time we'd finished eating it, the plane was approaching Freetown for the third time. The airport at Lungi, set among mangrove swamps, is located across a wide estuary from the city. To make it down safely you have to depend on the skills of an experienced pilot. Our friend from British Caledonian had been quite right not to attempt a landing in the thick morning mist. Air-traffic control was rudimentary. Only one or two local and international flights a day landed and took off from Lungi. At night, the runway was lit by a single strip of lights fueled by a generator; it was switched on only a few times a week for big airliners, and then only as the plane was making its final approach. Even in Freetown proper, electricity was a rare commodity. Outside the capital, it did not exist unless you produced it yourself.

We skimmed over the mangroves, suddenly bumping down on the runway. We breathed a huge sigh of relief. The plane taxied to a decrepit terminal building and lurched to a stop. Collecting our belongings, we got out quickly before our fellow passengers bound for London realized they had been put down in the middle of a mangrove swamp five degrees north of the equator. As we stepped out onto the runway we were immediately enveloped by a heat so intense and humid that it felt like walking through glue. As we passed through the opening into the airport building, Don and I were suddenly plunged into the chaos that is Africa. We'd never experienced anything like this. We found ourselves caught up, confused, and disoriented in a crush of people, all of them determined on being the first to get through the many hurdles that had to be overcome before you could hope to gain entry into Sierra Leone. Money needed to be changed, innumerable cards inspected and stamped. The officials were in no hurry, carrying on several conversations at the same time with various people around them, pointedly ignoring all the hands, cards, and passports that were being waved in their faces. With only one flight a day, this was their big moment, so they spun it out. Fortunately for Don and me, Austin Demby was there to rescue us from this human maelstrom. A Sierra Leonean working with our project, Austin had a reputation as someone who could get things done. It didn't take us very long to learn that Austin had a "cousin" everywhere we went in the country. Our path through immigration was suddenly smoothed with a perfunctory wave and a ceremonial stamping of many papers. Taking our documents, Austin gave us a series of instructions about where we were to stand and what papers we were to hand over. We complied obediently and gratefully. Miraculously, all our bags appeared—intact.

Austin, we learned, was a Mende (the major tribe in eastern Sierra Leone) and the son of a well-known chief. He is tall and handsome with an aristocratic bearing. Like all other local graduates, he had been educated at Forah Bay College in Freetown. Patricia Webb had recruited him straight out of school, and he turned out to be a marvel at fixing logistical problems, particularly if they were political. He also displayed courtesy and patience to a miraculous degree. Austin's cultivated manners and his amazing ability to find friends wherever he went rendered him indispensable to the project. Although I worked with him for several years, I saw him lose his temper only once. That was when someone tried to push ahead of him at a bank just moments before it was about to close. We had just spent two grueling days trying to procure diesel fuel when there was none to be had in the country, because the government had run out of foreign exchange.

"What do you want to do?" Austin asked when we finally made it through customs. "Would you rather go into Freetown, which will take us about two hours, or just go straight up to the project?"

I was exhausted and dazed by then, so I didn't know what the choices really entailed. How far was town from the project? I had no idea of the geography of the country or even where the Eastern Province was. If anything, Don was even more out of it than I was, since he still was suffering from jet lag after his flight over the Atlantic. Austin was too courteous to press me to make a decision. So I told him that proceeding directly to the site of the project sounded fine to me.

If I had taken a look at a map, I would have realized we were about to drive to the other side of the country.

After crossing the estuary separating the airport from the mainland on board the battered old ferry spewing out black smoke from its stack, we came out on a road and turned left. The road, Austin explained, was the only paved road that existed in the entire country. It basically consisted of two hundred miles of asphalt and potholes—more potholes than asphalt—before finally giving out onto dirt roads, much of the surface of which had been washed away in the rains. Stretches of the road bore names given to them by Bob Craven, the project director at the time: "Hemorrhoid Hammer" and "Hematuria Hill," for example. Covered with thick red laterite dust, the roadway was so badly rutted that it felt as though every bone in my body was being ground down until it became part of the dust that quickly invaded our hair, our teeth, every part of us. There was no air conditioning, so the only way to keep cool enough to breathe was to have all the windows open. The truck bounced and wound around the worst of the holes and ruts, often driving off to the side for distances where vehicles had made a fresh track in the bush marginally smoother than the road itself. Oncoming vehicles swerved drunkenly toward you on the wrong side of the road until you saw the pits and mountains they were trying to avoid. Goats, sheep, chickens, people with great loads on their heads, and occasional cattle added to the navigational hazards. I soon had a splitting headache. But rest was out of the question. The only prospect of attaining oblivion was by banging my head against the side of the truck hard enough to give myself concussion. I nearly did this a couple of times inadvertently as the truck bounded over an assortment of moon craters.

This ordeal by truck went on for the nearly eight hours it took Austin to reach our destination. By the time we lurched into Segbwema, I felt as if I'd been in a tumble dryer filled with red dust. I could barely move. And when I did, everything hurt. A shower was what I needed, but what I got was a bucket bath, and I soon found out how grateful I was. A single bucket is allotted for this purpose; it was the dry season now. When you needed water, you dropped the rope-secured bucket into an underground cistern that filled during the rainy season. The object was to get the bucket to fall below the surface of the water and fill properly before it floated uncontrollably, half filled. You then hauled it up carefully, so as not to spill the contents. I learned the routine very quickly: you start with your face and hair and work down to your feet. It was a miracle just to be free of all that caked red dust!

The Lassa project had been established on the grounds of the Nixon Memorial Hospital after the main lab facility had been transferred from the original site Joe had developed in Kenema. Years before, the Methodist-run hospital had enjoyed a reputation as one of the best in the country. Although it had seen better days, it still provided essential services, at a reasonable level of quality, to a large population and at minimal cost. The project had two labs: the platelet lab, which had just been improvised by Donna Sasso out of a bedroom in a hospital-staff apartment we also used as an office, and the main Lassa fever lab, which occupied a building across the way. The main lab had been built with funds that Joe had obtained several years earlier. With accommodations so scarce, most of us who worked at the project stayed in the director's house at the top of a hill opposite the hospital.

It didn't take me long to find out that almost everything in Sierra Leone needed to be improvised. Gasoline and diesel fuel were almost unobtainable, and we had to make special arrangements to buy imported fuel with hard currency. The country had been deteriorating since before Joe set up the project in the late 1970s. The telephone, which used to work (sometimes), was now entirely non-functional. There were a couple of handsets attached to the wall, but the wires were not connected to anything. The water system, although installed, had never worked. Nor was there any spare bottled gas for cooking. We had to use what we could scrounge to run the gas-powered freezers where we stored our precious specimens. I learned how to cook over a three-stone fire, one of the oldest cooking techniques known to man, dating back to Neolithic times at the least. It requires nothing more than some branches, a few twigs, and three large stones to balance your pot. Joe later showed us how to pop corn on one of these fires, a task for which we had the assistance of Bob Craven's lovely Labrador, Beans. Beans liked to participate in anything that we were doing, especially if it involved food.

But, that year, due to more than usually bad economic conditions, finding anything to cook was a problem. This was true even though the country was naturally verdant and fertile. But as a result of slash and burn practices, the original rain forest has vanished. Once the larger trees are felled and chopped up for export, the remaining growth is burned. The patches of thin soil left behind are used by subsistence farmers for growing cassava, coffee, and other crops. Rice is another staple grown principally in swamps. And there are many, many swamps. In such a subsistence economy, people seldom grow or catch more than they need to feed themselves and their families. At one point, we couldn't even buy bananas in the local markets—and we were living in the middle of a banana grove. In the Segbwema market stalls, it wasn't unusual to find only three tomatoes and five onions for sale at a single stall, carefully and tastefully displayed for purchase one at a time. Most people had to satisfy themselves with "chop," a pounded cooked leaf laced with a very small amount of meat, dried fish, or—if you were lucky—fresh fish from the swamps, spiced up with hot red peppers.

Despite poverty and shortages, I found Segbwema a happy and friendly town. People seemed to be able to defy the grinding poverty of their lives with good cheer. And when you needed your spirits raised, there was always palm wine, a heady brew taken from the top of the palm trees by special tappers. These men performed extraordinary feats, scrambling up the trunk of a tall tree, using only two bamboo hoops for security. To me, the wine tasted lethal. I preferred either the locally brewed Star beer or Coke. But beer and Coke required refrigeration. And that was another problem. The refrigerator was powered by kerosene. If the refrigerator wasn't working, which happened all too frequently, the bad news was immediately communicated: "Cold beer—no deh." So, each evening as the sun lowered and the temperature became bearable, we would repair to one of the local "bars," which consisted of an open-fronted hut or a small room with the all-important kerosene-fueled refrigerator. We'd sit outside on simple wooden benches or on high-backed chairs. People would drop by and pass the time of day—local people, Peace Corps volunteers on motor bikes, missionaries of various denominations driving dilapidated cars. Everyone was good natured and eager for a chat. The atmosphere was always convivial.

Bars came and bars went, but one of our favorites was Eddy's. It would later have the distinction of doubling as a lab for David Cummins, a visiting scientist from London. David spent much of his time, when he wasn't with Lassa patients, conducting some of his more critical experiments in Eddy's. These involved assessing the ability of platelets to aggregate in the bar's locally brewed beer. He even kept a running account of the data, which he posted above the counter. His activities produced a good deal of comment from the patrons, who weren't quite sure what to make of him, but who always enjoyed his presence.

I was new to Africa, and I was lucky to have the support of Donna Sasso from Joe's lab at CDC. Donna was an athletically built woman in her midthirties, a virologist who'd worked with Sheila Mitchell and me in the Level 4 lab in Atlanta, studying platelet function in monkeys infected with Lassa. The research we did had given us some important clues about why Lassa victims bleed and go into shock. Now that we had seen what could happen with monkeys, we wanted to confirm that the disease functioned the same way in human beings.

Donna was the sort of practical personality you need under the conditions we found. Fortunately, she was as strong as she looked, because strength was required to get the lab working. Everything in the lab depended on electricity—our light, our centrifuge, the platelet aggregometer I'd brought with me from England, but the only way to produce electricity was the generator that sat on the verandah just outside the lab. This was where strength came in. Every morning, it was Donna's job to start the generator. As it was an old machine, the only way to start it was by pulling a cord. When I tried, I couldn't even get it to turn over, but Donna could get it and just about anything else in the lab to work.

Every morning I rose as soon as it was light and made coffee on the three-stone fire. That was my breakfast. Donna, on the other hand, needed fuel: bacon and eggs and instant grits, which she bought in quantity whenever she had a chance to visit the American Embassy commissary in Freetown. When we were through, we would troop over to the hospital, collect our gear, and make rounds. Sometimes we'd see as many as fifteen suspected Lassa patients in a day. This was a far greater number than I'd been led to expect when I arrived. Donna had told me that we'd probably see about four or five new cases a day. Those new patients who could, sat on an old school bench outside the lab door and waited for the results of their blood tests. Those who were too weak collapsed on the bench or else were carried directly to the ward. The tests would reveal whether they'd developed antibodies to Lassa virus and measure their liver function. If the clinical exam suggested the patient might have Lassa fever, the decision to treat rested on the AST level. If the AST was over 150, the patient was given intravenous ribavirin.

While ribavirin was a boon for the patients, often leading to gratifyingly rapid recovery, the very success of the treatment presented a problem for my research. Because the patients rarely became as sick as the monkeys I had been working with, I had difficulty obtaining enough seriously ill patients to study. This was a minor drawback. There were still lots of patients, and plenty of sick ones at that. Needless to say, the patients could not have cared less about whether they might contribute to my research or not. They were just happy to make a complete recuperation. The word was out in town, communicated in the local dialect of Krio: "You getta Lassa fiva-o, you be fo go nah hospitool nah Segbwema."

We didn't just rely on word of mouth, though, to get our message across. The project instituted an education campaign to promote awareness of the disease, emphasizing the steps that people needed to take to protect themselves. A young Englishwoman named Cathy, recruited by a later director, Diane Bennett, took charge of the program, since she'd been trained in the use of theater in education. She developed a number of original dramas, shadow plays, and puppet shows as part of the campaign, many of them produced entirely by children. Typically, the protagonist in these playlets would become infected with Lassa fever because he'd failed to keep his house free of rats. He would have to be taken to hospital, where he'd be given ribavirin. Then he would make a complete recovery. Of course, everyone lived happily ever after!

The message was clear: you won't catch Lassa fever if you avoid contact with rats and keep them out of your house. But if you do fall ill, get help quick—not from the medicine man, but from the Lassa fever project.

Music, which occupied such a prominent place in local society, was also used to promote our cause. Cathy inveigled her husband, a musician, to lend his talent to the campaign. A unique presence in Segbwema, with his dreadlocks and West Indian good looks, he composed songs with a Reggae beat to drive home the anti-Lassa message. His idea of a good beat fit well with local tastes. Indeed, his style of music is thought to have originated in West Africa. Before long, the whole province was singing the lyrics to a song called "Lassa Fiva No Good-O." The song was a big hit at the local disco, which Austin and his brother had opened in Segbwema. It was so popular that tapes of the song were distributed around the country. Local musicians were recruited to lead processions through the streets. A large paper rat would be trundled along the way, then beaten with sticks, and finally consigned to the flames in a spectacular ceremony that was accompanied by fully robed and masked figures, everyone dancing and cheering.

The entire time we worked at the project, our protection consisted of gloves, gowns, and masks. But from experience we knew that that was all we needed. The important thing was not to stick your finger with an infected needle or get the virus into your mouth, eyes, or an open cut. We kept plenty of household bleach around as a disinfectant and were careful to use it on anything that might have virus on it. On the wards, the nurses employed the same techniques. In thirteen years, only two infections occurred among staff working with more than 1,500 confirmed Lassa cases. And those two were both accidents. One nurse splashed his eyes with a patient's blood, the other became exposed when a patient's vomit landed on a his sandaled foot and then leaked into an open wound. Both of them were treated promptly with ribavirin and recovered.

Once we had completed our rounds and taken our blood samples, we'd carry them to our platelet lab. The rest of the day Donna and I spent separating the serum and performing the necessary blood tests by hand. The platelet function tests were interesting. Normally, platelets are responsible for stopping hemorrhages. Oddly, Lassa fever patients had plenty of platelets, but they bled nevertheless. We suspected that the platelets must not be functioning as they should.

Our suspicion was borne out in our monkey experiments. Was the failure of the platelets the real reason for the bleeding? Or was it a general functional defect in cells that line the blood vessels that caused both bleeding and shock? To discover the answer, we had to separate the platelets and try to keep them alive, which meant, as we worked with it, we were keeping the virus alive as well. The separation had to be done very carefully. Once the platelets were harvested, we would place them in a machine and add chemicals that made them "aggregate"—clump together the same way they were supposed to do in the body, to stop bleeding. The platelets of a healthy individual would normally act to plug up breaks in the blood vessels, but the platelets taken from the sickest patients with Lassa fever would not clump. Something was very wrong with them. We worked for about six weeks to solve this riddle in the hope that we would obtain some good results by the time that Joe arrived a few weeks later. But then something happened that completely threw off our calculations and wreaked havoc on all our carefully planned studies. We were about to get a crash course in Lassa fever that was like nothing we'd experienced so far.

CHAPTER TWENTY-THREE

Jenny Sanders in Segbwema

Not long after my arrival in Segbwema, I got to know a number of the local expatriates. Apart from the missionaries, these were mainly young volunteers, either from the Peace Corps or VSO (Voluntary Service Overseas), the British equivalent. But unlike the Peace Corps, the VSO only recruited experienced individuals for very specific assignments. As a result, they tended to be a little older and usually proved more savvy than the Americans.

The Peace Corps volunteers usually lived in villages and worked on agricultural projects or else taught English. They were characteristically cheerful—and always grubby. Their clothes would be covered with reddish brown stains from the laterite dust, while their hands were dyed yellow from eating local food cooked in bright yellow palm oil. Not surprisingly, they had very little money, so they were always grateful for a good meal, a beer, or a Coke—anything you chose to share with them. They could always be counted on to turn up for a party. And we had many parties.

Joe had instituted a policy of acquiring the brightest Peace Corps volunteers for the project. When I first came to Sierra Leone there were three of them: a car mechanic named John who kept our trucks on the road; Susan Scott, a hospital administrator, and her husband, Andrew, an electrician who kept our generators running.

The VSO volunteers, on the other hand, were mainly skilled nurses working for the hospital. They were the cream of their profession in Britain. Most of them had come to Africa looking for adventure and the chance to make a difference in people's lives. Besides, they had little incentive to remain in England; if they were promoted at all, it was to administrative positions. They much preferred hands-on experience to filling out insurance forms. I was to get to know three of these nurses—Deirdre, Lesley, and Sheila. Of the three, Deirdre was the veteran, having been in the country for two years. Lesley and Sheila were new enough to Sierra Leone to still be coping with culture shock.

Deirdre happened to have a good friend and colleague named Jenny Sanders. Like Deirdre, Jenny had earned the highest British qualifications in nursing and midwifery (obstetrics). At the time, Jenny was working at Panguma Hospital, about twenty-five miles away.

Significantly, Panguma was situated just beyond the diamond mining area. With so many migrant workers living together in cramped quarters, sharing their space with countless numbers of rats, it had become a fertile breeding ground for Lassa. The hospital itself had acquired a certain notoriety because it was the site of the first nosocomial (intrahospital) outbreak of Lassa fever ever to have been reported in Sierra Leone.

The hospital's reputation did nothing to dampen Jenny's enthusiasm for her life in Panguma, though. And for good reason, too: no one had ever told her how members of the staff had become infected and died several years previously. As far as Jenny was concerned, life was terrific. She could play squash and swim in the pool of the local diamond mine club. She had her friends and her fiancé, Dominic, another VSO, who taught school in Segbwema.

Although Jenny routinely cared for the sick and delivered babies, neither she nor any of the other VSOs had been warned of the dangers of Lassa fever. It was as if everyone were in on a conspiracy of denial. This silence was even more absurd given the fact that information about Lassa fever was widely available. Anyone who needed assistance in regard to the disease had only to turn to us. But strangely, the VSO organization never asked for our help. A built-in prejudice might have accounted for this: British organizations usually reckoned that they could manage without American advice, particularly in an ex-British colony. As it turned out, the physician who was advising the VSOs was in his seventies and practiced in Belgravia, one of the most expensive and elegant parts of London; it was unlikely that he had much occasion to treat many infectious diseases, let alone an exotic disease like Lassa.

Sunday mornings at the project generally meant a day of relaxation at the bungalow, reading books in the shade of a large grapefruit tree. Little of importance could be expected to happen. Which was why it was so unusual for Bob Craven to receive an urgent message, summoning him to the hospital. What, we wondered, was so important that it demanded the presence of the director of the project?

After an hour he came back. But there was no sign of concern in his face. A gruff, taciturn man, he was seldom very communicative, so we weren't expecting him to say much. He did, however, inform us that a VSO nurse from Panguma had been brought in with a fever and that she was now resting in the bedroom Sheila and Lesley shared: it was Jenny. He went on to say that Michael Price, the British doctor at the hospital, was looking after her. Although Michael tended to assume that she was suffering from typhoid or malaria, he'd called in Bob because he also suspected that Lassa was a possibility.

When I went to see her, I learned that she herself thought she probably had malaria. I was drawn to her from the start; she was a good-looking, blonde English girl with an exuberant personality. In spite of the fact that she'd developed a fever and a headache, she told me that that hadn't stopped her from attending a party the night before. Acting on the assumption that she did have malaria, Michael began treating her with chloroquine. But she didn't get better. We began to realize that it was unlikely that she had malaria. That could only mean one thing: that she probably had Lassa.

The evidence, though, was inconclusive. Although Jenny didn't have antibodies to Lassa fever, that didn't tell us anything because an absence of antibodies is common early in disease. Nor did her liver function tests (AST) meet the somewhat arbitrary, but still very useful, standard that Joe had established as a guide to treatment. Because the results of the two tests were so ambiguous, Bob decided not to start her on a course of ribavirin. Strictly speaking, Bob's decision was the correct one.

But that didn't ease our worries one bit. When Joe had set these criteria, he was taking into account the time that it took people who lived in the area to get to hospital. Usually, they delayed as long as possible in hope that they would recover at home. Even the mission hospital's minimal charges were a major obstacle. Besides, it was quite customary to visit the witch doctor first. The hospital was regarded as the last resort. As a result, by the time they were admitted, patients were often in an advanced stage of the disease.

But Jenny was a different story. She'd only had a fever for a couple of days. If it turned out that she had Lassa, it couldn't have progressed very far, meaning that her tests might be relatively normal. The most important thing about viral infections is to start treatment as early as possible, before the virus has done any irreversible damage. Nowadays, we would treat someone like Jenny, as soon as there was any suspicion of Lassa fever.

The next morning, I was crossing the compound with one of the nurses, whose name was Coolbra. He had seen more cases of Lassa fever than anyone else I knew. He had also been watching Jenny with concern. I asked him what he thought. Waiting for his reply, I found myself watching the grass under my feet as I walked along the narrow path. I didn't want to see his expression.

"She has Lassa fever," he said. There was no doubt, no hesitation in his voice. It was what I had feared.

We continued to watch Jenny very carefully. Because the conditions in the wards were very primitive, her friends cared for her in Lesley and Sheila's apartment. We would see her sitting on the porch in the late afternoon each day on our way back from the lab, often with Dominic at her side. He was temperamentally anxious. He would have worried about her if nothing was wrong. Now he was frantic. Three days went by without anything happening. Jenny didn't get better but she didn't get worse either. And then, late on Thursday afternoon, our worst fears were confirmed.

Suddenly Jenny began to go into convulsions. Then she lapsed into unconsciousness. There could not be any doubt now: convulsions are one of the worst signs in Lassa fever. In fact, we knew of no patients who had survived after convulsions. This meant that the virus had now affected the brain. Jenny was started on ribavirin, finally, late that evening. There was nothing to do now but wait—and hope.

The next morning new tests were run on Jenny's blood. There was no ambiguity about them now. Her antibody test to Lassa fever was positive and her liver-function tests were highly elevated. Jenny wasn't just sick, she exhibited all the signs of a Lassa fever patient about to die.

On Friday night Bob Craven took over the watch. But everyone in the project was mobilized. When Bob needed rest, Michael and Don were ready to take his place. Donna and I provided lab support. Even those who lacked any special medical skills collaborated in the effort, insuring that there was always food available for the nurses. But it was Deirdre and her friends who were the real heroes. They set up a round-the-clock watch on four-hour shifts. Two nurses stayed with Jenny at all times. Their dedication was astonishing; it was the finest nursing care under any circumstances that I've ever seen. They turned her over so that she wouldn't get bedsores, they washed her and sucked out her airways so that she could breathe. They monitored her intravenous drip and made certain to give her regular injections of ribavirin. The equipment they had to work with was primitive.

All this time Jenny remained deeply unconscious, so much so that she didn't even respond to the pain from the injections. Nonetheless, the nurses talked to her continuously, convinced that somehow through it all she could still hear them and take courage from them to help in her struggle against the virus. But they all labored under the burden of a terrible knowledge that Friday evening. They all realized how close she was to the end.

To ensure that no one was in danger of becoming infected, we took the same rigorous, but simple, precautions we always used when we were handling Lassa fever patients and their secretions. First, we made certain to use masks, gloves and gowns whenever we were in the room with Jenny. Because human-to-human infection can be caused by blood-to-blood contact—though not by air—we were careful to avoid doing anything that might result in an accident involving needles or sharp instruments. And, as on the general ward, we made generous use of bleach.

Before anyone could enter her room we would brief him or her and carefully explain how Lassa virus could be transmitted. As an added precaution we gave the nurses and Dominic oral ribavirin—just in case they had been infected. The whole operation was testament to the years of work Joe and his colleagues had put in on the project, establishing how the virus was transmitted and how care-takers could be protected.

The living room adjacent to Jenny's jury-rigged ICU served as an all-purpose dining room, cafe, and lounge. It was where we convened to talk and exchange information or simply to relax and try to recover our energy for the next bout. We had to support Dominic, whose lack of medical expertise made him feel desperate and helpless. Sitting with Jenny, holding her hand, and talking to her weren't enough. He wanted to do more. But what else could he contribute? As it turned out, though, there was something more he could do for her.

Jenny needed blood to replace the blood that the virus was causing her to lose. The problem was that she had an unusual blood group: she was Rhesus negative. This meant that none of the donors could be African, because all Africans are Rhesus positive. So that Friday, Donna and I planned to cross-match every Caucasian in the hospital in the hope of hitting on a Rhesus-negative gene. First we tested Dominic. We were in luck—or so we believed. Dominic was Rhesus negative.

Because Michael was in a hurry to get some blood replacement into her, we didn't wait to start to transfuse Jenny with the blood we'd just drawn from Dominic. Since we would be sure to need more blood, we continued the process of cross-matching.

Late that same afternoon I returned to the lab to find Donna sitting on the steps outside the lab, next to the bench where the patients usually waited for test results, holding the broken tile she was using for cross-matching. This process entails mixing the donor blood with special reagents; the reagents that show a reaction agglutinate—that is, they clump together so that you can see coarse red granules. By checking to see which reagent has caused the reaction, you are able to identify the blood group. This is a simple, but usually reliable, method used throughout the world in places where there are no sophisticated blood banks. Donna had come outside to make use of what daylight remained to look for agglutination. I noted a look of anxiety in her face.

"What's the matter?" I asked.

She turned to look up to me. "Look at this," she said. "I was just testing my own blood and . . ."

"And what?"

"Well, these reagents we're using say that I'm Rhesus negative."

"Are you?"

"But, Sue, I know I am Rhesus positive!"

This was crazy. We decided to test more blood. Every donor without exception turned out to be Rhesus negative. It was like tossing coins fifty times in a row and having them all turn up heads. It was possible, but the laws of probability argued against it. After all, 85 percent of Caucasians and almost all Africans are Rhesus positive. Only one possibility suggested itself: the antiserum we were using to type the blood was outdated and no longer effective. We looked at each other and decided something had to be done. We lost no time. We hurried to the hospital and began to scour all the refrigerators to see if we could find newer antiserum.

In this instance, we were lucky; we found some reagents tucked in the back of an overstuffed fridge that appeared to be newer. Then we raced back to the lab and began to test the blood all over again. We began with Dominic's blood—blood that was even now being transfused into Jenny. This time the reaction wasn't Rhesus negative at all. It was Rhesus positive—and incompatible with Jenny's.

Donna and I rushed to Jenny's room and immediately took down the unit of blood. But we were too late: some of it had already been transfused. Transfusing Rhesus positive blood into Rhesus negative women does not produce an acute reaction, as other mismatched blood can. However, some woman do develop antibodies to the Rhesus antigen, which means that if they become pregnant by a Rhesus positive man, the pregnancy could be in serious trouble. But at this point pregnancy wasn't our primary concern. We didn't even know whether Jenny would survive the night. Almost certainly she wouldn't if she didn't get blood—and it would have to be Rhesus negative blood. But how was this going to be accomplished? All the blood we had on hand, now that we could test it accurately, was positive. We needed to find a Rhesus negative donor—soon. The question was where?

Blood wasn't the only thing that Jenny needed. Her breathing was deteriorating rapidly because her lungs were filling with fluids, and she was in urgent need of oxygen. Though the hospital was quite a good one for Sierra Leone, it had no oxygen supply. In fact it didn't even have an X-ray machine or any of the other basic facilities you would expect to find in a modern hospital. The hospital did have two unused oxygen cylinders, though. We appropriated them both.

But to get them refilled, they had to be taken to Freetown. The exhausting job of navigating the two hundred-mile stretch of hideous highway back and forth fell to Brian, the VSO director, and his driver. Once they brought the oxygen back we then had to confront another problem: How were we going to administer it? All Deirdre and her friends had were nasal catheters—tubes that you placed in the nose—but they were far from ideal because it is difficult to get much oxygen to the lungs this way. Jenny needed a ventilator. Actually, Jenny needed to be in a full intensive-care unit, but even if there had been an ICU anywhere in Sierra Leone, she was now far too sick to be moved. We improvised our own ICU as best we could.

By deciding to give her oxygen through the nasal tubes, though, we ran into yet another problem. Delivering oxygen in this manner meant that we couldn't use the candles, which were the only source of light when the generator was off, which was most of the time. Candles could ignite the oxygen and blow us all up. So we went around Segbwema and borrowed every flashlight and flashlight battery we could possibly find. Only then were we able to give Jenny the oxygen.

By late Friday night, Jenny had been unconscious for twenty-four hours. As she slipped deeper into coma, her breathing became more and more labored. Bob knew that she was developing pulmonary edema. Fluids were pouring into her lungs and preventing her from breathing. This was the development we feared, the same condition we had studied when we'd carried out our monkey platelet experiments. What we'd discovered was that the virus somehow stops the blood vessels in the lungs from functioning properly. Earlier that day, tests on Jenny's platelets had shown the same thing: they simply did not function.

Bob and I discussed entering Jenny into our experimental prostacyclin study. Certainly she fulfilled all the criteria for entry. She was in the final, fatal stages of Lassa fever. It was just possible that the drug, which had worked on patients with septic shock, would reverse the pernicious impact the virus was having and restore the function of her platelets. We decided we had nothing to loose and quite possibly a great deal to gain—namely Jenny's life. So we set up the infusions, monitoring her reactions every five to ten minutes to be sure that we had the dose and speed of the infusion right. And then again we were reduced to waiting.

As it grew close to midnight, Jenny's breathing had become so labored that she was grunting with each breath. There were only three people in the room with her now: Bob, Deirdre, and Lesley. It was almost as if they were synchronizing their breaths with hers, concentrating all their effort on taking in the next draft of air. Bob gently increased the speed of the prostacyclin infusion to see what effect it would have. After a few minutes she seemed to respond to it. Her breathing was becoming easier. Bob was wary. He wanted to see what would happen before he gave her an additional dosage of the drug. But when she began to grunt again, he speeded up the prostacyclin infusion. Again her breathing became easier. This decided him: he would maintain her on the drug as long as it seemed to help.

When I'd gone to bed on Friday night, I was sure that Jenny would be dead by morning. We all were. The next day, as usual, I rose at first light and went outside to the rock on the top of the hill to start the three-stone fire for breakfast. On the way, I met Bob coming in the opposite direction. He looked exhausted; he hadn't slept all night. He was tense and hyped up. Uncharacteristically, he was also very communicative. "It's a miracle," he said. "Jenny's alive. I think the prostacyclin did something."

Then Bob headed into the bungalow to get some sleep. After breakfast, Donna and I went down to see Jenny. Though she was alive, she still showed no signs of recovering consciousness. At least her breathing was easier, and her fever had ebbed.

Except for the continuous commute Brian and the VSO driver made to Freetown for a fresh supply of oxygen, we were living in virtual isolation in Segbwema. There was, of course, no telephone, nor any other means of getting in touch with Freetown except for the occasional contact we made with the U.S. Embassy by radio. Because the radio usually didn't work, it wasn't anything we could rely on. But we still needed blood. And it would have to come from outside Segbwema. There was only one other medium to communicate with the world, the same medium that Joe had used with such resourcefulness years before in Zaire: the missionaries.

They turned up, as expatriates do in this part of the world, emerging from the bush, unannounced, in response to the ubiquitous bush telegraph system. They came in all sorts of transport—in battered trucks and on motor bikes and bicycles, whatever moved. And they all arrived caked with red laterite dust. But none of them minded making the journey. Everyone expressed their concern for a dying young woman. Some of them knew Jenny personally, others simply identified with her plight. Donna and I were so intent on our task of finding a proper donor that we dispensed with all formalities. We greeted our visitors, took their names, then snapped on a tourniquet and pumped up the best vein, cleaned it, and inserted the needle. I was so preoccupied that most of the time I never even noticed what they looked like; it was only their arms that mattered to me, not their faces. Nonetheless, I could tell that they'd hailed from a wide variety of places. Their accents came in all flavors: American, English, Irish, Scottish, Canadian, German, Scandinavian. As I took their blood, I explained what we were trying to do and how Jenny's life was hanging in the balance.

Again and again we examined the blood, cross-matching the samples on the broken tile using the new reagents. Again and again we obtained the same results—positive, positive, positive. Only about 15 percent of Caucasians are Rhesus negative. Even so, we were convinced that sooner or later one Rhesus negative donor had to turn up.

And finally one did. A gift from God, perhaps. Whether it was some kind of divine sign—if only of God's sense of humor—or not, I have no idea, but the Rhesus negative donor appeared in the form of a priest from the Catholic mission in Kenema. He was Jenny's ultimate blood donor—and possible savior.

However, Jenny's condition continued to be critical. Though in some ways she was stabilizing, we still didn't know whether she would make it from one moment to the next. There were still some very bad signs. Jenny had now developed the massive swelling of the head and neck that afflicts some patients in the final stages of disease. The swelling was so pronounced in her head and shoulders that her features were all distorted; the pretty blonde she had been such a short time earlier had become virtually unrecognizable. At the same time she'd also developed decerebrate rigidity. This is a condition that freezes the victim in a contorted posture with the head extended backwards, the arms stretched behind the back, and the legs fixed rigid and straight and pulled back at the hips. It is a terrible sight. This unnatural posture results from the loss of all higher brain functions. Jenny was now so deeply unconscious that she no longer responded even to painful stimuli. Her friends, however, continued to nurse her and talk to her the whole time. They believed she could hear them. As before, we were reduced to watching and hoping.

And we kept taking samples of blood. No matter how bad things were, we never stopped studying the disease, hoping to learn more. But there was no escaping the fact that our nerves were becoming frayed, we were all on edge, sleepless, and drained. Somehow this ordeal had to come to an end. Day after day it went on, with Jenny hovering close to death. We looked for some sign that she was improving, that the disease had reached its peak and would now begin to ebb. Yet any improvement we detected was all in our heads. Nothing seemed to change. How long could we take it, how long could we go on? Already accidents were beginning to happen.

One day the blood specimen I'd just drawn from Jenny slipped out of my hands and broke on the floor at my feet. My gloves had become sticky because I had washed them in bleach solution. How careless! I thought, angry with myself for my clumsiness. I roused myself from the shock, and Donna and I looked at each other. Quickly, we poured bleach all over the spill, leaving the disinfectant to work for about half an hour before cleaning it up. The blood hadn't touched anyone, and no one was nearby when the tube fell, so I knew that there was no exposure. It was a scary experience.

At last, Jenny's fever had abated, which meant that we had won the war with the virus. Though we could take some satisfaction from that, there was no indication that she was about to recover consciousness. We debated what to do. Five of us were physicians, and, between us, we had considerable experience treating Lassa. Our greatest fear now was that we had brought her through her viral infection, but that the brain damage was so severe she would remain in a vegetative state for the rest of her life. We realized that we had a situation here that required the services of a neurologist. She needed to be assessed to see if anything could be done to make sure the damage wasn't permanent. Maybe she should be put on steroids. Given that there were no neurologists to be found in Sierra Leone, we knew that we had to get her to London. While she was well enough to be transported now, it would have to be done by plane—the whole way. The road journey would kill her.

As usual, the radio wasn't working, so there was no way to communicate with anyone in Freetown to ask for help. That meant that someone had to make the grueling trip to the capital, call on the British Commission, and try to make the arrangements in person. Bob and I volunteered to do this, since I knew the people in London whom we needed to contact and he was the project director.

As soon as we got to Freetown, battered and dusty as usual, we went at once to see the British high commissioner. He proved to be very supportive, assuring us that he completely understood what we needed and that he was very concerned about Jenny. With his assistance, we managed to get through to people at Queen's Square, the leading neurological hospital in London, and explain Jenny's needs as we saw them. The doctor we spoke to said that they were prepared to take her. But the hospital didn't have the final say. Only the senior consulting physician at Coppett's Wood Hospital did. I knew from experience that anything having to do with cases of hemorrhagic fever in the U.K. came under his supervision.

When I got on the phone with the consultant, he was quick to remind me that he personally had seen nine cases of Lassa fever, more than anyone else in England. He added that, having run the "bubble" unit at Coppett's Wood in North London, where all suspected cases of hemorrhagic fever in the London area had to be sent, he, of course, knew all there was to know about the disease. Naturally, he didn't need anyone to tell him about what treatment might be required for a Lassa patient. That I actually happened to be in West Africa, in the heart of Lassa country, working with people who had been studying the disease there for more than ten years, impressed him not at all. He was amazing.

"All of you have done a very good job," he told me in a condescending manner. "Keep up the good work, and look after her there."

I tried to keep my temper. "Sir," I said, "five highly competent physicians have examined her. She has fully recovered from Lassa fever, but she's suffering from complications of the disease. We are unanimous in our view that we do not have the facilities to give her proper care and that she desperately needs neurological assessment. She can only get that in a London hospital."

He was unmoved. He told us we were incorrect in our assessment that she was fit to travel by air. What he was saying, of course, was that we didn't know what we were doing.

He was adamant. "I'm afraid that I see no grounds to repatriate Miss Sanders. It would be inappropriate to expose British medical staff to the risk," he declared.

I remonstrated that Joe's years of work had shown that, even if she were still infectious, the risks were minimal if care was taken to avoid needle sticks and other direct exposure to blood. But he wasn't interested. He just wanted us to keep her where she was, even if she died, rather than take the risk of exposing someone in Britain to Lassa. The fact that she was now afebrile and that any risk, based on published data, was vanishingly small made no difference to him. Nor was the fact that she was a young, idealistic, highly qualified English nurse, who'd committed her skills to the poor of Africa. All these things were simply not relevant to him; he was abrogating all responsibility for her. It was devastating! I had failed. At that moment I was deeply ashamed for my country.

The deputy to the high commissioner, who'd listened in on this conversation, was appalled. Saying nothing, he left the room immediately after the call was over and sought out his boss. When he learned about Dr. Edmond's decision, the high commissioner, too, became enraged. Superbly so. Half an hour after I left the High Commission, I received a call from his deputy as I sat in the VSO office. (Occasionally the phone worked in Freetown.) The high commissioner then read me the text of the most elegant—but strongly worded cable— it has ever been my privilege to hear. The cable basically summarized the conversation I had had with Dr. Edmond, but it carried a note of reproof and protest that was impossible for anyone to miss. The high commissioner told me he would send the cable directly to the Foreign Office in London in hope that it would persuade someone with sufficient clout to overrule the consultant.

But if we expected any immediate results, we were in for a disappointment. Nothing happened. The British bureaucracy, it seemed, was hopelessly intransigent. What was acceptable for the rest of the world was evidently unfit for Britain. In short, the issue was this: at all costs British medical personnel had to be protected from exposure to the virus. Since in the same breath the good doctor at Coppett's Wood had stated that African medical personnel should continue to care for the Lassa fever patient, I can only assume that Africans were somehow considered less important.

I returned to Segbwema in despair. But when I entered our compound after the long dusty journey, I discovered cause for celebration. Donna came out to greet me. "Jenny is waking up!" she shouted. "She's responding to people's voices. She's even recognizing her friends who are nursing her."

It was the first good news I'd heard in a very long time. Donna also told me that she'd heard over the missionaries' radio that Jenny's parents were on their way to be with her. I was so relieved that they would now find a living daughter, not someone in a coma.

When her parents reached Segbwema, they were in a state of obvious distress. It was hard for them to hide their shock on seeing her. But I don't think that they ever understood quite how close a call it had been. On the other hand, there was little cause for optimism, either. Shortly after their arrival, she deteriorated again. This time she had contracted bacterial pneumonia. Things were so bad that she could hardly breathe anymore. There was only one solution: we would have to give her a tracheotomy.

A tracheotomy is performed by making an insertion in the windpipe and fitting in a tube. Once the surgery was done, Jenny's friends mobilized themselves again to provide her with care. The tracheotomy needed to be continuously aspirated so that the surgical opening would be kept free. To do this they continued to use the little foot-operated pump attached to a catheter, since they were unable to suck out the bronchi directly. Then she was given antibiotics.

Little by little she began to rally. And as the edema in her head and neck began to subside, a thin, fair-haired girl began to emerge. It was as if she'd become an apparition, a pale wasted double of the person she once was.

By this time Joe had arrived in Segbwema. When we told him about what had happened to Jenny, he expressed his amazement. He, better than anyone, realized how lucky she was to have survived. But there were two things about her case that particularly disturbed him. For one thing, it was apparent that she was needlessly exposed to Lassa in Panguma Hospital because no one had bothered to institute protective measures for the staff. And then there was the matter of the treatment itself. "In the future," he said, "we should learn from this. We must never delay giving any sick hospital staff exposed to Lassa patients ribavirin. In retrospect (which is always easier), Jenny should have had it right away when she first arrived in Segbwema." At this point, though, there was little more that he could do that hadn't already been done. Now our primary concern was getting Jenny to London.

Ten days after my abortive visit to Freetown, we received word that the high commissioner's cable had in fact made some impact in London. The British authorities had relented and reversed their original decision. They were now ready to allow Jenny to be repatriated. I was astonished. Obviously, I'd made more of an impression than I'd originally thought. However, Jenny's problems were not over. Actually, the worst part of her ordeal was still to come.

What none of us expected was that it would take a full-scale military operation to evacuate Jenny and take her back to England. The transport of a convalescent young nurse was judged to be such a major undertaking that it required the services of the Royal Air Force. Her departure turned out to be one of the most dramatic events in Segbwema's history.

Because we were concerned about the instability of her cardiovascular system, we wanted to make sure that her journey back to London was made as comfortable as possible. To spare her the grim, bone-shattering ordeal of riding down to Freetown, someone with connections had managed to prevail upon the president of the country, Siaka Stevens, to lend us his personal helicopter. It touched down on a local soccer field, an event witnessed by an enthusiastic audience made up of the entire town's population. However, the soccer field was still a mile away from the project site, and the road leading to it was in terrible shape. So now we somehow had to get Jenny to the soccer field.

As usual, we were forced to improvise. There was only one truck large enough to accommodate the mattress for Jenny, and that happened to be the rat wagon. This was the same truck that we used to transport the *mastomys* rats—alive and dead—that we trapped and then tested for the presence of Lassa. The irony of the association was not lost on us.

As soon as we reached the soccer field, the British doctor accompanying the mission went into action. The equipment he'd brought with him for this mission looked like something out of *Star Wars*. The two helicopter pilots, both Frenchmen, watched in astonishment as he picked up his respirator of the all too familiar gas mask design. Then they looked back at Jenny stretched out on the mattress. "Is it safe to take her?" they asked, suddenly apprehensive. Apparently no one had bothered to clue them in to her condition. We did our best to reassure them, and showed them that we ourselves were handling her closely, and did not consider ourselves at risk. However, our air force friend was not making a good impression, and this would not be good for Jenny.

So I stepped up to him and seized the respirator out of his hands. "Look," I said, "this is all you're going to need for her." I thrust into his now-empty hands a pair of gloves, the cannulae—the tubes she needed to breathe—and the hand pump to clear her lungs. The doctor didn't protest. I had a feeling he knew I was right. After all, he had just spent a full morning on our wards and had examined at least twice as many Lassa fever patients as any of the so-called experts back in London had seen in their whole lifetime.

When the helicopter took off, the spectators, who'd been eagerly following our little confrontation, burst into applause. Because everyone was familiar with Lassa and the devastation that it could cause, they rejoiced whenever anyone recovered. While we were fortunate in being spared the next leg of Jenny's journey back home, there was no question that she would have been far better off if we could have gone with her. No one should ever be subjected to the treatment she received.

Word of her impending arrival had already reached Lungi Airport. Apparently it was still believed that she was highly infectious. Why else would the helicopter pilots be ordered to land behind a hangar so that they couldn't be seen? You'd have thought it was a clandestine military mission. Then, as soon as the chopper was down, the Royal Air Force team leaped out of the transport plane, wearing gas masks and specially rigged protective red suits. When they got to the helicopter, they immediately installed Jenny in an isolation bubble before hustling her back into the transport plane.

Then, as the plane was warming up for takeoff, the intrepid crew quickly disrobed and tossed their suits out onto the runway. This was how they went about decontaminating themselves of a contagion that existed only in their minds. Apparently, they believed that while the suits were unsafe for Britain, they were perfectly okay for the people of Sierra Leone. Brian, the VSO director who'd witnessed this preposterous spectacle, was horrified and humiliated. But at least he had the presence of mind to recover the suits, which he bundled into the back of his Land Rover. Then he drove back to Segbwema to tell us what had happened.

After hearing Brian's account, we weren't sure whether to laugh or cry. But in any case, we were in complete agreement that we were going to have a party and celebratory goat roast. God knows, at this point everyone deserved a good time. The festivities were considerably enlivened by a costume show. Michael and two of the nurses decided to put in an appearance wearing the red suits the British crew had discarded on the tarmac. To complete the effect, they even put on gas masks. Needless to say, it made the consumption of a cold Star beer something of a challenge, but they managed.

Whoever had come up with the bright idea of putting Jenny in the so-called isolator was clearly more concerned about a nonexistent threat to the flight crew than he was with the welfare of the patient. Evidently no one had reckoned on how difficult it would be to care for a very weak patient, with an unstable cardiovascular system, locked inside a bubble. Even worse, the interior of the bubble wasn't air-conditioned; so Jenny roasted.

Her imprisonment didn't end when she arrived in London, either. In fact, she was compelled to stay in the isolator for sixty days straight, all because the British medical community and the public were thrown into panic by irrational fear of Lassa fever based on ignorance. Jenny was probably luckier to have survived her treatment by the British than she was to have come through her bout with Lassa. While it was true that she had a small amount of virus in her urine, it was a common problem in recovered Lassa fever patients, and one that could easily be dealt with outside the confines of an isolator. When Jenny was finally released from the isolator she was unable to walk because of the length of time she'd spent immobilized. Today, most of the world has finally moved away from the concept of the isolator, but Britain continues to go its own way.

To add insult to injury, the VSO received a bill for the equivalent of about $75,000 for this exercise. They had no insurance. It might have been even more expensive if they had followed Joe's advice and simply dispatched a physician to accompany her in first class on a regular British Caledonian flight. For if they'd done that, the British authorities probably would have impounded the plane and quarantined all the passengers.

Interestingly enough, to this day, Jenny has no recollection of her illness. She can only recall as far back as the day she arrived in Segbwema, complaining of a slight fever and a headache, nothing else.

CHAPTER TWENTY-FOUR

Sunset over Segbwema

While Jenny might not have remembered what had happened to her, Joe and I were determined that the international medical community didn't suffer from a similar lapse of memory. There were important lessons to be learned from the case. To that end, later that summer I gathered together several experts at the Central Public Health Laboratory building at Colindale, where I was then working, and got Joe to stop by on his way back to Atlanta from another visit to Sierra Leone. Joe gave his usual very good scientific account of his work with Lassa fever: how the disease is spread (or not spread); how to diagnose it, and how to treat it.

Then we called on other speakers who had been involved in various aspects of Jenny's case. We even had the pleasure of the company of the famous consultant who had denied repatriation to Jenny. The British air force doctor, with his high-tech respirator, came up to the stage to tell his version of the tale. And for an account of Jenny's sixty days of imprisonment, there was Dr. Stuart Glover, who had looked after her in her bubble. After I finished presenting the clinical case, Jenny herself came onto the stage to take questions from the audience. As she responded, I saw her again as the professional nurse she was.

I like to think that in the long term we may have had some influence on a new generation of infectious disease physicians. Certainly, they judged the evidence for themselves and concluded that the British government's policy regarding protective measures against hemorrhagic fevers was unnecessary at best, and dangerous at worst.

As interested as we were in getting our message across to the outside world, Joe and I were equally resolved to do the same in Africa—in the heart of Lassa country. He and his great friend Fakhry Assad, the architect of the Global AIDS

Program, hatched a scheme to hold a workshop in Segbwema in 1985. It would be an unprecedented event. On the face of it, the idea of convening an international gathering in rural Sierra Leone seemed impossible. How were we going to house and feed all the delegates, or find transportation for them? We had no hotels, no restaurants, no electricity, no running water, no airport, no vehicles. About all we did have were dirt roads. And plenty of Lassa fever. But Joe was insistent. "If we are going to teach them something useful about Lassa fever," he said, "then we have to show them Lassa fever."

Fakhry was behind us all the way. So we did it.

Taking charge of the considerable logistical problems were our two Peace Corps recruits, Susan and Andrew Scott. They persuaded the Catholic Mission at Kenema to provide the workshop participants with food and lodging. It was this mission that had given us the priest who had donated Rhesus negative blood for Jenny. However, we still needed to arrange for a bus that would take our guests over the dirt road to Segbwema each day, and then find a way to feed them lunch.

With her customary resourcefulness, Susan found Mary for us. Mary was indisputably the best cook in Segbwema and had long been known for the tasty food she offered in her "chophouse." Concocting dishes from the best local Mende rice and fresh fish from the swamps or the nearby river, Mary prepared meals that even won plaudits from people accustomed to the international cuisine of Geneva and Paris.

The workshop began in Freetown, at a hotel that more or less guaranteed some electricity, at least some of the time. While the majority of participants came from all over Africa, there were also a number of delegates from Europe as well as one Soviet, rumored to be a KGB operative, who was working at WHO. Etiquette demanded that participants in the workshop pay a courtesy call on the U.S. Embassy, whose staff had always been extremely supportive of our mission. As we emerged from the embassy, we noticed that our Russian friend was behaving very strangely. Suddenly he froze and stared down the hill toward the bay.

He muttered, "I have to go down the street—at once!" And then he ran off.

We all looked after him, just in time to see him scamper into a doorway underneath the bright red flag with the hammer and sickle.

"Oh dear," said Fakhry, "he just made a terrible faux pas. He went into the U.S. Embassy before he went to see the Soviet ambassador. If they saw him come out of here he could be in trouble." We all laughed.

After two days of lectures in Freetown, we were ready to go upcountry. We roused everyone at 5 A.M. and herded them out onto the street to wait for the buses. The buses were a contribution of the government of Sierra Leone. We'd made contingency plans in the event that they didn't show, but in this case, we were gratified to see that something did work as it was supposed to. As the buses lumbered over the wretched road on the way north, our colleagues from East Africa gaped at the passing view in amazement. They had no idea as to just how deplorable conditions had become in Sierra Leone. They were used to seeing poverty and underdevelopment in their own countries, of course, but nothing like this.

Once we'd got the delegates to Segbwema, we divided them into three groups and set them to work. One group made the rounds of the wards to see Lassa fever patients; another went into the field to see village housing conditions and to catch rats; and the third stayed in the lab to learn the techniques for diagnosing Lassa. In the evenings, we reconvened on the hill above Kenema to eat meals cooked for us over an open fire and conduct a postmortem on the day's activities. During these sessions, the delegates developed the WHO guidelines for handling hemorrhagic fevers in Africa. These were simple, cheap, and sensible, and were based on the techniques we knew had worked in Sierra Leone. As workshops go, it was certainly a very unusual one, and extremely productive besides.

And what of Jenny Sanders, who unwittingly had been one of the inspirations for the workshop? Jenny went back to her work soon after she recovered. Joe and I had occasion to visit her and Dominic about six months after her bout with Lassa. The two were now married and were living in a flat near the Crystal Palace in the south of London. We shared a takeout pizza and talked. Jenny was a bit apologetic, but whenever she tried to recall what had happened to her during her long siege nothing came to mind. Maybe, I thought, it was all best forgotten. Jenny and Dominic now have three children and live in the West of England.

Sierra Leone marked a major personal turning point in my life. In the course of my work at the Lassa project, which extended from 1985 to 1990, I developed a deep affection for the country. I would visit once or twice a year—sometimes with Joe—and stay for extended periods each time. But on each succeeding visit, I became more aware of the changes that were taking place. Few of them were for the better. When I returned in 1990, I could see at once that Segbwema had undergone a transformation. But I didn't understand just how profound that transformation was until the evening of my first day back in the town.

I was sitting in a bar—not at Eddy's (Eddy was long gone by this time)—on the main street early in the evening. Usually the street would have been crowded with people coming back from the fields and doing their shopping. Any number of little children—*peekins*—would stop, stare at those of us who regularly collected at the bar, and call us *pumwe*—"whitey"—until their parents shushed them. Wherever you looked, you'd see animals roaming free—stray dogs, chickens, and goats. For the drivers of the barely serviceable vehicles that made up Segbwema's rush-hour traffic, there was the constant challenge of avoiding these creatures, not to mention the *peekins* and the potholes. But on this particular evening, the street was eerily quiet. Even the dogs and the goats were subdued. The children seemed to have vanished entirely. Those few people who were in sight had grave expressions. If they looked at us at all, it was with wary eyes. Something seemed about to happen. There was a heaviness in the air that had nothing to do with the impending rains.

Suddenly, an army truck barreled over the hill, lurching past the police station, which sat in the shadow of a tree full of yellow-bellied weaver birds chattering madly as they alighted in their strange pendulous nests. The truck came to a stop opposite the bar where my friends and I were seated. An officer jumped out of the truck and ran to the police station. Something urgent seemed to require his attention. Meanwhile, several raw army recruits waited in the back of the truck. Wearing ill-fitting camouflage suits, they looked out at us with fearful faces. In their hands, they clutched machine guns with their clips in place. An unsettling silence had descended on the street. Everyone was watching the truck and waiting.

After a few minutes, the officer came back to the truck and got into the cab. Then the truck set off down the hill until it had gained sufficient speed to start the engine. "Statah motah, no deh," we assumed. The silence persisted for another moment or two, then everyone seemed to release a collective breath and come alive again. We finished our beers without saying another word and left. That evening, as I often did, I sat alone on the high rock outside Austin's house, warmed by the heat of the sun, my favorite place. I watched the sun set in a vast bright red ball over the distant misty hills of Panguma. Egrets flew home to roost over the rice paddies, and birds called to each other from the tall palm trees. As it grew dark, I could just catch the lights of a truck bouncing erratically over the road that led to Kenema away on a distant hill. I went inside quickly because of the mosquitoes.

We later learned that there was good reason for the people of Segbwema to be afraid. Only a few days previously, at about midday, the town had been emptied entirely of people. Rebel soldiers, mostly young teenagers armed with machine guns, had come over the Liberian border and killed people at random in the marketplaces of Kailahun and Koindu, both of which were situated perilously close to Segbwema. Virtually all of Segbwema's population had hidden out in the bush until they were assured that the rebels had been pushed back and the danger was over. At least for the time being.

The conflict ignited panic among the foreign community. All British medical personnel in the area were ordered to evacuate. This meant shutting down the hospital, the only source of medical care for the region. The last operation performed by a British surgeon there was an amputation. The patient had been shot in the arm in Kailahun that same day and there was no way for the surgeon to save it. Even as he was cutting, his mind was on other things—like getting out. Though the border was now sealed, there was no way for Sierra Leone to contain the civil war that had engulfed neighboring Liberia.

During our stay, Joe and I began to hear rumors about a home-grown "rebel" leader, who was said to be delivering ultimatums to the government. But what was really going on? We had no idea. Who were the rebels? I asked. No one seemed to be able to answer me, which is why we waited anxiously for Austin Demby to come back from Daru junction, where an army base was located. Austin had friends in the army—he had friends everywhere—and if anyone could pump them for information, he could. When he returned, he looked very tense. Things were quiet for now, he told us, but there was little likelihood that they would stay that way for long. He hinted that we might want to cut short our stay in Segbwema.

Nonetheless, while we agreed that it was no longer as safe as before, we still had to hold a large party at the main bungalow—it was a tradition, after all. It was expected of us, and we owed these people a lot. We decorated the bungalow with palm fronds and other vegetation we pulled out of the bush. Naturally, plenty of beer was stocked up and plenty of it was drunk. And no party would have been complete without a goat roast. Two goats were procured for the event. This presented certain difficulties for me because they were tied to a tree for several hours before being slaughtered. Since I was around them, I began to get to know them, which made it that much harder for me to contemplate their imminent execution, let alone consumption.

That party marked the end of our time in Sierra Leone. Neither Joe nor I would see it again. Now the whole of the Eastern Province is in rebel hands. People have been indiscriminately slain. Forty percent of the population has been turned into refugees. We can only assume that Segbwema has been overrun by rebel forces and that the site of the Lassa project has been completely looted. With the end of the rat-control program, which Joe had instituted, there can be little doubt that the rats have returned. And with the rats will come Lassa virus. The crowding and the refugee situation will make things much worse. But this time there is no Lassa fever ward, no Lassa fever doctors, no Lassa medicine, and no cries of "Lassa fiva, No Good-O" as our truck bounced by.

CHAPTER TWENTY-FIVE

Of Mice and Peanut Shells

Adding to the horror of artillery rounds, poison gas, and machine-gun fire that menaced soldiers in the trenches during World War I was a strange disease marked by kidney failure and bleeding. It appeared again during World War II, sweeping through the ranks of German troops campaigning in Norway and Finland. No one was sure what it was, but it seemed to thrive on war. It could have been leptospirosis, a disease caused by a long, thin bacterium carried in rats. On the other hand, it could have been caused by one of a family of viruses, the hantaviruses—carried by rats—which cause Hemorrhagic Fever with Renal Syndrome (HFRS) in patients. We now know that these viruses are widespread in Europe, particularly in the areas where many of the battles were waged in both world wars. The trenches were teeming with rats.

In 1915, when the strange new disease was first described in connection with an outbreak among British troops in France, it was called "renal dropsy." A similar disease was also being seen about the same time among civilian populations in eastern Russia near the port city of Vladivostok. When it again struck, this time among Japanese troops rampaging through Manchuria in the 1930s, it was called Songo fever. Establishing itself in China in the 1940s, it became known as Epidemic Hemorrhagic Fever (EHF) and has continued to advance inexorably south ever since. Also well known since the 1930s is a similar disease in Scandinavia, called *Nephropathia epidemica*, which, while apparently closely related, is much milder than the Asian disease.

But it took the Korean War for the disease finally to earn its place in the annals of medical literature. During that conflict, it attacked nearly three thousand UN soldiers, including many Americans, killing four hundred. With each new incarnation, the disease acquired a new name. This time was no exception. It now became known as Korean hemorrhagic fever. Again, it was believed that rodents were involved in its transmission, but there was no concrete proof.

Efforts to discover the causes of HFRS extend at least as far back as the 1930s. Some of these attempts involved extremely dubious experiments. In one of them a "filterable agent"—that is, a virus—was said to have been given to human "volunteers" in the Soviet Union. The Japanese documented similar experiments, injecting the virus into prisoners they seized when they occupied China. In the 1950s, a scientist named Myrhman in Scandinavia took a more heroic approach when he drank fifteen milliliters of infected urine to discover its effect on the human body. When the urine failed to produce any discernible results, he injected himself with five milliliters of blood from an infected patient. That he suffered no ill effects suggests that, luckily for him, his patient no longer had virus in serum or urine.

In 1978, Karl Johnson and a Korean colleague Ho Wang Lee returned to Korea to investigate. Assuming that rodents were responsible, Karl and Ho Wang took serum from patients who had recovered from Korean hemorrhagic fever and then tested it to see whether it would react with slices of renal tissue from field mice captured in Korea. Their hypothesis was correct: they isolated a virus and called it Hantaan, after a river in Korea. The virus that caused the disease was carried by the striped field mouse, called *Apodemus agrarius*, distinguished by the stripe of blond fur that runs along its spine. Inspired by Karl's pioneering studies, the race was on to isolate viruses. George French, working out of the military lab at USAMRIID, in Frederick, Maryland, used the same technique employed by Karl and Ho Wang to isolate very low levels of the virus. Instead of relying on mice, he developed a cheaper, easier, and more practical tissue culture system. By 1981, when I started to become interested in Hantaan virus, we still had not been able to grow it in sufficient quantity to be able to identify and characterize it. To be sure, Hantaan had been isolated in mice and in tissue culture, but that was as far as researchers had got. High concentrations are necessary to determine the type, shape, size, composition, and closest relatives of a virus. So the next logical step was to acquire the capacity to grow the virus in sufficient quantities to work with it and identify it. Only then could we develop a routine diagnostic test that was simpler than the more cumbersome one then in use. At this point, confirmation of human infection was still performed by testing a slice of tissue taken from an infected field mouse kidney with human serum from a suspected case.

A possible solution to the problem suggested itself. Paul Price and Karl Johnson had cloned a line of tissue culture cells at CDC, which they called Vero E6.

In cloning, a single cell is made to divide again and again, so that many cells are produced, all containing exactly the same genetic material as their common ancestor. This cell line had turned out to be excellent for growing other hemorrhagic fever viruses like Lassa and Ebola, so it was a natural candidate to try with HFRS. Our goal was to generate enough virus particles to be able to take a look at them under an electron microscope. With that instrument, we could fix virus particles on a special grid and shoot electrons at them, which would produce an actual image of the virus. Once we knew its shape, size, and structure, we would be able to get a good idea of what kind of virus it was. But we needed to grow it to a concentration of at least a million viruses for each cubic milliliter of fluid (that is, about twenty drops). This is a lot of virus. We also needed the same high level of virus in order to study its molecular characteristics. A good tissue culture method might also allow us to isolate new strains more easily. If all went well, we would ultimately be able to create better diagnostic tests. In all, there was a great deal at stake.

I sought out Karl to sound him out on my experiment. He was about to leave CDC to take up a position at USAMRIID. "Joe," he replied, "I wouldn't waste my time," he said. "We tried to get Hantaan to grow in ordinary Vero cells, and it wouldn't, so there's little reason to believe that it will grow in E6 cells."

In spite of his pessimism, I thought we should still try. Donna Sasso, the technician who also worked in Sierra Leone with the Lassa project, took on the responsibility of preparing the cultures for tests we could use to detect whether the virus was growing as we hoped. The virus we would be using came from rodent material that Karl and Ho Wang had isolated. At first, the going was slow, and I was becoming discouraged. But then, a couple of days after we'd begun, Karl dropped by the lab.

"I was wrong about that experiment," he admitted. "Of course, you should try it. Just because Vero didn't work doesn't mean we should shut the door on E6. It could be different."

I took heart from his words. No one was more hard-nosed, intellectually honest, or demanding than Karl. He could be just as critical about his own work as he was about anyone else's. He was everything you would want in a mentor.

For all our efforts, though, the virus remained decidedly uncooperative, refusing so much as to recognize the existence of our cell line. Then the experiment was thrown off by bacterial contamination, forcing us to return to the freezer for fresh rodent tissue containing small amounts of virus. It was incredibly frustrating. Each time Donna would place fresh virus in the culture, it would pull a magic act, vanishing almost instantly. Not only wasn't the virus growing, it wasn't even hanging around. Normally, we would change the fluid that nourishes the cells every two or three days, but Donna decided that there was nothing to lose by leaving the fluid in longer and longer, just to see what would happen. Would the virus appear in larger quantities? We also decided to increase the virus stock from what we'd started with in hopes of speeding things up. You have to be very careful to add just the right amount of virus to tissue culture. Too little, and nothing happens. Too much, and the virus can actually interfere with its own replication.

In the same way that a farmer knows when to start picking apples off the trees so that he can get them to market while they're still fresh, a researcher has to know exactly when a virus is ripe for plucking. Timing is everything. The process requires a green thumb. Luckily, Donna had one, and she was also blessed with patience and persistence.

Most tissue culture becomes ragged within five to six days, but because Vero E6 is especially resilient, we took the risk of letting the cultures sit without interference for as long as two weeks. Ordinarily, we would have transferred the material onto new cells (a process called "passaging the virus") within a matter of days. Our reasoning was simple: we decided that this was a virus that needed time to take hold, and we didn't want to wash it away by changing the fluid. The experiment went on for weeks while we struggled to find the conditions and circumstances under which the virus would grow. Donna hung in there, even though there was every possibility that we would never succeed in growing anything.

Each time we examined the infected cells, we would notice that a little more yellow fluorescing material had apparently attached itself to them. This was what we'd been waiting for, a sign that we were actually getting more virus particles. Encouraged, we kept transferring this virus to new cells, hoping to avoid any contamination in the process. As the virus grew, so did our expectations. We were now approaching the time when we would be able to put the cells under an electron microscope. If we were lucky, we might actually see the virus and, in so doing, become the first people in history to set eyes on it.

Finally, about six months after we had started, the day came when we felt that we had taken the process as far as we could. We decided to remove the infected material from the flasks of tissue culture and then place it in a fixative called glutaraldehyde, which would kill the virus, so that we could work with it safely. The fixative would have the additional advantage of stabilizing the virus structure, so that we could see it under the microscope. When all this was done, we gave our precious sample to Erskine Palmer, our electron microscopist.

Erskine was a slight, soft-spoken man of few words, who had contributed greatly to the study of virus structure. We couldn't have found anyone better to help us with this project. He was the man we would be relying on to find out whether our experiment of half a year had succeeded or not. Once we'd given the material over to Erskine, I felt as much trepidation as I did handing in a final exam. Suppose the virus never showed up? All our work would have been for nothing.

The process of preparing the material for examination under the electron microscope requires two to three days. It has to be fixed with a special chemical, stained with a heavy metal (which coats the virus, so that the material shows up clearly when bombarded with electrons), and, finally, embedded in a hard plastic. Once the plastic has set, the material then must be cut into very thin slices, so that it can be observed under the microscope. What you eventually look at is the heavy metal dye shadowing the structure of the virus, not the virus itself.

After three days of nail-biting tension, we gathered in a small dark room where Erskine was getting the electron microscope ready. There were five of us altogether, including Erskine, Donna, and myself. Some microscopists would have cleared the room so that they could concentrate without distraction, but Erskine knew how much this meant to us. We stood behind his chair, looking over his shoulder at a green screen. Whatever appeared on the screen was what the electron microscope would reveal.

Several minutes went by as Erskine adjusted the vertical and horizontal holds with two large black knobs on either side of the microscope.

Watching with less-than-trained eyes, we were hoping to see one of two things: either the telltale symmetrical signs of a virus particle, or else the outlines of a virus envelope studded with minute projecting antigens. We knew that the envelope might come in one of any number of shapes, so we weren't sure what to expect. Any sort of rounded object that appeared would have to be studied in minute detail. We had only a vague idea what type of virus it would be. Possibly it could be an arenavirus, like Lassa, since it was a virus that lived in rodents. We continued to stare at the EM screen with Erskine until our eyes were bleary with the effort. We all fell silent as it became obvious that we were going to see nothing like a virus. With five sweating bodies packed into the tiny room, it was beginning to steam up. We all strained to see something.

Nothing.

Cells, garbage. No virus particles. Not one.

Back to the lab. What to do? After all, we knew there must be something there: Through the light microscope, using fluorescent dye, we could see the viral material shining like a beacon in a New England fog. But, we had to remind ourselves, you can see bright fluorescence with other viruses at concentrations of only 1,000 to 10,000 virus particles in each milliliter of fluid. We had to have at least a million virus particles if we were to see a single virus under an electron microscope!

So we had no alternative but to grow the virus up in larger quantities, and then use an ultracentrifuge to spin the virus particles to the bottom of a test tube. An ultracentrifuge spins at rates as high as 100,000 revs per minute, in contrast to a regular laboratory centrifuge, which spins at 5,000 to 10,000 rpms. Using an ultracentrifuge would concentrate our virus by as much as ten to a hundredfold. Perhaps that would give us sufficient quantity.

The downside? To begin with, it was a process that was going to take a whole lot more time. And there was something else to consider: the danger. High speed centrifugation of concentrated virus is a particularly risky business because of the sheer number of virus particles involved in the process, and the high energy of the centrifuge, which can produce aerosols full of infectious material if a vial breaks. So, we took it into the Level-4 lab. Donna was protected by her space suit and a good deal of experience. She was willing to take a calculated risk.

Our hopes were hanging by a thread, but having gone this far, we were determined not to give up. Several more weeks passed until we had our "pellet," an almost invisible tiny blob of material at the bottom of a tube: the end result of the process of spinning in the ultracentrifuge. Once the pellet was recovered, we fixed it in glutaraldehyde. Once again, it was Erskine's turn.

It took three more days for him to prepare the pellet for the electron microscope. Then, when he was ready, he called us in to have a look.

It was early afternoon as we once more gathered around the EM in the dark room, our eyes now fixed on the screen that projected the image Erskine was scanning. He began his search, spinning the knobs of the microscope as he systematically charted the tiny grid on which he'd planted our material. You can get billions of viruses on a pin point, let alone a pin head. If we only had a few, they would be easy to miss. Erskine would need to inspect every square nanometer of the specimen, trying to spot a virus capsid (envelope) or the spikes protruding from a viral membrane. Viruses range in size from 20 to 250 nanometers in diameter, and since a nanometer is 10^{-9} meter, the search takes a lot of time. Periodically, he would pull out the screen so that he could see better himself, which left us nothing to stare at but the darkness. All we could do was guess at what he might be seeing.

Erskine is excruciatingly methodical and scrupulous, both essential qualities in an electron microscopist. A myriad of shapes and shades will appear in any preparation, most of which have little to do with what you are searching for. But each squiggle and amorphous form has to be scrutinized to make sure it is *not* the virus particle you are interested in. Sometimes only a part of a virus particle might be visible because the rest of it has been damaged. It takes an expert to spot the real virus, even when it is there, and what Erskine was doing now, as we stood in the darkness, was eliminating phantom images, shadows of distorted cells, and cell fragments, all masquerading as bits of virus.

After a while, we could see that Erskine's attention was riveted on something the rest of us were unable to discern. I held my breath. I could sense that he'd just caught a glimpse of our holy grail and was now about to get close enough to photograph it. Then he raised his head.

"There are certainly virus shapes here," he said, speaking slowly. "We are probably on to something."

I could hear our collective pulses racing. Or maybe it was just my own.

"Come on, Erskine," I entreated. "Let us see." I was on tenterhooks.

But Erskine was not a man to be hurried. He needed first to be convinced himself. As chief arbiter among ghosts and shadows, he would let us know when he was ready to go public.

Several minutes more went by. Then Erskine clicked back the screen, allowing us to see the image for ourselves.

"Peanuts?" I exclaimed.

This was the first time that anyone had ever laid eyes on this virus.

Erskine gave me a somber look.

"Actually, I think this is a bunyavirus. I can't be absolutely sure, but at this point I would place it in that family."

Bunyaviruses are enveloped viruses. The virus acquires its envelope by pillage, looting the cells it infects for the part of the material it needs to manufacture itself. Then the virus uses its loot like a coat, wrapping it around its nucleocapsid, which is the stuff of the virus itself. The envelope is usually rounded, but it can be pulled around during the process of preparing it so that it can be seen under the electron microscope, and stretched into many different shapes, including one that resembles a peanut shell.

The shape should tell us what kind of virus it was. If Erskine's assumption was correct, and it did turn out to be a bunyavirus, then it would be the first rodent bunyavirus ever found. That would put Hantaan into a virus group (or genus) all its own. Sera from patients with HFRS had already been tested against all known bunyaviruses, and the existing bunyaviruses did not react against any of them. This meant that the structure of the virus Erskine was examining and its chemical makeup were different from those of any known bunyavirus. But then there are a lot of bunyaviruses—all animal viruses—and new ones keep popping up all the time.

So part of the mystery was solved. We now had more of a handle on an infection that caused serious disease and death in many parts of the world. The tissue culture system that Donna had developed with such patience, over such a long period of time, has proven so effective that even today it remains the standard method for isolating Hantaviruses and for preparing diagnostic reagents. The new genus was therefore also named after the river in Korea, close to where Ho Wang Lee and Karl had found the mouse from which they isolated the original virus. Donna's system has allowed the isolation of thousands of viruses from around the world. The Hantaviruses have now been forced to give up some of their secrets, and with a greater understanding of the genetic composition of the virus, researchers were subsequently able to develop vaccines.

Over the months that followed this vision in a small, dark room, we prepared the cell lines and the purified virus for distribution to colleagues at USAM-RIID, in Belgium, France, the U.K., and Japan, where similar work was going on. Ironically, if these developments had occurred ten years later, we would never have been able to be so generous. Instead, we might have had to guard the method we used to isolate the virus by patenting it, notwithstanding the fact that we had refined the process in a U.S. government laboratory. But 1982 was another era, and the spirit of scientific cooperation and collaboration was still flickering, however faintly, in many laboratories around the world.

Eric Dournon was an infectious disease physician who also ran a small lab in Paris. Although it was grossly underfunded and understaffed, it was highly productive, so productive, in fact, that Eric became the first scientist to have searched for and found *Legionella pneumophila* in France. During a visit to the Level 4 lab at CDC in 1980, I suggested that he must try to chalk up a second triumph by identifying HFRS in France. I was interested in finding out whether there was any contemporary equivalent of the renal dropsy disease reported in the trenches of Picardy during World War I.

Eric became a special friend. He was a slight, dark-haired, dark-skinned Frenchman, endowed with Gallic good looks, and he was a true adventurer. He was later to teach me to scuba dive in Corsica. His teaching method was simple and to the point: he first gave me ten minutes of verbal instructions and then he threw me in the water. He said that it was the same way he'd learned. A complicated man of great intelligence, he would chain smoke while fishing, more interested, I suspect, in the time that it gave him to think than in the possibility of luring a fish to its extinction. In fact, he often caught nothing. In any case, we both agreed that fishing was a philosophy, not a sport.

After talking it over, Eric and I concluded that he should be looking for a milder form of the disease in France. We agreed that it would probably be something similar to the *Nephropathia epidemica* described in Scandinavia. NE was also now known to occur over large areas of Russia. It was caused by a slightly different virus from Hantaan, attributed to a type of rodent called a bank vole or *Clethrionomys*. This cute little red creature, which looks like it could have come straight out of Beatrix Potter, lives in banks and hedges all over Europe. So Eric needed to keep his options open; he would be searching for a disease that was similar, though not necessarily identical, to the more severe disease seen in Korea and China. When he left Atlanta, Eric was enthusiastic about his new quest. He liked teasing out new diseases, and HFRS represented a significant challenge.

About four months later, Eric sent to me a set of sera from a man who had been hospitalized in Paris with acute renal failure. According to his account, the patient was an officeworker who owned a cottage in a small village near Rheims, right in the heart of champagne country. He'd fallen ill after visiting his cottage for the weekend. When questioned, the patient said that he'd been chopping wood in a barn that had been shut up for some time. Almost three weeks to the day from his weekend foray, he developed fever, chills, and intense muscle pains, particularly in the lower back. He was seen by a physician and hospitalized.

Then he stopped passing urine. The kidney specialist happened to be a good friend of Eric's, so he had already been primed to be on the lookout for a case of HFRS in France. Naturally, he got in touch with Eric and arranged to obtain the appropriate specimens from the poor man, whose only error was having spent a quiet weekend in the country. Now he was the first laboratory-confirmed case of HFRS in France—a dubious distinction. (Fortunately, he did make a full recovery.)

When the specimens arrived in Atlanta, Donna tested them against our new reagents. As soon as she had the results, she came in and showed them to me.

"We've got it. The Frenchman is positive," she said. Another first for Eric.

This was just the beginning. Eric and a friend, a fellow physician, wanted to find out more. So they set out to see the incriminated barn to determine for themselves what was going on there. They decided to trap some rodents and then test them for the presence of infection. They entered the barn without any protective clothing. Hey, they must have thought, it was just an old barn, and, after all, this is France. Put on a gas mask, and the neighbors would think you were crazy: *complètement fou.*

But Eric and his friend knew what they were doing. At least they thought they did. After setting some live traps (designed to catch rodents alive, so that you can take specimens from them), they searched around for evidence of rodent infestation, remaining in the barn for about twenty minutes to half an hour. The next day, they returned to the barn to collect the rodents from the traps. No luck. The empty traps didn't necessarily mean that there were no rodents, just that none took their bait. Eric and his friend returned to Paris with nothing to show for their trouble—except that, exactly twenty-one days later, Eric's colleague developed fever, severe muscle pains, and headache, marked by intense pain behind the eyes. Perhaps, he thought, it might be *grippe*—the flu—but when his symptoms persisted for several days, he decided to contact Eric.

It didn't take them long to figure out that he might well have been infected with Hantaan virus. It turned out that he had HFRS all right, but it was a mild form, and he made a full recovery. As Eric and I had surmised, the French disease was not unlike the Scandinavian disease, *Nephropathia epidemica*, which was also usually mild or moderate, and very rarely fatal.

In the spring of 1982, I happened to be returning from travel in West Africa. Because so few flights were direct from West Africa to the U.S., I was obliged to go through Europe. That gave me a chance to stop off in France for a couple of days to see Eric and help conduct a systematic study of the barn. This time we would try harder to outwit the rodents. Given what Eric's friend had gone through, we were understandably a little nervous about the expedition. Stirring up the dust might be all you had to do to acquire the disease, and there was evidence from Russia and the Far East that natural infection with Hantaan could be acquired by aerosol—through the air—in sharp contrast to other viral hemorrhagic fevers. We decided to err on the side of caution and put on protective face respirators. Then we went about securing a supply of live traps and other materials for catching and processing rodents.

Now, we knew we had to be discreet. We could be certain that our appearance in gas masks would alarm the locals, and there was also a possibility that they might call the police, who, though we were doing nothing illegal, would probably take a dim view of our strange activities. We elected to maintain a low profile.

The night before, we drove from Paris to the village and found lodging at a charming inn. We then sat down for a serious French country meal, which we washed down with the local champagne. As field investigations go, this one had some style about it. At dawn, we drove out to the barn and parked. Like spies engaged in a top-secret mission, we quickly and quietly got on our gear, grabbed the box with the rodent traps, and slipped inside the barn to investigate and lay the traps. We saw no sign of anyone having been there since Eric's last visit. Just as well, we thought. We remained in the barn about forty minutes.

When we had finished, Eric peeked out to see if the coast was clear. No one was in sight. Not that we expected to see anyone. It was only 6:30 on a Saturday morning. We shed our protective clothing and placed it in a bag for incineration. Then we washed off our gloves with special disinfectant and put them into another container for disposal. Once all that was done, we drove back to the inn for breakfast. We hoped to impress the other guests as health fanatics who'd just returned from our early morning constitutional. Indeed, refreshed after our clandestine mission, we set off to visit some of the local health officials as well as local physicians. We were naturally interested to learn whether they knew of any cases of the disease in the area. Several thought that they had seen patients with the disease, but they weren't sure. Later, once we'd established surveillance and carried out antibody surveys, we determined that, in fact, HFRS is not uncommon in that area.

That night, at about dusk, we slunk back to the barn. This time, I went in to collect the traps while Eric kept watch. Great, I thought as soon as I walked in. There was at least one live rodent in the trap. This meant that I would have to process it immediately, right there in the barn. I slipped back out for the rest of the gear, and then settled in to work. When I obtained the specimens I wanted, I popped them into the dry ice container we'd brought with us.

Not bad for a day's work, but we needed more. So we reset the traps, and decided to return again early the next morning to find out what we had. Then it was time for another excellent meal. This had some advantages over epidemiology in the African bush.

At dawn the next day we went through the same routine, except this time Eric went in the barn and I stayed outside to keep watch. On this occasion we trapped two more rodents. Tests subsequently showed that one of them had antibodies to Hantaan virus, so we now had confirmation that they did in fact carry the virus. Although we ourselves didn't succeed in isolating virus from these rodents, other researchers were more successful. The European virus isolated by colleagues in Finland, now called Puumala after the town near where it was first isolated, turns out to be even more finicky than Hantaan virus and grows with great difficulty in tissue culture.

What made the area where we were working so interesting was that this part of France contains remains of some of the earliest human settlements in Europe. I couldn't help wondering what effect this disease might have had on the Roman Legions, two millennia before it crept into the trenches of World War I battlefields. We felt more confident now that much of the renal dropsy described so elegantly by the physicians of the early twentieth century must have been HFRS. Nothing happened; clearly, our precautions had paid off, or we were just lucky.

Over the next few years Eric was the first one to identify Lyme disease in France, and then became increasingly involved in the care and study of AIDS patients. Then he died—prematurely and suddenly—only ten years after our visit to the barn. It was a great loss to French science, to his family, and to me.

Four years later, I was back on the trail of virus-carrying rats, this time in China, where there are two kinds of HFRS viruses. One, called Seoul virus, is a relatively mild disease, spread by house rats and concentrated mainly in urban areas. The other, which afflicts mainly rural areas, is far more severe, with a death rate of about 5 to 15 percent. Admittedly, it is a less acute disease than other hemorrhagic fevers and only rarely causes bleeding. Nonetheless, when bleeding does occur, it can leak into the brain. It doesn't take much of that to cause irreversible damage to nerve cells and, inevitably, death. This is the same Hantaan virus that my lab team had finally succeeded in isolating for the first time.

The small field mouse with the pale stripe of blond fur along its spine known as *Apodemus agrarius*, which is responsible for Hantaan, is semiaquatic. It flourishes on the banks of creeks and rivers, as well as in irrigation ditches. It especially loves rice fields. Just like Lassa virus, Hantaan infects baby mice, which then carry the virus for life with no ill effects. So China was full of happy, well-fed field mice, peeing out this virus in their urine all over the rice fields of this extremely populous country.

The result, I was told, was that HFRS was officially listed as the second most important viral disease in China. Only hepatitis was considered worse.

In late 1985, Sue, working with several Chinese colleagues, had investigated HFRS in a mountainous region in Zhejiang province where the disease was a major problem. But we needed to follow up on her original studies. We arrived in China in the autumn of 1986, one of the two peak seasons for HFRS, the other being spring. Both fall and spring follow the two rice harvesting seasons. It seemed reasonable to suspect that there was a link between the harvests and the outbreaks, yet, curiously, the symptoms and severity of the disease changed from spring to fall. No one had as yet come up with an explanation to account for the difference.

Sue and I were joined in our investigation by two Chinese colleagues, Professor Wu, an infectious diseases physician of the First Medical University, and Yi-Wei Tang, an epidemiologist who had previously worked with Sue. Our search for rats began in Shanghai as we boarded a ship that would take us south along the coast. Let Sue continue with the story.

The dockside was chaotic. We were led through the confusion and ushered to a "first-class" cabin. I took one look at it and thought: If this is first class, what can second and third class be like? Two bunk beds were squeezed in a space about ten feet square. The bedding didn't appear to have been changed any time in the last three months. The only other piece of furniture was a small table. There was barely any space to pile our boxes, one of which contained a platelet aggregometer and another a small automatic hematology machine for the hospital in Tiantai.

In all my travels throughout the world, I have never found toilets so completely unusable. They were unisex, though I can't imagine anyone of either sex wanting to go anywhere near them. The stench was so nauseating that you had to avoid that whole area of the ship where they were situated. Unspeakable matter was even flowing out of the restroom doorways. I immediately restricted my fluid intake. When I climbed into my bunk that evening and lay down, I very quickly realized that I was not alone. Something was crawling on me. The bed was full of cockroaches. How was I supposed to get to sleep? I'd been given a sleeping pill at CDC for use in case of severe jet lag. Coma, I decided, was the only way to get through the night.

The next morning, we disembarked at a small port city on the east coast of Zhejiang province. In my post-sedated confusion, I never did discover exactly where it was, or even what it was called. The four of us were led to a car, and then driven many miles over rugged roads, climbing higher and higher into the mountains. Our first destination was Tiantai, which occupies a spectacular position high in the mountains of Zhejiang. In 1987, cars were a rarity in the town, so we had to compete for space with bicycles, carts, tractors—and people. Lots of people. Since we were barely able to move, they came over to our car and pressed their noses against the windows and stared at us in amazement. In Tiantai, people almost never saw a Caucasian. So, I thought, this is what it feels like to be a new zoo exhibit.

We were taken to the hospital and given caps, masks, and gowns to put on. Members of the hospital's staff then escorted us to the medical wards.

"All the patients on the first floor have HFRS," explained our guides. "Which stage of the disease would you like to see?"

HFRS is classically divided into five stages. Naturally, we said we would like to begin at the beginning, with stage one. But just as we were about to start our tour, we were asked to move to one side. A man, his body contorted in a strangely crumpled but upright posture, was being hurried along the hallway, supported under each armpit by an attendant. This was China: no luxuries like gurneys or wheelchairs.

We entered the first room. "This is stage one," our guide said.

All the patients here bore the characteristic signs of the initial phase of the disease: fever, flushed cheeks, swelling around the eyes. When we looked inside their mouths and under their armpits, we could see signs of petechiae, a scattering of tiny red spots caused by minute bleeds resulting from the damage to their platelets.

In the next room, the patients were in stage two: all of them were in shock, including the man we'd just seen being hustled down the hallway. Thankfully, he was now in bed. In the next room, we found three patients, all stage three. They had no kidney function.

Finally, we were taken to see patients in stages four and five. These patients were in recovery. Some of them—those in stage four—still had disturbed kidney function, while the lucky ones, in stage five, were beginning to stabilize and get better.

I had never imagined anything like it: one hospital floor virtually devoted to a single infectious disease. But I gained considerable respect for what the Chinese physicians were doing. Their understanding of the disease allowed them to diagnose it quickly and take measures to minimize the number of deaths.

That night, we were treated to a spectacular meal, served by Buddhist monks in an ancient monastery. They were vegetarians, and though every dish was given the name of a bird or animal, they were all composed of vegetables artfully disguised as slices of pork, legs of chicken, breast of duck, whatever. We all dug in. Joe, who is something of a gourmet, was trying everything. Walking back to our lodgings, we passed through a courtyard that adjoined the kitchen. The door was open.

I glanced in to see two rats scurrying across the floor.

Ah, well. Buddhists will not harm animals, including rats. The one consolation was that, unlike West Africa, we knew we were not likely to be offered rat meat. We just hoped that everything we had eaten had been cooked well enough to destroy any Hantaviruses that might have contaminated it.

From Tiantai we went on to visit Jiande, further to the west. The town sat on the banks of a river that was created by the construction of a massive hydroelectric dam. The dam also accounted for the existence of what was called a lake of a thousand islands. Actually, they weren't islands at all, but the tips of hills, which had been submerged when the area was flooded.

The next day, we went on a tour of nearby villages where many patients had fallen ill with HFRS. This time our guide was Xu Zhi Yi. A specialist in viral diseases, particularly hepatitis and HFRS, he'd worked with CDC for many years and published some important studies. A delightful and cultivated figure, he could speak impeccable English, even though he'd learned it entirely in

China. Like many of our Chinese colleagues, he had suffered a great deal during the Cultural Revolution.

Evidence of the recent harvest was everywhere apparent. No matter where the eye settled, the ground was covered with drying rice. When there was no more room left in the fields or on the paths, the farmers simply laid the rice out to dry on the road, forcing cars to navigate around them. Even when you walked inside of the houses, you'd find rice piled on the floor, and if you looked really close, you could see tiny footprints in it. Rats. We walked down to the river and crossed a rickety bridge. On the opposite side lay a small cluster of houses where we'd been told many cases of HFRS had occurred.

As we continued walking along the bank, we passed workers cheerfully ladling night soil into baskets, which they then carried to the fields to use as fertilizer. Reaching the last compound, which was made up of two houses and a few sheds, we introduced ourselves to the occupants. But the only members of the family available to talk to us were an old man and his two grown grandsons. They told us that they were married, though we saw no sign of their wives or children. "Five of us have been in hospital with HFRS," the grandfather acknowledged matter-of-factly. HFRS was such a common phenomenon that apparently no one considered such a large outbreak in a single household to be anything unusual. We looked around the house. There were no surprises, just plenty of rat holes.

We asked the grandfather if he had ever made any attempt to get rid of the rats. He rattled away in local dialect. We waited for Xu to translate.

"He says he did," Xu said. "He put down poison, but it only killed the pig and the cats."

Xu knew exactly what the old man was talking about. "That's the problem," he explained. "You know, a few years ago, I was working in Anhui province. We had a big epidemic of HFRS. So the government decided to poison the rats. There was only one way to do it. We placed poisoned rice everywhere there were rats. We colored the rice bright orange to identify it. We stationed a soldier over every pile. The soldier's job was to chase away children, pigs, ducks, dogs, rabbits, cats, anything that wasn't a rat. There are plenty of people in China. So there are always enough spare soldiers."

Back at the local Anti-Epidemic Station (the Chinese equivalent of a State Health Department), we discussed how we could go about catching live rodents so that we could speciate them and isolate virus. It was the only way we would have of finding out what the causes of the spring and autumn outbreaks were, and why the disease was different in each case.

As the meeting went on, the decibel level began to increase. To a non-

Chinese speaker, Chinese sounds like it's being shouted most of the time and that a heated argument must be going on. When I asked for a translation, it turned out to be astonishingly succinct, given how long the conversation had been going on.

"It may not be possible to get the rats," someone said. "Dr. Ma says he cannot do it." Qou-Rong Ma, a chubby, round-faced man, was the field investigator, whose specialty was catching and processing mice and rats.

What was the problem? Was it difficult to catch rats?

No. That was not the problem.

Were there not enough rats to catch?

No. That wasn't it either. There were plenty of rats—we'd seen as much for ourselves.

Could they be caught alive?

Yes. Absolutely.

"Do you have enough traps?" Joe asked, growing more puzzled.

"Yes," said Dr. Ma.

Traps were not a problem. Well, did they think it important to catch mice and rats? we asked.

Yes, they agreed that was very important.

This rather fruitless dialogue went on for quite a while. Dr. Ma usually didn't talk much, but now he became positively voluble. Suddenly, he sprang up from his chair and disappeared into another room. When he came back, he was holding a rat trap. It was a live trap, consisting of a wire cage. It looked perfectly capable of doing the job. So far so good.

Then our Chinese colleagues started calculating: one trap every five meters, in so many fields, on so many nights. There was a standard scheme you needed to follow in setting out the traps.

Can't you do that? Are there enough traps? we asked, still puzzled.

Yes, of course, they said. No problem.

Well, what *was* the problem?

It only took a quick bit of mental arithmetic to make us understand. In most of our investigations we normally used collapsible traps. You can load a great stack of collapsible traps on a truck. In China, though, the wire traps were big and had to be piled on top of each other. They did not collapse. Moreover, they were pretty battered, so that they would make for an uneven and rickety pile. It would take several truck loads to transport the traps necessary to do the job. The expense and logistics made the operation completely impractical.

We were happy to be of service. The CDC, we informed them, would be able to supply them with more than enough collapsible rat traps to do the job.

The traps proved to be the key. The following spring, when they were deployed in the fields, they yielded a number of both large brown rats and mice. But there were more of the former than the latter, and when tested, it developed that only the rats harbored the virus in their urine. In October and November, however, when the traps were set, it turned out to be the mice that were the culprits, not the rats. That was why the two outbreaks were different. If the Chinese hoped to come up with an effective vaccine, then they would have to make it work against both strains. Once this was known, the Chinese researchers began to go to work. Clinical trials are already under way. In the United States, Connie Schmaljohn at USAMRIID, the pioneer and the accepted scientific leader in the molecular structure of Hantaviruses, has developed a genetically engineered vaccine that also shows promise. Connie worked closely with her mentor, another virologist, Joel Dalrymple. Joel was another of the hunters of these viruses—again, a heavy smoker, who died suddenly and prematurely in 1990. Tobacco is undoubtedly more dangerous than hemorrhagic fever viruses.

The development of a vaccine would go far to stem HFRS, of course, while the introduction of ribavirin as a treatment should do much to reduce the mortality rate of the disease. We had got the virus to reveal itself and had pursued it into the Champagne country of France, and then cornered it in Zhejiang. It would not be the first virus we had to chase halfway around the world. But not every virus would surrender its secrets so easily, and no one knows this better than Joe. He will tell you about the time Lassa fever moved to Chicago.

CHAPTER TWENTY-SIX

Lassa Moves to Chicago

It was late in the afternoon, January 13, 1989, and Azikiwe was sitting in his Chicago office, reviewing a blueprint. The phone rang. It was his wife, Veronica. It was unusual for her to call at work, particularly at this hour, when the children would already be home from school. With six boisterous kids, Veronica was kept very busy. It was clear from the sound of her voice that she was upset.

"Azikiwe," she said in her lilting West African English, "it's your mother. Valerie called to say that she suddenly took ill and died."

The blood seemed to drain from his head. Was he hearing her right? His mother hadn't been ill. There had been no warning, nothing. Azikiwe had just been thinking about bringing his parents over for a visit. They had never seen the United States. What a pity that his children barely remembered their grandmother. The oldest boys, Ogbejele and Oyakhi, had a vague recollection of the old lady, but the other four children were just too young.

Realizing he could not afford to surrender to his grief, he picked up the phone and called his boss to explain that he would need some time off. Then he booked a seat on a flight leaving for Lagos, Nigeria, via New York, the next afternoon. He dreaded this trip, and not just because of all the emotions that it was certain to stir up inside him. Even under optimal conditions, the trek was a hassle. He detested the airport at Lagos, one of the most corrupt and inefficient in the world, and then he would have to endure a long journey over treacherous two-lane roads with crazy traffic and police road blocks all the way. But he steeled himself and left.

The journey proved every bit as excruciating as he'd imagined it would be. At least he emerged from Lagos airport with all his baggage. He was bringing gifts for his family, and he had pretty much resigned himself to seeing them disappear. They had not. But then he had to face a six-hour bus ride to Benin City, from which it would be another two hours, again by bus, to Ekpoma, near his native village. In theory, the bus was spacious and air-conditioned. But it was so packed that people were sitting four abreast in rows intended for three. Other passengers were standing in the aisles. Of course, the air-conditioning didn't work. The bus driver drove like a man possessed. Swerving incessantly to avoid smashing into oncoming vehicles, he kept his foot on the gas almost the whole way. The ride was so jarring that several passengers actually fell ill, but with the stoicism bred of the struggle to survive in rural Africa, no one complained.

The routine police road blocks made for an additional nuisance. Money had to change hands before the bus would be allowed to continue on its way. Azikiwe felt like a man blessed just to be off the bus in one piece when he reached Benin City. He made the final leg of the journey in a smaller Nissan bus filled with women returning home from market. There was hardly room for them, let alone for the produce they carried: woven banana-leaf cages of chickens and ducks, sacks of fermenting cassava flour, and tins and bottles of pungent, deep orange palm oil. Many of the women carried dozing babies, which they strapped on their backs in colorful cloth swaddles. Only babies could have slept through all the noise. Noise was part of the experience of travel in rural Africa. Everyone was always talking and gesticulating, trying to make themselves understood over the rumble of the motor and the squealing of frightened animals.

When Azikiwe finally alighted in Ekpoma, he had to shake out his limbs and recover his equilibrium. He was exhausted, but he was unexpectedly exhilarated to be close to home. As he began to search for someone to take him to his village, Ishan, he realized that Ekpoma had grown since his last visit. It seemed curiously unfamiliar, even though he had been there a thousand times. For several minutes, he failed to recognize anyone. He was quite disoriented, a bit uneasy, even frightened. After wandering around for a while, he finally spotted an old friend who agreed to give him a lift on his motorbike. They left his bags in a secure place. One of his brothers would drive back and fetch them.

Arriving at his parents' house, he was overwhelmed by conflicting emotions. Instinctively, he looked for his mother. Then it sank in: he would never see her again. But he had the consolation of the rest of his family. How wonderful it was to be among them once more. After exchanging greetings with everyone, they all moved into the *bafa*, the traditional thatched pavilion, with open sides to catch the cool breezes, used for family or village gatherings. He found himself gazing at his father. He looked a different man, Azikiwe thought. How much he had aged in the four years since he had last seen him.

When Azikiwe woke, it was the small hours of the morning. He was still on Chicago time. He sat up in bed, alarmed. Something was wrong. Then he realized what it was: the silence. The village was dead quiet. No cars, no engines, no clock ticking, no dogs barking, nothing. He got up the from his woven-mat bed and slipped outside, stepping across several sleeping forms. It was still dark. The African night sky was wondrous, clear, and filled with stars that shone so brightly he couldn't believe it was the same sky that he was used to seeing above Illinois. The night air was not so heavy as that of the day, and it even brushed cool against his skin.

He sat down on a low stool near the *bafa* and reflected. He thought about his family and childhood friends. About how most of his educated friends had followed his example by abandoning the life of a subsistence farmer for work in the big cities. Probably none of his friends had been as financially successful as he'd been. But in his disquiet, he thought: to have cut himself off from his roots like this—had it been worth it?

He had never wanted for food or for shelter as a boy. He'd had plenty of adventures while he was growing up. So what was it that drove him away? Boredom? Desire for a better life? He certainly led a comfortable existence in the States, he worked for a top engineering firm, he had all he wanted . . . but even so, he felt uneasy, as if he had left undone something important and undefined.

The next day, Tuesday, was consumed in preparations for his mother's funeral and the period of mourning that would follow. As the head of the family, Azikiwe also had to see to it that important funeral rituals were carried out. This meant that he had to get in touch with village elders, wise men, drummers, and a juju man (the local shaman, magic man, and traditional healer). Without them, the rituals could not be properly performed, and without these rituals, there would be no way to ensure his mother the smooth transit to the next world that was Azikiwe's responsibility to provide. Of course, all the elders and the juju man would have to be adequately compensated for their contributions. Then Azikiwe would also have to see to it that enough food was stocked away, because the mourning period would last for several days, and many relatives who lived far away would have to be put up for the duration.

The funeral next day required the participation of all members of the family, including the children. Throughout the proceedings, the body remained under the watchful eye of the juju man, who was there to guarantee that the spirits permitted the proceedings to go ahead without a hitch. In Africa, it didn't matter which particular religious faith you believed in; no one ever questioned the existence of the spirit world. It was as much a reality as the things you could touch, the scents of the frangipani, and the whistle of the wind in your ears. In this sense, Azikiwe was truly African in his ability to remain a committed Christian, while simultaneously honoring the power of the spirit world.

The formalities took most of the day. It wasn't until late in the evening that Azikiwe at last had the opportunity to ask his family about his mother's final illness. Surprisingly, no one seemed anxious to talk about it. His sister was rather vague, and his uncles equally unforthcoming. Even his father seemed uncertain. Azikiwe couldn't quite figure out why he in particular was being so reticent. What was the problem? After all, his mother was an old lady—in the demographic context of rural Nigeria, at any rate—and her death couldn't have come as a total shock.

Azikiwe spent the next few days with his family in the expectation that he would leave before the end of the month. But just five days after the funeral, his father started to complain about feeling cold. He said that his back was sore and that he had a headache. Azikiwe went to the local village dispenser to obtain some medication. He was given chloroquine, a drug widely used at the time for treating malaria. Given the prevalence of malaria, it was the therapy of first choice for almost every fever in much of Africa and was doled out much as aspirin is in the West. His father took the chloroquine, but didn't get any better. He began to complain of a severe sore throat and nausea. Soon he was unable to eat, or even to swallow very well. His fever soared. Azikiwe noticed something else even more disturbing. His family begin to recoil in fear from the old man.

It was only then that his sister told him that his father appeared to be suffering from the same illness that had struck down his mother.

The terror that his father's illness provoked wasn't restricted to members of his family. His neighbors were responding the same way. The silence that Azikiwe felt deep in the night had taken on a new meaning. He now heard rumors of juju at work. No one spoke openly. If you said too much, it would touch you next.

While it was true that Azikiwe had grown up here and had absorbed the village culture, he was also a Western-trained engineer. He had been trained to think like a scientist—and, to some extent, like an American. Certainly, he didn't look on disease as a curse. But now he was torn between two worlds, one based on science and rationality, the other on the unknown or the spiritual realm, with all its rituals. In the world in which Azikiwe had come of age, the juju man was master; it was he who was responsible for maintaining order and harmony. Every phenomenon could be explained by juju. If your animals died, it must be juju, and it was necessary to find out who had cast the spell and stop him.

Was Azikiwe's father under a spell?

Apparently, his mother hadn't been the only one who died recently from this strange illness. People, they said, were getting sore throats and dying all up and down the street. No one had any explanation. Why would you die of a sore throat? Nor did anyone know what to do about it. In spite of all his Western training, Azikiwe, too, could see where juju might be the only way to account for what was happening. How else to explain this strange illness? For as long as Azikiwe could remember, there had been stories about the presence of witches in the village. Some people even claimed to know who they were. It was possible—just possible—thought Azikiwe, that there really was a witch at work.

Meanwhile, his father continued to deteriorate. He no longer talked, but only lay in his bed, mute in his agony. Perhaps, Azikiwe thought, this was his father's response to the loss of his wife. Maybe he only wanted to join her in death.

On January 28, his father died.

Azikiwe now had to remain in order to see to his father's burial, but immediately after the second funeral, he left Nigeria; his job and family responsibilities in America demanded his return. With a heart that beat in pain and a mind full of confusion, he made the preparations for his departure.

Azikiwe was greeted by Veronica at Chicago's O'Hare Airport when he returned on February 1. It was a tearful reunion. He'd lost both his parents in a manner completely incomprehensible to him. Neither the religious support he'd gained from his church nor the spiritual upbringing he'd had as a child offered a satisfactory explanation or consolation. The world of spirits and spells receded, but the sense of loss remained as sharp as ever.

Things hadn't been easy for Veronica while Azikiwe was gone, either. She and two of their children had been bedridden with the flu. Indeed, half of the community had been struck by this epidemic, which was still going on. But maybe things would be better now. After a restless sleep, Azikiwe got up the next morning and reported back to work.

A couple of weeks later in Atlanta, I was preparing an EPI 1, the document that offers a brief description of any epidemiological investigation the CDC has agreed to undertake. It is drawn up upon notification of an outbreak, prior to sending a team into the field. A brief factual account of the situation, the EPI 1 indicates why the investigation is being launched and what it hopes to achieve. The document also includes the names of those involved in any aspect of the outbreak. The EPI 1 for February 15, 1989 reads as follows:

On February 15, 1989, Joseph B. McCormick, M.D., Chief, Special Pathogens Branch, Division of Viral Diseases, Center for Infectious Diseases, Centers for Disease Control, received a telephone call from Robert Chase, M.D., an infectious disease practitioner in Winfield, Illinois, about a case of suspected Lassa fever in a forty-three-year-old Nigerian-born man who had recently returned from a trip to his hometown of Ishan, Nigeria. Upon review of the patient's history, physical findings, and laboratory results, Lassa fever was considered to be highly likely.

Azikiwe had no sooner returned to work than he began to feel feverish. It was now February 3. He reasoned that he was so fatigued and emotionally drained that it was no wonder he felt unwell. Or maybe he'd contracted the same flu Veronica and the children had come down with the week before. His mind was on what had happened in Nigeria. He decided to leave the office early and go home. Yes, he thought, it was probably the flu.

But something nagged at him.

His children and wife had been ill with fever for two or three days, but then they had started to improve. Azikiwe's illness was different. His fever mounted as the days went by, accompanied by an excruciating headache. Aspirin barely touched it. He also began to have a sore throat; he was lucky if he could swallow a spoonful of soup. His children would join him in the evening, since he was unable even to sit at the table, and would try to encourage him to take some food. Sitting at his bedside, they even shared food from his plate.

Veronica and the older children were not unmindful of what had transpired in Nigeria and were naturally worried. A few days later, on February 7, Azikiwe began to complain of an intolerable pain behind his eyes. His fever was still high. Veronica decided that enough was enough: he needed to see a doctor. Bundling him into the car, Veronica drove him to the HMO clinic. The doctor who examined him found that he had swollen tonsils and lymph glands, and some degree of abdominal tenderness. His white blood count was low, but this was thought to be a symptom compatible with influenza. So Azikiwe went home with a diagnosis of flu and a prescription for acetaminophen, which was supposed to alleviate his fever.

On the morning of the eighth, Azikiwe summoned the energy to return to work and somehow managed to make it through the day. But the next day he again left early, after putting in only an hour or so at his desk. As much as he might have wanted this to be the flu, he knew in his heart that this was something worse. Much worse.

He went back to the HMO clinic. In addition to the fever and a painful throat, he told the doctor that he had a bitter taste in his mouth. At no time, during this visit or the earlier one, did Azikiwe mention what had gone on in Nigeria. Nor did the doctor ever ask him whether he'd recently been abroad. Besides, they were in the middle of a flu epidemic; why worry about zebras when herds of horses were thundering through the clinics? Still, Azikiwe's illness was something of a puzzle, since it had persisted for much longer than the flu usually does, and it had become much more severe than would ordinarily be expected in an otherwise healthy forty-three-year-old man.

On the occasion of this visit, the doctor did observe something that he hadn't noticed before. Azikiwe appeared to have pus in his throat. This time, the doctor diagnosed a strep throat, even though the strep screen was negative. Azikiwe was given penicillin and sent home.

He became much worse. On February 12 he developed bloody diarrhea. Then he began to complain of severe pain in his ribs and back. Veronica noted that he was coughing up thick sputum. She could no longer even get him to drink water, because his throat was so painful. She did not know what to do next.

When Veronica brought him back to the same clinic, he was running a fever of 103°F, which had been present for nine days straight. His systolic blood pressure was low, just over 100, he had marked swelling in the neck, and still more pus on his tonsils. His abdomen was just as tender as before. Veronica made certain to mention to the doctor that he had blood in his stool. Although she was panicked, she tried to bear in mind what Azikiwe had told her. American doctors weren't like they were in Nigeria; they knew what they were doing, so she shouldn't think that they would fail to find the cause of his illness. Now, after these three visits to the HMO, she wasn't so sure.

But once again, neither she nor Azikiwe thought to say anything about the deaths of his parents the previous month. The physician diagnosed strep pharyngitis and hemorrhoids. He continued to treat Azikiwe with penicillin.

A blood test was done to check on a number of things, including the level of his liver enzymes. Remarkably, although his enzymes proved to be sky-high, no one seemed to take note of this. Now Azikiwe was manifesting every symptom necessary to establish a diagnosis of Lassa fever. It was no longer tenable to diagnose his condition as strep throat or even a complicated flu.

Desperate, Veronica took him to another clinic. Not that it did any good. The ear, nose, and throat specialist who saw him diagnosed tonsillitis and doubled the dose of antibiotics. Still, no one asked him about travel.

Again the couple returned home. Completely distraught, Veronica sat by her husband's bedside, wiping his brow, looking after his every need. At least she wasn't alone. The minister at their church was a great support; so were several members of the congregation, who came over to help prepare the meals and look after the children. To Veronica, it must have made more sense to turn to the church and to God, since Azikiwe had now been to four different doctors without seeing any improvement in his condition.

Azikiwe began to slip into a fitful sleep punctuated by periods of incoherent rambling, mostly in his native language. His wife tried to speak to him, but he seemed not to hear her. She collapsed in tears, no longer able to contain her grief.

On the night of February 14, Veronica took her husband to emergency room of the DuPage County Hospital. The physician who saw Azikiwe in the ER didn't know what to make of this patient. Here was a man who had had high fever for nearly two weeks, who had lost fifteen pounds, according to his wife, and who was obviously confused and very sick. His sore throat had failed to respond to various antibiotics, and now, in addition to bloody diarrhea, he was also suffering from severe nose bleeds.

In spite of Azikiwe's incoherence, the physician could see that he didn't have either jaundice or hepatitis, both of which can cause hallucinations and dementia. By now, it was late at night. The physician admitted the patient, put in an IV to try to restore fluids, and ordered some tests to be run immediately. When the results came back a few hours later, the physician was puzzled and surprised by the level of the liver enzymes. Ordinarily, such a high level might have indicated hepatitis. Yet it was already clear that the patient wasn't jaundiced. Early the next morning, another physician, Robert Chase, the infectious-disease consultant of DuPage Hospital, came to see the patient. He was the first one to think of asking Veronica about whether Azikiwe had been traveling any time recently. It didn't take him long to establish that Azikiwe had recently been in Nigeria. He immediately realized that he needed outside help and called CDC.

It was a Thursday, and I was sitting in my office working on a paper when Dr. Chase came on the line and began to describe his patient's symptoms.

"Is there anything that can do this sort of thing in Nigeria?" he asked.

"Absolutely," I replied. "Lassa fever. This sounds like a classical case."

After fourteen days of agony for Azikiwe, at least an answer was in sight. But I recognized that the prognosis was not a good one. I told Dr. Chase that Azikiwe was well beyond the stage in his illness when ribavirin, the drug that had proven so successful with Lassa in West Africa, might reasonably be expected to save his life. But, I said, there was still a chance: provide him with all the life-support care possible and hope that it might pull him through the acute crisis of his infection. It might also give the ribavirin time to work against the virus. In West Africa, a patient this far advanced would die—no question about it. But no patients in West Africa ever had the support offered by a modern intensive care unit. It might just work.

"Is it safe for us to intubate him and put in a Swan-Ganz catheter?" he asked.

We had updated and published our recommendations for handling such cases in the United States just a year before, basing them on our experience with Lassa patients in Sierra Leone as well as other published data on hemorrhagic fevers. This case presented us with our first opportunity to apply our new guidelines. I assured him that he could safely go ahead and intubate and catheterize the patient. I provided him with detailed instructions about how to proceed with Azikiwe's care without endangering the staff. Although it might be too late, Dr. Chase agreed that he would obtain intravenous ribavirin and begin administering it to the patient as soon as he had it in his hands.

I told him that he wouldn't have to deal with the situation alone. I assured him that I would be there later that day with a team from CDC. I put the phone down and dialed the manufacturers of ribavirin. They promised they would get the drug to Chicago as fast as humanly possible. Sue was in Senegal, so I called Cuca Perez, the technician who worked with her.

"Put the lab together, Cuca," I said. "We leave this afternoon."

We needed five hours to put together our mobile laboratory, arrange transportation to the airport, rush home for a bag of clothes, and make the necessary arrangements with the DuPage County Health Department before we could leave for the airport. There were several issues at stake. We had a rip-roaring case of Lassa fever right in a Chicago suburb. A lot of people were going to be interested, and most of them would want to know whether we were at risk of seeing more cases. In one respect, though, we were lucky. With the AIDS epidemic, a heightened awareness had developed among the medical community that was almost totally absent a few years previously. Most physicians and other health-care workers knew about the risks of infection from blood-borne viruses and excretions. When it came to treating patients, attitudes and practices had changed dramatically. People now automatically put on gloves whenever they handled blood and secretions, and they were taking far greater care to avoid needle-stick injury. Their caution extended even to those patients not diagnosed with AIDS, for the sensible reason that no one could be absolutely certain, without a reliable test, that the person might not be positive for HIV. It was quite possible, I reasoned, that doctors at the hospital, as well as those who'd first examined Azikiwe at the HMO clinic, had escaped contracting Lassa themselves because they'd followed these standard precautionary measures.

Just as Azikiwe's physicians were getting ready to put him on life support, he began to develop a problem often seen in cases of severe Lassa fever: Adult Respiratory Distress Syndrome (ARDS). This is what Jenny Sanders had had in Sierra Leone. Simply put, Azikiwe couldn't get enough oxygen to his blood because his lungs weren't allowing enough of it to pass into his circulatory system. Viral hemorrhagic fevers like Lassa cause extensive damage to the capillary bed, the tiny network of blood vessels that supplies oxygen to all the organs and the tissues. This is particularly true in the lungs, because the blood vessels become leaky and fill up with fluid. It's almost as if the individual is being held under water. To assist his breathing, Azikiwe was placed on a respirator. At the same time, a Swan-Ganz catheter was inserted in order to monitor and manage his failing heart. The willingness of the anesthesiologist to intubate Azikiwe and put him on life support, including a respirator, all on the basis of a recommendation from a physician he had never seen, was a testament to the confidence invested in those of us who worked at CDC.

The measures I had proposed, though, weren't enough. Azikiwe had been on life support for no more than two or three hours before he went into cardiac arrest and died. The ribavirin had not yet even arrived from California.

Just as I was leaving my office to go to the airport with our team, Dr. Chase called to tell me of Azikiwe's death. This changed our mission. A team was no longer necessary, so Cuca repacked the mobile laboratory. We wouldn't need it unless we had a second case. If more cases did occur, we would be prepared for them. Undoubtedly, several people had been exposed, so further cases were certainly possible. I took with me only one person, a young medical officer named Gary Holmes. The trip to Chicago was to be Gary's first experience with hemorrhagic fever—and a rather dramatic introduction it was, too.

Our objective was to set up a system of surveillance for secondary cases. Then there was another question to deal with.

"What should we do with the body?" Dr. Chase asked me over the phone.

I recommended that he arrange for a postmortem liver biopsy and blood specimen analysis to be done, so that we could be certain of the diagnosis. I also told him to ensure that whoever handled the body wore the appropriate barrier protection of gloves and gown, and at all costs avoided accidents with sharp instruments. I further suggested that the body be embalmed, since cremation was unlikely to be culturally acceptable. Though embalming was likely to kill any residual virus, I still had some anxieties. While whatever happened to the body afterwards was up to the family, I did have one suggestion to offer: at the funeral, they should make do without an open casket.

Once in Chicago, Gary and I talked to anyone who had been involved in the case. Little by little, we began to piece together Azikiwe's story. It became apparent that something might be happening in Nigeria, but we would have to get to that later. Right now, our priority was to ensure that we identified everyone who had been in contact and determine the degree of contact for each one. Anyone who had come into close contact with Azikiwe was put under surveillance for three weeks, time enough for the virus to make itself known if it were present. We decided that the only people with a high risk of infection were his wife and children. So his family was given oral ribavirin.

Two days later, we were in Azikiwe's home talking to his family when Veronica received a phone call came from Azikiwe's sister Valerie in Nigeria. She had already been notified about Azikiwe's death. But this wasn't what she'd called about. Since Azikiwe had left Nigeria, other members of the family had been stricken with the same ailment. Another sister, who was twenty-eight, and an eight-year-old cousin had become ill, too. While both of them had recovered, Azikiwe's thirty-six year-old brother, a physician himself, was stricken. He had died almost at the same time as Azikiwe had in Chicago. Valerie told Veronica that the family was making efforts to contact other relatives who'd attended the funeral to discover whether any of them had also fallen ill. But it wasn't easy. Some of them lived far away, and it would take time to get in touch with them. It was a nightmare of suffering for the family.

It seemed to me that we had to get much more information about what was happening. We needed someone on the ground in Nigeria who was an experienced professional. Luckily, I knew our man: Oyewale Tomori, better known as Wale. Wale had worked with us at CDC and was now a professor of virology at the University of Ibadan in Nigeria. Given the unreliability of the Nigerian phone system, I counted it as a minor miracle to get through to him without delay. When I related Azikiwe's story to him, he promised to get out to Ekpoma right away to see whether he could find out what was going on. This was the start of a complex investigation for which we had practically no precedent. The next part of the story belongs to Sue.

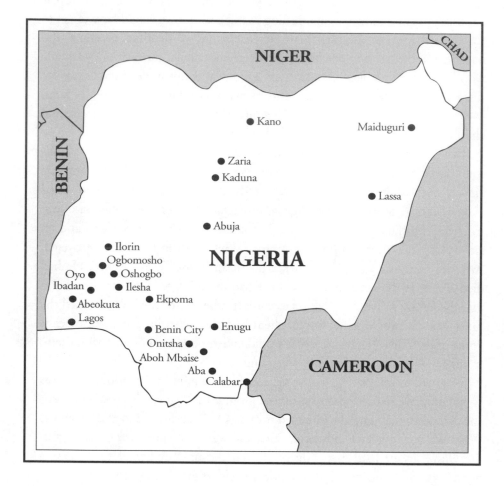

CHAPTER TWENTY-SEVEN

Juju

Joe and I emerged into the unrivaled chaos of Lagos airport, bleary eyed after twenty-four hours on the go. In all our years of travel, we'd never seen an airport that displayed such total anarchy. In the midst of our confusion, we began to realize that we were being met by two competing parties. One was a representative of our CDC colleagues working in Lagos, and the other was composed of two Nigerians. We had no idea who they were, but they certainly knew who we were, and insistently tried to corral us.

They had tickets for us, they said, adding that we must fly at once to Enugu.

Why Enugu? We weren't sure, but we knew that the reason we had finally received a formal government invitation to come to Nigeria was because of a physician with connections in high places who was based at the Medical School of Anambra State, in Enugu. Otherwise we might never have got into the country. In the six weeks that had passed since Azikiwe's death, we'd been trying to get to Nigeria to find out where he'd become infected—so far, to no avail. So we were obviously grateful to the man who had expedited the process. Whether the two fellows who met us now had any connection with the physician in question was impossible to determine.

"By the way," one of the Nigerians said, "where is the ribavirin?"

Ribavirin, it soon became clear, was an obsession for these gentlemen. They loomed over us while we slumped down on the edge of the nonfunctioning baggage belt, trying to gather our wits. They wouldn't take no for an answer.

"You have to come with us," they asserted.

Then, before we could reply, they repeated, "Where is the ribavirin? Where is the ribavirin?"

They wanted ribavirin. Now. If we wouldn't go with them at once to Enugu, then why didn't we just give them the ribavirin? Where was the ribavirin?

I suppose that we should have expected such a bizarre welcome to Nigeria, after the events of the previous two days. We'd been besieged by phone calls. Calls from Nigerians in high places in the United States, calls from friends of Nigerian friends or relatives of somebody important—whoever. We lost track. One thing was certain: someone in Nigeria was scared of Lassa fever.

And about time, we thought to ourselves.

We had been in touch with Wale Tomori. He confirmed our worst fears: he was finding case after case of Lassa fever. There were rumors of many deaths. So we knew that there was an outbreak of Lassa, maybe more than one, but we weren't sure exactly where. Although we'd brought ribavirin with us, we weren't about to give it to our Nigerian welcoming committee. For one thing, we had no idea what was going on, much less who these people were. For another, use of the drug for Lassa was not FDA approved, and we were authorized to distribute it only as part of an experimental protocol. So we slipped away from the two men, accompanied by our CDC colleagues, promising them that we would bring ribavirin to Enugu tomorrow.

The words "Ribavirin, ribavirin, please give us the ribavirin," were still reverberating in my ears long after I left the airport.

Before anything could be done about investigating the outbreak, though, we first needed to sit down and discuss the situation with representatives of the Nigerian government. Dealing with the government here is a chastening experience. Nowhere else in the world have we found negotiations to be quite so convoluted. We made the rounds at the Ministry of Health, going from one office to the next. Either one of two things happened: the person we were supposed to see wouldn't be there, or else he would be there and insisted on giving us a lecture about how we were to go about our business. Once he was done, he would remember to add that actually, come to think of it, he wasn't the one we should be speaking to at all; the person we wanted was in another department, but first, of course, we would have to make an appointment. We soon realized that an appointment was no guarantee that anyone would show up.

Occasionally, we'd find some official who would assure us of his cooperation, expounding with such eloquence that I doubted he could be trusted.

Of course, nothing would come of it. We waited, but nothing happened. There were many excuses, but the people giving them to us seldom bothered to make them plausible. It soon became clear that no matter what anyone said, there was only thing they were really talking about: money.

A government official promised to provide us with a vehicle and cover some of the field expenses. We didn't believe him, but, then, I supposed, anything was possible.

Eventually, we decided to leave Lagos for the interior and see what was happening for ourselves. But first, we were interested in finding out whether anyone else Azikiwe had come into contact with—either his friends or his family—was infected. The answer was to be found in the laboratory of the chief virologist of Lagos, whose name was Nasidi. Wale had found the family after our phone call, and had obtained information and some blood samples from them. He'd then taken the samples to Lagos and delivered them into the hands of Nasidi, who also happened to be a good friend of his. Nasidi had received his training in the Soviet Union. When he returned home, he not only had a medical degree, but a Russian wife as well. Although he was a practicing Muslim, he took a pragmatic approach to both religion and life. And he had a very sharp sense of humor. What he didn't have were any reagents to test Wale's samples, so he had to wait until our arrival to process them.

In spite of our jet lag, we unpacked the reagents and performed the tests. We all stood around in anticipation as we waited to read the slides. It was Nasidi who examined them first. We stood beside him, and noted down the results against his list of Azikiwe's family members and friends. Then, wordlessly, Nasidi stood up and allowed Joe to have a look. No one had read more Lassa tests than Joe. Nasidi started jumping around in obvious delight.

"That's it!" he cried. "All the positives are from those who gave Wale a history of Lassa-like disease."

It was the correlation that pleased him, not the plight of the victims.

"Then that must have been just about everyone you took a sample from," Joe said, glancing up at me, "because almost everyone here is positive."

The next day, with Nasidi in tow, we were on the road again, bound for Ibadan, a distance of about two hours. There we were hoping to find Wale, so we could hear his account of the outbreak in person. In the back of our truck we carried a liquid nitrogen container for any samples we could gather, as well as gloves and materials to bleed people. That was all. Though Nasidi had a long string of promises of government support, I very much doubted that we'd ever see any tangible evidence of it.

At least we had the use of a vehicle with diplomatic plates, lent to us by John Nelson, the director of the Child Survival Program. In fact, we were entirely dependent on him for in-country support. Without him and the assistance of the U.S. Embassy, we wouldn't have got very far at all. Once again, the U.S. State Department proved to be highly effective. The police had a bad habit of erecting roadblocks at periodic intervals, stopping traffic so that they could extort money. Since they were well-armed, no one was about to protest too strongly. But they were sufficiently impressed by our diplomatic plates to wave us through without subjecting us to harassment.

When we arrived in Ibadan, we immediately went to look for Wale. We found him in great shape. He said that he was doing a lot better than Nigeria was.

"The country is going to ruin," he said. "A beautiful country, a rich country, and the people who are running it are killing everything."

Tribalism was rampant, corruption a way of life, the oil wealth was disappearing, presumably into secret bank accounts in Switzerland and the Cayman Islands, and all over the world Nigerians were acquiring a sordid reputation as drug couriers and con artists. Not a happy state of affairs.

In the days when he worked with us at CDC, Wale had made himself famous by singing riotously while wearing his space suit. His songs were spirited, but there were times when some of us would have preferred to have gone about our business in silence. The technology of the lab didn't allow for this, however, since we were all connected by air hoses. Whether we liked it or not, Wale's Nigerian tunes were piped into everyone's ears.

Wale kept us in suspense about our real destination as he filled us in about his first visit to Ekpoma, near Ishan, where Azikiwe's parents had lived.

"As soon as I heard from Joe about the engineer dying in Chicago, I thought I would go and see what was going on in Ekpoma," he said. "When I got there, I found devastation. Both parents were dead, so were many of his relatives. It was awful. So I found as many family members as I could and took blood samples. It seems that they all became infected about the time of the funerals, probably at the funeral itself. The outbreak there now seems to be over. Some of the dead man's relatives fled in terror to Port Harcourt on the south coast. So I think we will have to go there, too, to find out what happened to them, but we have to go to Enugu first."

Enugu. It was the same place that the Nigerians at the airport wanted to take us. Why Enugu? we wanted to know.

"I'm afraid," Wale said, "that the Lassa epidemic has spread."

How could he be so sure? we asked.

While he was investigating the situation in the village, Wale went on, he happened to attend a meeting at the University at Enugu, in adjacent Anambra State. The subject was HIV. While AIDS hadn't made a significant impact on Nigeria yet, there was good reason to assume that it soon would. Cases of AIDS had already turned up, some in Anambra State itself. In fact, a physician at the meeting informed Wale that there were two AIDS patients in the local hospital. Wale was told that, if he wished, he could take a look at them.

The two patients—a man, Dr. Ikeji, and a woman, Dr. Anamba—were desperately ill with high fevers, and they were in shock and bleeding. Wale was told that they were both surgeons who had worked in the same hospital. Wale is trained as a veterinarian, but he looked at the dying surgeons carefully and reached his conclusion.

"There was little I could do," he said. "They were beyond hope. But there was something else I knew at once. What they had wasn't AIDS. No way was it AIDS. It looked like Lassa to me."

So he proceeded to inform the staff that they were mistaken about their diagnosis and that they needed to take all possible precautions to avoid becoming infected themselves. Then he drew some blood samples from the two patients, got back in his car, and headed straight for Lagos.

"The samples were sitting on the floor of the car. I didn't want to have an accident and then have them fall off the seat and break open, so that's why I put them on the floor. I had two full vials of Lassa fever with me, and I kept looking at them, scared stiff that they might break as they rolled about."

We later isolated one hundred million particles of Lassa virus per milliliter from those samples. It was among the highest virus concentrations we have ever seen in human blood.

Wale now revealed that we were expected in Enugu. Not only that, but authorities there had even gone so far as to arrange a conference for us. Well, I thought, now I can understand what the two were doing at the airport.

We decided that first we would go to Enugu and try to track down the source of infection of the Lassa victims there. Then we would return to Ekpoma, the closest city to Azikiwe's village.

When we arrived at Enugu hospital, we learned that the two surgeons Wale had seen were now dead. We were greeted by Professor Nwokolo. He was the physician who'd cared for them in a private clinic before they were admitted to the hospital. He was now a very worried man. In fact, everyone who worked at the hospital was very worried. They were all convinced that, at any moment, they were going to come down with Lassa and die. As we began to talk to Professor Nwokolo, it dawned on me that this was the man with connections in high places, this was the man who'd obtained our invitation to the country, this was the man who'd sent two emissaries to the airport for the ribavirin.

He wanted it for himself.

We talked to everyone we could, taking careful histories and blood samples. We then hastened to explain to them that the incubation period for Lassa was almost over and that if they were going to get the infection, they would have already fallen sick. We were able to conclude that Enugu Hospital had no further cases of Lassa fever. We were all relieved—until we heard about the third dead surgeon.

Where had he come from? From the south, we were told, in Ibo territory, Imo State. The two surgeons had come from another part of the same state. Were there any samples of his blood? No. But some people knew the names of the hospitals where he and the other surgeons worked. It was complicated, but we collected as much information as we could and headed south for Imo.

Our next stop was Owerri, Imo State's capital. There we met with the state Minister of Health. In Africa, you cannot go into a government office and simply ask questions. Elaborate greetings are essential and certain formalities must be adhered to. Great ceremony was attached to these encounters, especially since we were a joint Ministry of Health/CDC team. And no ceremony in Ibo country can take place without cola nuts.

Cola nuts contain great dollops of caffeine, which explains why, in times gone by, messengers and travelers could sustain themselves with them, and little else, over long distances. More importantly, cola nuts are the currency of hospitality. The Ibo people revere the cola nut. It is the object of special presentation ceremonies, which have to be performed in order to ensure friendly relations. People will even talk to the cola nut, which they invest with great powers. However, only men can take part in these cola nut rituals; women don't get to eat them—or talk to them. This was no loss to me.

So it was only after the cola nuts had been duly addressed, appreciated, and consumed—by the men—that we could finally get around to asking the minister whether he was aware of any cases of Lassa. Yes, he said, he knew of the physician named Ezirike who had died in Enugu. He came from a semiurban area called Aboh Mbaise, located just outside of Owerri. The death of this third physician was shrouded in mystery and rumor, however. Depending on whom you believed, he was either a victim of a business conspiracy—poisoned by rivals who ran a competing hospital a mile away—or else he was a victim of witchcraft. Whatever the explanation, you couldn't prevail on taxi drivers to take you there. They'd put their feet to the gas pedal and floor it until they were well out of sight of the village where the hospital was located. When we reached it on our own, the hospital was deserted.

Though the hospital was new, it was tiny and miserable. Two dark rooms served as the wards, holding about twelve beds in all. The operating theater was nothing more than a small concrete room. There was no equipment. I had little difficulty imagining what it must have looked like just two or three weeks before. A few patients in iron beds. A few young girls as nurses. Very little in the way of drugs and equipment, and no sense of safety or good medical practice. Surgery would be performed in primitive conditions. Nothing moved except the flies and the mosquitoes, and the lizards chasing them over the walls and into the light fixtures.

As we sat outside on a bench under the eaves, Ezirike's widow emerged from a nearby house. She was sullen and angry, and she refused to talk. But the dead surgeon's father soon put in an appearance, and then one of his brothers. They proved more communicative. Their stories were embellished with further dark hints of juju, chicanery, and intrigue. They had no doubt that their family was being targeted by nefarious elements.

While Joe and Nasidi began to explore the surrounding village to see what evidence they could find of a Lassa fever outbreak, Wale and I returned to the hospital to go through the contents of the physician's tiny office. As hot and humid as it was outside, it was worse inside, so we propped open the door for some air. The office was infested with mosquitoes, making it a magnificent breeding ground for malaria. I resigned myself to being eaten alive. We went to look for the charts to find out the names of the patients whom the physician had been treating. There were no charts. No outpatient records, no inpatient records, no operating records. Nothing. Only one set of papers existed for each patient: the drug charts. We started to go through them. Each chart we looked at turned out to be a detailed list of every drug each patient had received. It was only when we came to the end of each chart that we understood the reason why only the drug charts had been maintained so meticulously while no other records had been kept at all. These charts were not medical documents, but financial papers. The more drugs a patient was given, the more the hospital could charge him.

Nonetheless, there was information to be learned from these records. We decided to use them to reconstruct a clinical picture of each patient. We had the dates of admission and discharge, and the dates of death. Ezirike had only four or five antibiotics available to him. It was clear that he went by a set regimen. A patient with a fever received a certain set of drugs. If the fever persisted, Ezirike switched over to a second set of antibiotics and, for good measure, added chloroquine on the off chance the patient might have malaria. If a patient vomited, he'd give him an antiemetic; if he experienced any pain, he'd give him an analgesic. Despite his limited stock, he administered drugs in amazing numbers and combinations. He would prescribe up to six injected drugs and as many oral ones, including vitamins and other substances of marginal benefit to his patients. It was a good way to make money.

Finally, we discovered that the patients were being given iron and blood transfusions. That gave us pause. Did that mean that they had started to bleed? Then we found that the patients were being given anticonvulsants. Lassa patients have seizures in the final stages, and anticonvulsants control seizures. And if all else failed, Ezirike would give steroids in a desperate and futile attempt to bring back their blood pressures as they went into shock and died.

Sometimes we'd turn up a note in the margins that confirmed our assumptions that we were dealing with Lassa, such as "bleeding per rectum," or simply, "convulsions." We felt as though we were translating a Rosetta Stone, deciphering one ancient script in order to interpret another. In most cases, the notes would trail off, and the bill would be tallied up.

We found some notes about those who had not paid. Apparently payment was left up to the patients' families. The patients themselves were dead.

For two days, we sat in the little room, our legs munched by mosquitoes, studying these charts. By the time we'd finished, we were able to piece together the terrible story of this little hospital.

Seventeen patients had died of an acute febrile illness with shock, convulsions, and bleeding. Many had had terrible sore throats. We were able to trace the infections as they spread from patient to patient. There was one week in February (about the time Azikiwe had died in Chicago), when several patients had died within hours of one other. It was at this point that the handwriting in the charts had changed. It wasn't a nurse who was keeping track of the drugs anymore, it was Ezirike himself. We imagined that he must have been frantic to save his patients, resorting to whatever medications he had left at his disposal, trying anything in any combination to see whether it would have any effect.

But nothing worked; nothing was ever going to work. He couldn't save his patients, and in the end, he couldn't even save himself.

We traced the story even farther back, to January, when Ezirike's nephew, a student about nineteen years old at Enugu University, had come home for a few days. This was the same town we had just visited, the same town where Ezirike had gone to die. Apparently the nephew had been afflicted with a sickle-cell crisis, which comes from a condition called sickle-cell anemia, a commonly inherited disease in West Africa. The name of the disease is derived from the shape of the patient's red blood cells. Normally, these cells have the appearance of little round red-brimmed hats under the microscope. In sickle-cell anemia, they take the form of a sickle, or a crescent moon. The boy was admitted to his uncle's little hospital. Like every other patient, he'd received lots of injections. Unlike most other patients, he quickly got better.

About a week after being discharged, however, he developed a fever and a severe sore throat. He was readmitted—and given a great many more injections. So were the other patients, regardless of what was wrong with them. All the evidence indicated that syringes and possibly even IV drips were shared. They were, after all, expensive.

This time, the boy got worse. His chart—dry and technical as it was, cluttered with the names of antibiotics, dosage levels, and sums of money owed—was a cry from beyond the grave. It told of increasing desperation and anguish, of a futile search for something, anything, that would halt the inexorable progress of the virus. When one drug didn't work, Ezirike had tried another and then another. The virus continued its attack. The boy began to vomit and bleed. He went into shock. Convulsions, finally. Then death.

About a week later, a patient who had been in the hospital at the same time as the boy, but who had been discharged fully recovered, was readmitted with a fever. The pattern repeated itself over and over again. Somehow, Ezirike must have believed that he could control the situation. Maybe it was pride, maybe it was fear; more likely it was ignorance. He just didn't realize the nature of the beast that was now loose in his tiny hospital. Whatever his motivation, he delayed for three weeks before appealing for help. Seventeen patients had to die and he himself had to become fatally infected before it finally sank in that perhaps he couldn't handle the crisis on his own.

After reviewing the charts, we concluded that the poor nephew was probably infected in the hospital, possibly by an injection or an IV drip, during his first admission. As to who the index case was, we were never able to say for sure. The records were inadequate and staff fear too great to piece the story together accurately.

To see whether the infection had spread beyond the doomed hospital, we went around to other health facilities in the vicinity. We talked to the doctors and nurses and pored over charts, looking for evidence of additional cases. We asked about patients who'd recently died, and we examined the records to try to determine the cause of their final illnesses. We bled the staff to see whether they had become infected. The virus had extinguished itself from the human population of the hospital by killing most of its victims and chasing other potential victims away in fear.

As the trail grew cold, we decided to survey hospitals in the rest of the Owerri area. One visit was to a small but well-run private clinic. The doctor, who was the owner, had practiced in the U.S. Midwest for some years. He sat up immediately when we told him about our investigation.

"Yes," he said, "I think I know what you are looking for. Come with me. I have a patient upstairs you should see."

We went up the narrow staircase. The doctor escorted us into a private room where we found a man in his late thirties lying in bed. He was bundled up under the covers. He was febrile and very weak but not bleeding. His throat ached horribly, and, when we examined it, we noticed yellowish exudate on the tonsils—all signs of Lassa. He also complained of severe abdominal and back pain—again symptoms compatible with Lassa fever. He told us that his business kept him on the road. Perhaps that is where he acquired his infection. We took some blood from him. Before leaving, we showed the staff how to nurse him safely without infecting themselves. It was only when we'd got back to Atlanta and put the specimen up for culture that we were able to establish for certain that the man indeed had Lassa. (Fortunately, his recovery was uneventful, and none of the staff treating him came down with the illness.)

We found another case in the main hospital of Owerri: a young woman who had aborted a dead fetus. Lassa is very severe in pregnant women, and the baby usually dies before it is born. The mother does better if she is in the first six months of pregnancy, or if the fetus aborts. Maternal death soars dramatically in the third trimester, particularly if the uterus is not emptied. This woman was quite ill, very much alone, and scared. She would not speak to us.

As soon as the nurses understood what we'd come about, they were thrown into panic. Now that they realized what the young woman might have, they wanted nothing to do with her. They wouldn't go anywhere near her. Her family, too, must have shunned her, because they had already vanished. She was completely alone in the world. Since she was lying on a miserable cot on the floor, we suggested that she be transferred to a general ward, where she could be better looked after. When no one else offered to help her, Joe and I carried her ourselves. But then we still couldn't find any nurse to care for her.

We did our best to convince the nurses that they were in no danger at all if they took some simple precautions and made sure not to come into direct contact with her blood. They listened to us and said they understood. But I'm afraid that they still weren't persuaded. I had the sense that once we were gone they would continue to have nothing at all to do with her.

Again we took serum. Again we sent it to Atlanta for culturing.

While all this was going on, wonder of wonders, the government vehicle we'd been promised in Lagos actually turned up. Wale was expecting someone to bring us money for our operational expenses. But there was no money.

Where was the money?

No one knew. The driver swore ignorance. Wale railed at him, but it did no good. So we sent him packing back to Lagos.

We next turned our attention to determining how the two surgeons, Doctors Ikeji and Anamba, had become infected. We knew that they'd come from Aba, a busy market town in the south of the state. We drove to Aba and located the chief medical officer of the town. He said that he was delighted to see us. We told him we wanted to go to the hospital. No, he said, first we had to see his superior. Though we weren't much interested in seeing him, we complied.

After we were introduced to the superior, a heated discussion ensued between him and the medical officer, the upshot of which was that lunch tokens were to be distributed to all of us. We protested vigorously that we did not want lunch. We wanted to get to the hospital. After much debate, we prevailed, but in the process we succeeded in sacrificing the good will of the medical officer. His only interest in our mission, it seemed, was the free lunch.

We drove around Aba, but at first were unable to gain access to the hospital where the surgeons had worked. Not to waste time while Nasidi went off to initiate tortuous negotiations for access, we turned our attention to a survey of all the other hospitals and clinics in the town, talking to the nurses and physicians and taking their blood, searching for evidence of Lassa fever. While we failed to find any other suspected Lassa cases, we did find ribavirin. That suggested that people were on the alert for Lassa. When a surgeon showed the drug to us, we asked where he'd purchased it.

"The market," he said casually. "Where else?"

We looked at the packaging. It was made in China.

The market in Aba was mobbed; people seemed to bring a great deal of passion to their haggling. Whatever you wanted was available for a price: plastic pots, cooking pots, rush mats, drums, rice, onions, newly butchered meat swarming with voracious flies. And ribavirin. If they didn't have what you wanted, then the vendor would urge you to come back in half an hour. It was impossible to figure out how they could mysteriously procure the article you were after in so short a time. When it came to drugs like ribavirin, some obliging entrepreneur would produce for you whatever you asked for in no time at all, with perfectly faked packaging to make it look like the real thing. Fake drugs are a major industry in many developing countries.

We made another attempt to gain access to the hospital that we believed to be the source of the infection for Aba. It was located at the end of a narrow lane heaped with mud and refuse and filled with potholes. When we finally managed to reach the hospital, we discovered the doors bolted. The hospital was empty; everyone had gone. We were seeing the same thing all over again.

Wale and Nasidi networked. Next day they somehow succeeded in rounding up the brother of one of the dead surgeons. He, too, like the people in Aboh Mbaise, was convinced that the whole thing was a plot by competitors who would resort to anything—even juju and poison—to shut down the hospital and kill his brother. As a result, the family had refused to allow anyone to enter, including even Ministry of Health officials. They believed that, once they opened up the hospital, the conspirators would somehow take advantage of the situation. It was all that Wale and Nasidi could do to persuade him and his relatives that we had no interest in harming them Finally, the brother relented and took the padlocks off the door.

Unlike the hospital in Aboh Mbaise, this one had opened for business only two years previously. Its clientele was mostly poor and drawn from the nearby market area. Because it kept its rates low, the hospital was always filled. The building was constructed like a prison, with balconies facing out over a central covered well and concrete-block rooms radiating off of it. Inside, there were two tiny operating rooms, each about eight by ten feet in area. One contained a converted gynecological couch, which doubled as the operating table. A large porcelain sink sat in a corner. Light was supplied by a single flyblown fluorescent strip hanging off its wires from the ceiling. On the floor we noticed a couple of gas burners with cooking pots on them. Presumably these were used for sterilization. Hanging over a rack were a few tattered surgical gloves. Everything wore a look of desolation.

Once we'd actually got inside the hospital, the brother was very cooperative. He even invited two physicians who'd been employed at the hospital to join us. Once they began to speak, they couldn't stop. Their stories came spilling out. Oh yes, they began, there have been several deaths at this place . . .

The first victim was the hospital matron. She died only a few weeks before, in early January. She was an active, healthy woman, with no history of illness. But without any warning, she developed a fever and a sore throat, and failed to respond to the usual treatment. Her death followed quickly. It sounded like Lassa to us.

There were others like her, including a nurse and one of the patients.

And what about the two surgeons whose deaths we were investigating? They were why we were in Aba to begin with. At least here we had the advantage of having plenty of documentation available. We began pulling out the records of recent surgical procedures as well as charts for all patients admitted in the last few months. What we were looking for was a single event that would link the two surgeons, something that would have occurred about ten to twenty days before their deaths. Since they had died on the same day, and indeed had been transferred to Enugu together, we assumed that they'd been infected together. The female surgeon, Anamba, had done most of the operations. That made sense: operations were a highly probable source of infection.

That still left the third surgeon, the doctor who ran the hospital. Had he become infected performing surgery as well?

"No," replied one of the doctors, shaking his head. "He did not like to operate. He did not go to the theater. He stayed on the wards and looked after the patients there."

What about the rest of the staff? Was anyone else ill at the time the two surgeons died?

"Oh, yes, " the doctor said. "One of our nurses was very ill, but she has gone back to her village. No one knows where it is."

This evoked our curiosity.

"What did she do here at the hospital?"

"She was a theater nurse," the doctor explained. "Her name was Peace Uba."

Now we had a clue. Perhaps there was some link between the nurse and the surgeon. When none of the operating notes yielded any information, we asked to see the register of the operating theater. What we were looking for was an emergency operation that involved both Peace Uba and Anamba. Then we found it. In the middle of February, just about twenty days before the surgeons' deaths, the register indicated that an emergency procedure had been performed. We looked for notes in hope of further elaboration, but could not find them. However, as soon as we began to interview the staff, their memories came flooding back. Oh yes, they said, they remembered the patient. He was a young man who had been in the hospital for some time. He was admitted from another hospital with what was called a "failed appendectomy." He'd been getting better without any special treatment, except for the usual injections of antibiotics and other drugs. Then, after more than a week, he developed a fever again. When his condition continued to deteriorate, his doctors assumed that it must be a recurrent abdominal problem and rushed him back into the operating theater to open him up.

The more we probed, the more details we obtained. Everyone, it seemed, recalled the case because of the way the patient had bled on the operating table. It was uncontrollable. No matter what anyone did, the bleeding simply could not be stopped. There was blood everywhere. Anamba was having so much difficulty that she called for Dr. Ikeji, the hospital head.

We checked the register again. This was the only time that month that he had been in the operating theater at all. Now we knew how he had become infected. Despite Dr. Ikeji's help, the young man died later that night, after returning to the ward. We asked for the name of the nurse who was assisting at the operation. It was Peace Uba.

After reviewing the young man's case carefully, we reached two conclusions: that he had had Lassa and that he'd most likely been infected in the hospital, probably by a shared needle. We needed to know more. We decided to round up the staff and interview and bleed them to see whether they were also infected.

The next morning, when we reached the hospital, we discovered that its central hall, ordinarily the gloomiest of places, had been quite transformed. To our astonishment, we were greeted by over two hundred nervous, giggling girls, most of them in their late teens or early twenties. While we were told they were nurses, they described themselves as students. Joe and I questioned the girls and took notes, Nasidi bled them, and Wale sorted and stored the specimens.

Every girl answered our questions in much the same way. They admitted that they had little in the way of education or professional training, but even so, they still performed all the jobs that nurses usually do. They all said their age was eighteen. They gave injections, distributed drugs, cared for patients, and cleaned up after them. But when it came to the young man we were interested in, practically none of them remembered him. I was getting tired, dehydrated in the heat, and depressed.

Then I talked to another girl. I asked the usual questions. Have you been ill in the last four weeks? I asked. And if so, what kind of illness did you have?

"Yes," the girl replied shyly. "I had a heart attack."

She was also eighteen.

"A what?"

I was dumbfounded. A heart attack in an eighteen year old? As far as I could see, she looked in the peak of health.

"Describe it to me," I said.

"I had a pain here." She put her fist to her chest.

I was suspicious. One of the symptoms of Lassa fever is chest pain, which may be due to pericarditis, an inflammation of the sac surrounding the heart. I asked her to tell me more. Was she admitted to hospital? I asked.

"Oh yes," she said. "I was in the same bed as Peace Uba."

I caught my breath.

What did she mean, the same bed? It wasn't clear whether she meant the same ward, or whether it really was so crowded in the ward, that they'd had to share a single bed. But another thought occurred to me before I could pursue this.

"Did you have anything to do with the operation on the young man who was bleeding?"

"Oh yes," she said. "I cleaned up the cloths."

After hearing her story, I nodded to Nasidi, who slipped the needle into her vein and took blood for Lassa antibodies.

Later that evening, we sat in the hotel lounge in Owerri and sipped cold beers. One question preyed on all of us: Where was Peace Uba?

Nasidi made up his mind to find her. The next day he began his search. It didn't take him long. Around lunchtime, he reported back to us.

"I know where to find her," he said. "Let's go."

He wouldn't tell us how he did it, but his sources turned out to be reliable. Peace's family, he said, were simple folk, subsistence farmers, surviving in the African bush on what they could grow. Peace was their big hope—an educated girl, a girl who, until Lassa had struck, was looking forward to a bright future as a nurse. Nasidi let slip that she was supposed to be beautiful.

We had to drive for many miles, deep into the bush of southern Nigeria. How Nasidi knew how to get to where we were headed was beyond me. At last we came to a small farm and stopped the truck. We climbed out and walked down a grassy bank to a little house. Nasidi knocked on the door. For several moments he just stood there. He pressed his ear against the door. I could see by his expression that he believed someone was inside. Then the door opened and several people emerged They huddled in conference with their visitor. Some agreement seemed to have been reached. Then Nasidi came back to fetch us.

"She is here," he said, "and her family assures me that they'll cooperate with us."

Peace Uba didn't appear immediately. Instead, we were treated to a warm-up in the form of several members of her family. We asked them the routine questions and drew their blood. I suspect that they hadn't had so much excitement in years.

Finally, the object of our search came out herself. She was a tiny, nervous girl, but very pretty. She had gone to some trouble to make herself look nice, which undoubtedly explained her delay. In any case, she walked a few oddly lurching steps in our direction and then shyly took a seat beside Nasidi.

He was grinning from ear to ear. He couldn't help himself. Nasidi liked pretty girls a lot.

But it soon became clear that something was very wrong. When he turned toward her to speak, she didn't respond. She kept staring at us without so much as glancing at Nasidi. He looked nonplussed. What was this? Were his charms failing him?

He spoke again, touching her lightly on the arm. She gave a start. There was a nervous expression on her face. Her smile, which only moments before we'd found so fetching, seemed empty.

Nasidi told us he had heard she was stone deaf. It looked as though that was the case.

Deafness is a complication from Lassa, and it can be total and permanent. Then we asked her to walk. Slowly, she rose from her seat and took a few steps forward. She began to stagger; it was a classic ataxic gait, meaning that her legs were no longer responding to the commands of the brain, with the result that she had no sense of balance. This was another terrible complication of Lassa. While the deafness might persist for the rest of her life, it was possible that her unsteady gait would disappear in time. After taking her blood, we completed our notes and spent what remaining time we had trying to reassure Peace's family.

It was sad. We had reconstructed the story of how Lassa had come to Aba and how it had infected the three surgeons and the two nurses who had worked in its hospital. But while we had established the diagnosis, we hadn't succeeded in penetrating the mystery that the disease presented. Just how much of it was there in this part of the world? From the sampling of the population we took, it seemed relatively uncommon. But how was it transmitted? What was its source? And then there was another question: When would it reappear and who would it carry off next time?

In our search for Lassa, we'd traced a circuitous route, beginning in Enugu, where we found the disease had come and gone; then continuing on to Owerri, to eat cola nuts with bureaucrats; resuming our investigation in Aboh Mbaise, where one surgeon and sixteen patients had died in a single hospital; going on to Aba, where the virus had killed two more surgeons and emptied yet another hospital; and finally ending up in a small village to find Peace Uba.

Now it was time to see what we could find out about Azikiwe's family, which was what had prompted our investigation to begin with.

The first place we looked for his family was Port Harcourt on the south coast. Wale had heard that several relatives had fled here after the funeral. However, we failed to find any of them. Maybe they'd gone into hiding; in any case, they didn't want to be found. So we decided to head to Ekpoma and Ishan. Our route took us north, toward Benin City. On our way, we came to a fairly large town at a major road junction called Onitsha. It rang a bell.

Then it occurred to me what it was. In 1974, three cases of Lassa fever had been recorded in Onitsha. One was a nineteen-year-old Nigerian boy, and the other two were German missionary physicians. The first German had become infected after he cared for the boy. He bled profusely, had multiple seizures, then lapsed into coma and died. The second missionary also became infected. He was taken to Enugu teaching hospital, the same hospital where the surgeons from Aboh Mbaise and Aba had been treated. He fared better than they did, however, and went on to recover.

Mysteriously, investigators failed to discover anyone else in the area who'd come down with Lassa and then recovered. Was this some rare virulent strain that killed nearly all people who were infected, insuring that there would be virtually no survivors? The disease had many mysteries, and this was one of them.

We passed through Onitsha without stopping—we had no time—and went directly on to Benin City, where we had an appointment with a top health official. Shown into a large office, we were introduced to the official, who made sure that we understood that he wasn't just some ordinary bureaucrat. No, he was a prince. Then we went through the motions that had by now become familiar to us. First, he listened politely while we tried to explain just how serious a problem he had on his hands, and then he gave us every assurance of his government's cooperation.

Then nothing happened.

We ended up setting out for Ekpoma alone. Later we learned that the chief minister of the state had gone on television to let the local people know that the Lassa fever outbreak was due to juju.

This was something we were getting used to. Not surprisingly, witchcraft loomed large in Ekpoma. Wherever we went, we could sense the fear of witchcraft. The house where Azikiwe's parents lived in Ishan was locked, and we could not get in. However, Wale had been there earlier, and he had obtained good histories and specimens from surviving family members. So we already knew that they'd all had tested positive. What we needed to do now was find out how many other cases there had been in the vicinity of the house. We needed resources to trap and bleed rodents. Catching a rat is not a major challenge; most people can do it. What made it so tricky was that we needed to catch the rats alive, so that we could take blood and liver samples. And we had to be sure that whoever handled them knew how to avoid getting infected with Lassa virus—which meant that we would have to do it ourselves.

We then embarked on a survey of the local community. It was important to find out how much Lassa fever there was around. The city of Ishan was set out so that most of its inhabitants lived on the main street. To each house was attached a lot of about one third to a half an acre, where families could grow enough produce to meet their needs. As we went from house to house, we encountered the same suspicious responses. No one was willing to talk. Even a simple smile or expression of welcome was hard to come by. Eventually it became clear to us that nothing was going to happen unless we had the sanction of the chiefs. Our trouble was that it was difficult to figure out who exactly the chief was. In the past, the identity of the chief would be well established, and his word was law. But in Ekpoma, the situation was much more fluid; migration, modernization, and increased communication with the outside world had altered the attitudes of communities, eroding the power and prestige of the traditional tribal hierarchies. Now, when we asked, it appeared that there was more than one chief, and it was uncertain at any given time under whose jurisdiction a certain family fell. It was even less clear whose orders carried any weight. A chief might assure us that he was the one in control, but we never had any way of confirming the truth of this.

It didn't leave us with much of a choice. Without any official sanction, we could only continue to go from door to door, asking questions and trying to get blood samples. Usually people are very helpful, even when you just turn up unannounced on their doorstep. They will even answer amazingly personal questions. This experience was different. We could see fear in the looks that greeted us. Our questions met with grudging responses when they were answered at all. And there was no question of drawing blood samples. As soon as one person refused, everyone else around them—and there were always people around—followed their example. We did very badly.

The whole business was sinister. Our every move was tracked; we were watched from behind closed doors and through parts in flimsy curtains. We felt like interlopers, like sources of the contagion ourselves. Since Azikiwe and his family had most likely contracted Lassa at the father's funeral, we were anxious to find out as much as we could about how funeral rituals were conducted. After all, there was every reason to suppose that these practices could only spread the virus. But we couldn't find out anything; the rituals were conducted in such secrecy that no one dared talk about them. Wale, however, did turn up some tantalizing information.

When people died, the bodies were taken to a local morgue, where they were kept refrigerated until the families had enough time to gather together for the funeral. Once the clan has assembled, the body was then returned to them. While we had some evidence that blood contact must be a part of the funeral ritual—how else to account for the engineer's exposure to the disease?—we had no idea how this contact had occurred. Wale said that he'd heard a rumor that the hearts of the deceased were removed. Was this true? And if it was, what happened to the hearts? And what happened to the people who handled them?

In spite of an almost total lack of cooperation, it became increasingly clear that there had been a great many deaths. Invariably the symptoms of the disease that carried them off were the same: sore throat, fever, and bleeding. No other virus but Lassa (or Ebola) could produce symptoms like these.

We decided we might do better if we split into two teams. Wale and Nasidi went off in one direction, while I linked up with a microbiologist from the University of Benin. Meanwhile, Joe was still off searching for the index case, who was in hiding. I'd expected Wale and Nasidi to be away for a while, but they reappeared so quickly that I knew something had gone awry. One look at them and I could see that they were terrified by whatever experience they'd undergone.

"What is it?" I asked. "What's happened?"

They threw anxious glances behind them, as if they still weren't certain they were free and clear.

"Machetes," Nasidi managed to get out.

"What?"

"We were chased with machetes," Wale clarified. "They didn't even give us time to find out what we wanted."

It didn't matter; by this point everyone in Ekpoma knew what we were about.

Even after we'd succeeded in locating Azikiwe's family, they were no more communicative than anyone else in the area—with one exception: Azikiwe's sister, Valerie. According to her account, the suspected index case was a cousin, who was in her late teens or early twenties. She'd fallen ill in late December. During her illness, she'd come into close contact with her aunt, Azikiwe's mother, as well as other members of the family. In January and February, two other cousins—a six-year-old boy and a forty-three-year-old woman—took sick and died. What Valerie could not tell us was whether these two victims had had any contact with the rest of the family. It doesn't appear as if Azikiwe knew about their deaths when he returned home for his mother's funeral.

We decided to try to find the suspected index case. It soon became apparent that this wasn't going to be easy. We were told that while she'd recovered from her illness, she was now considered a pariah, stigmatized as a witch, because she'd brought so much trouble on the family. Some members of the family had even beaten her, causing her to flee.

And where was she now? Valerie heard that she'd taken refuge with a distant and more sympathetic family member. But where exactly she was hiding out no one seemed to know.

With the tenacity of detective-movie gumshoes, we managed to ferret out her whereabouts. Apparently, she was holed up in a nearby village at the home of her uncle. When we reached the village in question, we did succeed in locating the uncle, only to discover that it was the wrong uncle. The uncle we wanted was living in another village not far away. So we continued our odyssey.

In the next village, we resumed our search. You have to rely on the directions people give you; there are no addresses and few street names. This time, we found the right uncle but not the girl. He told us that she wasn't there. Understandably, she had no wish to be exposed. Her life could be in danger if she were—at least that was what her uncle said. I suspected that he, a wizened old man in his late sixties who spoke a little English, might not be telling us the truth. So we persisted. Wale patiently explained our purpose, assuring him that we only wanted to speak to her about her illness and possibly take a blood specimen.

Finally, he succeeded in convincing him. The uncle smiled warily and then led us into his small house. We found ourselves in a crowded sitting room. The uncle invited us to sit. A few moments later, his wife entered the room. She proved a much tougher customer. She said that there was no way that she was going to allow us to see the young woman. She seemed quite adamant.

Wale wasn't about to give up, not after we'd come this far. Once again, he was obliged to make our case, explaining why it was so important that we have a chance to speak to the girl. I could see by the expression on his face that we had won the uncle over. After a several minutes, the uncle and aunt retreated into a corner to deliberate. A compromise was reached. The uncle said that, while we would be permitted to talk to the girl, we could forget about taking a blood sample. Wale, being the diplomat he is, agreed to their conditions. Better, he thought, to accept incremental progress than none at all.

A few minutes went by while we waited. Then the girl was brought in. She was thin and pale. It was clear that she was terrified. She kept casting furtive glances around the room, all the while avoiding our gaze. When we were able to calm her down, we began to question her about her illness. The symptoms she described suggested that she'd had a milder case of Lassa fever, which undoubtedly explained why she had recovered. She admitted that she'd been in contact with several family members, though she didn't seem to remember all of them. After some hesitation, she went on to tell us about the ordeal that had followed her illness. It was with difficulty that she was able to recount how certain members of her family had hounded her and beaten her until she'd had to flee for her life.

Bruised and bleeding from her wounds, she made her way through the bush until she reached the safety of uncle's house. Not content with the punishment that had already been inflicted on the poor girl, someone in her family had gone to the juju man and had a curse placed on her. Now she was even more afraid. She was a virtual prisoner. She couldn't leave this house and despaired of ever escaping.

After talking to her for several minutes, we believed we'd succeeded in convincing her that we weren't acting as agents of her family or any juju man. Wale decided to up the ante and ask her if we could take a sample of her blood to test for antibodies to Lassa. Grudgingly, she agreed. But just as Wale was about to plunge the needle into her vein, she shrieked and fled from the room. We had to start all over again. It took us another hour to bring her back into the room and calm her down before Wale was able to take a small specimen. When the blood specimen was tested in Lagos, it proved to be highly positive for antibody to Lassa virus. The type of antibody we found is called IgM, which signifies a relatively recent infection. What the test couldn't tell us, though, was where she may have been infected. Rodents were ubiquitous, and people commonly caught and ate them, so there were thousands of opportunities for the virus to be transmitted.

It wasn't only machete-wielding villagers who were responsible for bringing our work to a crashing halt. We were told that the U.S. Agency for International Development (USAID) would provide us with the resources to conduct surveillance and control programs. Somehow, though, nothing happened. We learned later that the USAID representative, who had assured us of support, later mentioned in confidence that no money would be allocated to Lassa fever work. Apparently, Lassa did not enjoy a high priority when it came to the distribution of U.S. funds to developing countries. This was not a triumph for the U.S. State Department.

For the next two years, we sought funding, writing proposal after proposal to undertake the studies that we believed were essential if we were ever going to find out to what extent Lassa had spread in southern Nigeria. We wanted to carry out rodent surveys. We wanted to know in particular how Lassa fever had managed to get into downtown Aba, a city of one million, when Lassa was supposed to be an exclusively rural disease. We also needed to find out the role that rituals like funerals were playing in the spread of the disease. If Lassa was spread mainly as a result of high-risk practices—like exposure to blood in funerals or the reuse of needles in hospitals—then we had a very different situation from what it would be if the disease was being spread by a natural infectious process. Only by understanding the way in which Lassa is transmitted could we hope to curb it and protect the population. This is exactly the sort of thing that epidemiologists are trained to do. But while we'd acquired some success in combating viruses, we were not quite so lucky when it came to fighting superstition or the mindless attitudes of bureaucrats, who seemed to have better notions of what to do with "their" money than help save lives.

As to Wale, he remains our good friend. He is working at the WHO office in Zimbabwe. Each year he writes the same message: "Another Lassa outbreak. Lassa is alive and well in Nigeria, but nobody will take any notice."

These epidemics seem to be regarded as nothing out of the ordinary, no more cause for alarm than a thunderstorm. After all, Nigeria also experiences annual yellow fever outbreaks that carry off hundreds and, sometimes, thousands of victims. And yellow fever has been vaccine-preventable since the 1940s.

CHAPTER TWENTY-EIGHT

Ebola in Virginia?

It was November 30, 1989. Fred Murphy, now director of the Center for Infectious Diseases at CDC, came to my office in Building 15.

"Joe, I just got a phone call from General Russell at USAMRIID. Peter Jahrling has isolated a virus identical in appearance to Ebola virus from monkeys housed in a private holding facility in Reston, Virginia."

Jahrling is one of the principal researchers of hemorrhagic fevers in the United States. But Ebola in Virginia? In a beltway suburb of Washington, no less? Luckily, I was sitting down. This was going to take a bit of getting used to. I played for time.

"Why were they even looking?" I asked.

"Apparently they were working up an outbreak of simian hemorrhagic fever (SHF) and saw filovirus in tissue culture," Fred said, adding that no one was more surprised by the discovery than Peter himself. We all knew Peter very well; it was he who had given ribavirin to monkeys some years before and shown that it could successfully treat Lassa fever. More recently, he'd been involved in the investigations of outbreaks of SHF in monkey-holding facilities in the U.S. SHF is a severe disease in monkeys and is supposed to exist only in Africa and possibly in India. Once SHF begins to spread, it can wreak havoc. Aside from the fact that SHFV (Simian Hemorrhagic Fever Virus) is a large DNA virus— apparently with no relatives—little else is known about it. We at CDC were not involved in these investigations, nor should we have been. SHFV is a purely veterinary problem, and does not cause disease in humans. In fact, it apparently does not even normally infect humans.

Peter had been called to Reston to help with yet another SHF outbreak. On the face of it, the investigation seemed pretty routine. After collecting several specimens from monkeys that had been infected and died, he put them into tissue culture and then examined them, certain that he wouldn't find anything unusual. But his electron microscopist saw something that caused him to respond with alarm. He called Peter over to have a look at the photo he'd just taken of one of the specimens. The image contained a form that resembled a big snake.

Peter understood immediately what had alarmed the microscopist. The snakelike form looked like a type of virus known as a filovirus (from the Latin word for thread). More to the point, it looked like one of the deadliest viruses known to man. It looked like Ebola.

As if that weren't enough, these monkeys hadn't come from Africa. They had recently arrived from the Philippines. There wasn't supposed to be any SHF in Asia, let alone Ebola. We were invited to come to USAMRIID to discuss the situation and share whatever insights we might have. After all, the CDC had more experience dealing with Ebola in the field and in the lab than any other agency. Moreover, CDC was charged with the responsibility for investigating any public-health risks to humans. By this point, Karl Johnson was no longer at USAMRIID, where he'd served after leaving CDC in 1982. He had left in 1985 to work independently in California. It was too bad; his expertise in hemorrhagic disease would have been invaluable.

On the way to Washington, Fred and I mulled over several burning questions. How had the virus got from Africa into Asian monkeys, which had been shipped from Asia through Europe? Was someone smuggling African monkeys into Asia? Had they come into contact with a source of the virus from Africa en route? Monkeys are imported in the freight hold of passenger planes. They are boxed up in wooden crates and shipped on large pallets. The less reputable dealers often send them in pitiful conditions. But these monkeys had flown KLM—luxury class, compared with many other airlines.

On the other hand, we speculated, suppose this virus was not from Africa after all, but was in fact a new member of the filoviruses—from Asia? As far as anyone knew, there were only two members of this filovirus family: Ebola and Marburg. Ebola (which has two distinct strains, from Zaire and Sudan) had been found only in northern Zaire and southern Sudan, and Marburg seemed specific to the area around northern Lake Victoria in Uganda and the Mount Elgon cave area in western Kenya. One unexplained case of Marburg disease had turned up in a hospital in South Africa. The victim had been touring in Zimbabwe. But that was it. That was the sum of what we knew at the time about the origin of filoviruses. In spite of heroic efforts on the part of many researchers, no one yet knew where they came from.

Marburg presented an interesting precedent. When the first cases occurred in Marburg, West Germany, in 1967, it was the first time that a filovirus infection had been identified. Investigators determined that the virus had originated in monkeys recently imported from Uganda. Thirty-one laboratory technicians, veterinarians, and animal handlers and their close contacts were infected. Seven died. It was later learned that the monkeys responsible had been dying off at a much higher rate than was expected, even taking into account the stress of the journey. Over a period of about three months, fully one third of the animals were dead, usually dying on the order of two or three a day.

We had to keep our minds open. This could be the same kind of story.

Interestingly, we never considered the possibility that this new discovery might have been the result of laboratory contamination. That wouldn't have been an unreasonable explanation, but we had too much respect for the quality of Peter's work. If Peter said that he'd grown Ebola from a monkey, then Ebola is what it was.

Once Fred and I landed at Dulles, we rented a car and drove to Frederick, Maryland. As soon as we arrived at Ft. Detrick, we made our way to the head offices of USAMRIID. An urgent meeting to discuss the situation was about to begin. All the top brass of USAMRIID were assembled, including General Russell, the head of the division, Peter Jahrling, and C. J. Peters. We were introduced to Dan Dalgard, the veterinarian from Hazelton Laboratories, the company that owned the monkeys. I was pleased to see that a representative of the Virginia State Department of Health had also been invited to attend.

Peter Jahrling began by presenting his findings. He related how Dalgard had sent some specimens taken from sick monkeys at the Hazelton facility in Reston. Apparently, the veterinarian had already encountered cases of simian hemorrhagic fever in his monkeys and had assumed that it was more of the same. It certainly looked very similar. In fact, he was right. The monkeys *were* infected with SHFV. Peter's technician, Joan Rhoderick, proved once again why an observant lab tech is so vital. She had noticed that the cells in one of the flasks were being blown away; something was killing them. Peter sniffed the open flask for the telltale odor of bacterial contamination. Deriving no immediate satisfaction from this test, he gave the flask to an eager young intern named Tom Geisberg and asked him to prepare the material and practice his newly acquired skills with the electron microscope. The outcome of this exercise has been described by journalist Richard Preston in *The Hot Zone*. It sure looked like Ebola, and, suddenly, Peter Jahrling remembered he had sniffed a flask full of the stuff.

Was this some convoluted, in-your-face move by Ebola? After so many years in eclipse, to show up suddenly within a cab ride of one of the world's most sophisticated laboratories, not to mention the seat of government?

There was no doubt in my mind that, whatever this virus was, it was going to get some attention. A number of the military contingent sitting around the table were men of few words, many of which now were expletives.

The room was full of excitement and ego. Curiously enough, no one at the meeting, except for me, had ever actually seen a patient with Ebola. As I listened to them, I recalled kneeling on the floor of the hut in Nzara, taking blood samples from Ebola patients by the light of a kerosene lamp. That was the real thing. But here, in Washington?

Our brief was to develop a coherent plan of action that would give priority to any issue involving public health. This was no small matter. To achieve this objective I would need to work closely with the Virginia State Health Department and provide whatever assistance they needed. That is what the taxpayers fund CDC to do.

It was decided at the meeting that USAMRIID would continue to deal with the animal problem. It was up to them to establish to what extent the virus had spread among the animals in the facility and to devise a strategy to prevent the disease from spreading beyond its walls. While the military dealt with all things simian, I would work with state health authorities to limit the risk to the human population. That was how the responsibilities were divided up. I was later surprised to read in *The Hot Zone* a suggestion that CDC was preparing to take over the entire operation. Neither Fred nor I had any such idea, nor did we ever express such a view. It wasn't my style, for one thing—and, in any event, we didn't have the resources to pursue problems involving monkeys. The army was welcome to them.

Of course, all of this was going on in the shadow of the most media-conscious place in the world. Up until this point, we had been unable to obtain any significant funding out of Washington to support projects on viral hemorrhagic fevers. We had to sit by and watch as large sums of money were allocated to less definable afflictions like chronic fatigue syndrome.

"What we really need are some *Mastomys* rodents with Lassa virus running around the Capitol," we'd joke. "That might get some attention. Congress might finally figure out that these viruses are real."

Suddenly, what had been a joke seemed perilously close to becoming reality. We were acutely aware that once this news broke, we would have the media breathing down our necks. The knowledge that there were likely to be reporters with mini-cams and microphones converging on us only added to the tension in the room.

The possibility of media attention only meant one thing: it was essential to keep a low profile and a calm approach to avoid panic. I'd encountered the phenomenon with the Lassa case in Chicago, and some suspected cases of Ebola in New Jersey a few years earlier. The secret to being effective is to remain balanced and composed, figure out the possible risks, develop a plan to limit them, and then calm people down by explaining exactly what was going on. I had always dealt with the press by giving them the facts as I knew them, without trying to gloss over what I did not know. So even when I was anxious, it was essential to present a calm and collected demeanor.

The first thing that was done after the meeting broke up was to put out a press release, which was issued jointly by USMARIID and CDC. It stated simply that an Ebola-like virus had been identified in a group of monkeys in a research facility in Reston, Virginia. Our plan of action required us to identify human contacts at Hazelton, evaluate and contain the monkeys, and find out whether there were monkeys elsewhere that might be infected. To do this, we needed to survey other animal importers and, perhaps in that way, figure out where the Reston monkeys might have acquired the virus. This task fell under the jurisdiction of the Division of Quarantine at CDC.

For some time, the Division had been concerned about the circumstances and conditions related to the monkey trade, so they moved with alacrity and very effectively. Our plate became full from the beginning. The whole operation quickly evolved into a major collaborative effort.

The individuals whom the Virginia State Health Department assigned to the problem proved to be superb diplomats. Not only were they professionally competent, but they consistently remained rational, which wasn't easy under the circumstances. I had far less difficulty than I would have expected explaining to them how Ebola was and was not transmitted. It might have been tempting for them to place anyone who had come near infected tissues or monkeys into a high-risk category. In reality, this would have been unnecessary and could have caused needless anxiety for those individuals, but it would have been more expedient for the state health officials: the if-in-doubt-lock-it-up approach. The state officials, however, chose not to panic.

Unfortunately, some members of the press exploited the crisis atmosphere and used it as an opportunity to seize the spotlight. In so doing, the message we were trying to convey—that our experience in the field and in the lab had given us the knowledge we needed to contain these viral infections—was sometimes distorted. One notable exception was Brent Blackledge, a staff reporter for *The Journal*, the local newspaper of Fairfax County. Not only did he produce several good stories, he also served his community by giving them accurate information and playing down the slim potential for catastrophe.

I set about assessing human risk with my colleagues at CDC, including Sue, the only person who had substantial experience with Ebola infection in monkeys. In tandem with a CDC epidemiologist, Steve Ostroff, and the people in the Virginia State Health Department, we drew up a list of potentially exposed people as well as a questionnaire that would allow us to assess the level of their exposure. We had to identify anyone who had had any contact with the animals or their tissues.

Here things did not go quite so smoothly.

When I went to Hazelton Laboratories, I found that a meeting had been called for all employees who might have had any contact with animal tissues, except for those in the animal facility, with whom we dealt separately. I hadn't arranged for the meeting; it was the company's idea. Presiding was a veterinary pathologist from the military who was working with the monkeys at the facility. She not only had no idea who I was, she had little idea what Ebola was, either. Since the meeting seemed so well organized, I thought it best not to interfere. In fact, I said nothing until the end, when I was asked to do so.

That doesn't mean I wasn't tempted. I had to listen to a description of the disease that only vaguely resembled reality. But I decided that there was little to be gained by standing up and contradicting her in front of all these people. That would make it seem as if none of us knew what we were doing, and create the impression of a situation out of control, which would, not unreasonably, cause concern. When I was finally given a chance to speak, I told them that it was highly unlikely that they would be infected. There was practically no possibility that the virus could have somehow jumped across the facility from the animal room and infected a secretary in the coffee room. You certainly couldn't even contract Ebola by just being in the same room with a piece of infected monkey tissue. The only individuals at risk were those who'd had very close contact with sick animals. I assured the gathering that we were going to identify those individuals and watch them very carefully.

Naturally, people wanted to know what would happen if they became ill. I told them that, in the event that someone did become ill, they would be treated with the best medical care available in Fairfax County Hospital. I emphasized that the hospital staff was prepared to handle such a situation. After all, I said, we had established that, even among close contacts of sick and dying Ebola patients in Africa, the transmission rate was only about 10 percent. This risk increased among those who nursed very ill patients for longer periods, but it was still quite low. The major risk was from a cut or needle stick with an instrument that had infected blood on it. Those employees who thought they had been in contact had handled tissues wearing gloves, in good laboratory conditions, and there were no reported accidents involving cuts, scratches, or needle stick. On the other hand, I had seen the rows of dying patients in the Sudan, and had witnessed the terrible consequences of transmission in Zaire. So I didn't want to seem too laid back. The challenge was to use my knowledge and experience to balance the risks.

Possibly the most reasonable person I worked with at Hazelton was Dr. Dalgard, the veterinarian responsible for the monkeys. I could sympathize with his dilemma. He was being pulled in three different directions. He had to be concerned for his animals, but at the same time he had to think about the risks they might have posed to humans. Nor could he ignore the possible repercussions the incident might have on the financial health of the company that paid his salary. In any event, the employees remained calm. At least they didn't panic or riot or, worse, go home and call the newspapers or their senator. Instead, they worked with us to organize themselves into risk groups according to the level of their contact with any tissue or animals.

Throughout all of this, there was a good deal of discussion about whether Ebola could be spread through the air. This hoary monster rears its head whenever Ebola emerges. All the epidemiological evidence of human disease, including the three outbreaks of 1976 and 1979 and the first Marburg outbreak, however, argues against it. Although many cases have been treated in poorly ventilated huts, the only people who ever contracted the illness from another human were those who were in direct contact with the patients, specifically with their blood, urine, vomitus, or diarrhea. If Ebola were spread by aerosol, Roy Baron and I might well have been infected in Nzara. When an Ebola infection broke out more recently among a group of villagers in Gabon, it was the result of handling a chimpanzee that had apparently died of the virus and had been brought home by some youths. Again, it was the handling of the infected animal and the preparation of fresh meat for eating that were undoubtedly the causes of the outbreak. Indeed, it was comprehension of this and the quick and astute handling of the situation by Alain Georges and his team from Le Centre Internationale de Recherche Medicale de Franceville that contained the outbreak and prevented any further cases.

The problem comes from the difficulty of proving a negative: you can show that something does happen, but it is difficult to show that it does not happen—ever. We have clear examples of aerosol spread with diseases like smallpox, flu, and measles, but there are significant differences between these diseases and Ebola. First, they are primarily human pathogens; this means that the human is the principal source of transmission, and the respiratory tract is a principal target. These diseases are basically "wired" for aerosol spread in humans; it is their main strategy for getting around. Second, because diseases like measles have developed such an easy and rapid method of dissemination, their rates of attack are much higher than those observed with Ebola. In the airless huts, where most of the African cases were nursed, other airborne viruses, and even bacteria such as TB, are very easily transmitted; in those cases, many are infected. Not so with Ebola.

It is true that experiments have been conducted at USAMRIID to show that aerosol spread of several hemorrhagic fever viruses is possible. But doing so requires the use of a muzzle system attached to the face of guinea pigs and monkeys, which delivers a large dose under pressure. In this experiment, all the animals were infected. Obviously, this is not a natural system. The only lesson you can draw from this experiment is that if you push it hard enough, you can probably produce a pneumonia with anything you like, including the nonpathogenic bacteria that live on the skin, such as *Staph epidermidis*. It is significant that this experiment did not use any controls with such nonpathogenic organisms to demonstrate this point.

Given the evidence, therefore, I was convinced that aerosol spread, though not impossible, was going to be an extraordinarily rare event. It was certainly not my number-one concern.

Once we'd come up with a comprehensive list of all potential contacts and categorized them according to their level of exposure, as high risk, medium risk, and low risk, we then established appropriate levels of surveillance. We made sure to maintain daily contact with the individuals on our lists in order to ascertain their overall well-being and to check whether they were running a fever.

It was also my job to ensure that we had prepared suitable provisions and an appropriate place for any potential patients to be treated. We were operating under the same guidelines we'd used in Chicago. Patients would be admitted to a regular hospital capable of giving them the care they needed, provided that barrier-nursing techniques were followed. There were several reasons for this approach. First, a patient with hemorrhagic fever needs the highest-quality medical attention. This is very difficult to provide unless the patient is easily accessible to the hospital facilities, especially intensive care. Second, there are few Level-4 high-containment facilities available for patients. Valuable time will be lost in getting patients to them, and additional people will be placed at risk of contact while they're being transported. In any event, a patient with hemorrhagic fever endures travel very poorly because their cardiovascular system becomes unstable, putting them at risk for shock and bleeding. The trip can kill them. Third, a staff assigned to a containment facility may not be able to maintain the highest standards of tertiary care unless they practice it every day, something that is difficult unless the facility is located in a big city hospital.

As evidence to support my contention, I need only mention that, some years ago, patients with Lassa fever were unknowingly admitted into general wards of London hospitals; some of the patients weren't even diagnosed until after they'd recovered. But even without barrier nursing, not a single member of the medical staff became infected. Azikiwe had more Lassa virus in his blood than almost any patient we have ever measured, and yet no one with whom he came in contact in a tertiary care hospital near Chicago was infected.

I spent considerable time with Dr. Allan Morrison, the epidemiologist and infectious disease expert at Fairfax General Hospital, and with the staff of the intensive care unit. They listened carefully to my assessment of the risks they were facing and explanation of what precautions they should take. To prepare for any eventuality, Cuca Perez had come up with the mobile lab, which had facilities to test Level-4 pathogens. That way, we would have the capacity to examine the blood and serum from any case where Ebola infection was suspected and make a diagnosis without delay. Cuca was given the job of doing the tests, should any be necessary.

We didn't have long to wait.

On December 4, an animal handler at Reston went to work in the morning with a fever and began to vomit. In the highly charged atmosphere of the monkey holding facility, that was all it took to set off alarms. The handler was rushed to the hospital and admitted to the intensive care unit. The staff carried out routine admitting procedures, using the protective techniques we had established. I examined him and took a history. It was clear to me right away that if he had Ebola, which I doubted, it was not the classical form of the illness from Africa. One of the characteristic signs of Ebola is that it starts with the sudden onset of fever, muscle pains, joint pains, and headache. It lets you know at once that you are sick. This man didn't have anything like that.

But there was the remote possibility that he had an infection with another form of Ebola we knew nothing about. So we had to be prepared for a different profile of the illness. Within twenty-four hours, however, Cuca had confirmed my impression with a string of negative tests. No Ebola. By this time, the man no longer had fever. If I had to guess, I'd say that the animal handler had suffered from an anxiety attack. Given the circumstances at Reston, I can't say that I blamed him.

CHAPTER TWENTY-NINE

The Asian Bombshell

I became aware of unusual activity in Joe's office, next to mine in the little office complex of the Special Pathogens Branch. It was the evening of November 30, 1989. Fred Murphy was in with Joe. It was apparent from the sound of his voice that Fred was excited. I was naturally curious to know what was going on. I stuck my head round the door. When they saw me they asked me to come in.

"General Russell has just called Fred from USAMRIID," Joe said. "Peter says that he's found what looks like Marburg or Ebola virus in some sick monkeys in a facility at Reston, just outside Washington, DC."

I knew Peter Jahrling well; he'd spent years conducting research on Lassa and Ebola in the Level-4 lab in Fort Detrick. But Ebola in Washington! Come again?

"He must be seeing things," I said.

Then I stopped to think.

Well, it wasn't really impossible, given what little we knew about Marburg. If it happened once before, it could happen again. Fred, who is a big man, with a decisive voice, elaborated on what General Russell had told him over the phone. Apparently they had been culturing tissue from a dead monkey and found it full of a filovirus.

"Where are the monkeys from?" I asked. I was waiting for him to say Uganda. That is where the Marburg monkeys had come from.

"The Philippines," replied Fred.

The Philippines?

"But," I said, "the only cases of monkeys with this virus have all been from Africa. There shouldn't be SHFV in the Philippines."

Both Fred and Joe agreed. They had been struggling with that, too. But we all knew Peter well enough to recognize that we needed to take his word seriously. If he said he'd seen a filovirus in a Filipino monkey, then he probably had. On the other hand, he'd also told Fred that he had confirmed that the monkeys definitely had SHFV. We could only conclude that the dead monkey had had a mixed infection, caused by two viruses, neither of which should be seen outside Africa or India: SHFV and a filovirus.

And the monkeys were sitting in a Washington suburb.

Fred and Joe went to USAMRIID the next day to meet with the army and civilian personnel involved. When they returned to CDC, they sat down to develop the CDC's strategy. Joe decided that he would go back to Reston, accompanied by Stephen Ostroff, to set up disease surveillance and contingencies for safe care of potential human cases. Once Joe had determined on his next step, he turned to me and said, "Sue, We have no idea where these monkeys got infected. We need a plan to track their journey and search for any clues for contacts with Africa."

This was going to be complicated, I thought. We began to make phone calls to people we knew, who might have some idea about monkey traffic in various parts of the world. They included Jim Meegan at WHO and Mark White, who was running the CDC-sponsored Field Epidemiology Training Program in Manila, as well as researchers in Germany and the Soviet Union, who might be able to confirm rumors about dying monkeys that were coming out of their countries. Mark was particularly important to us because the infected monkeys had come from the Philippines. Joe persuaded him to do what he could to investigate the monkey facilities in Manila and see what he could learn about the animals and their caretakers. Mark had the benefit of having a few vets on his team of epidemiologists; this was just the sort of work they would love. It was also, as luck would have it, a difficult time in the Philippines, when rebels were fighting in many rural areas, making travel difficult for Mark.

I also held extensive conversations with the New York shipping manager at KLM, who was very helpful. It turned out that the monkeys from Manila had been shipped during October 1989. As we developed our sources, we discovered that there were four monkey dealers in Manila, but that the sick animals had come from only one of these. They had been shipped in the holds of regularly scheduled KLM flights and had spent a night in transit at an animal hostel at Amsterdam airport. We heard another story, which we couldn't confirm, that some monkeys that had arrived in the U.S. had been subjected to overheating during the flight. Were these the same animals? As it turned out, they had come in a different shipment on another airline. The overheating had nothing to do with the filovirus monkeys. Under ordinary circumstances, about 5 percent of imported monkeys may die during the flight or soon after arrival because of the stress of the journey. But we were now talking about mortality rates of 20 to 50 percent. There was another complication: we discovered that the animals destined for Reston had spent time at yet a second animal hostel, at Kennedy Airport in New York. Now we had New York involved as well as a suburb of Washington.

In the hallway, I happened to speak to a colleague who had considerable experience in the military. He'd heard what I was getting involved with.

"I have to warn you about something, Sue," he said.

Now what? I thought. I *said:* "What is it?"

"Mark my words: This outbreak will change the whole character of your branch. Nothing will be the same when it's over."

I couldn't imagine how this could be—why this outbreak would be so different from any other I'd investigated. I thought little more about it, and turned instead to the business at hand. I called Peter at USAMRIID in Fort Detrick, Maryland.

"Sue, it's crazy here," he said. "We're overwhelmed with dead monkeys. I've never seen anything like it."

He filled me in as best he could, providing me with some leads about the source of the monkeys. He said the vets were dealing with the monkeys themselves, and that he was grateful that he no longer had any responsibility for them. We agreed to keep in touch.

The African connection. I was sure that there had to be an African connection. But how to find it? Joe and I discussed various strategies. It didn't take us long to decide what I had to do.

"You'd better get yourself an invitation to go to Amsterdam," Joe said. "Someone needs to get there and see whether anything happened to those monkeys in transit."

This assignment meant long and difficult transatlantic phone calls with special political problems. Immediately, questions were raised about jurisdiction. The Dutch authorities were alarmed by the possibility that monkeys with Ebola virus might have passed through their airport. I got into contact with Jim Meegan at WHO in Geneva. Jim was an army virologist then working as WHO's hemorrhagic fever specialist, but he was also an academic virologist, specializing in the laboratory diagnosis of arboviruses, which are spread mainly by mosquitoes. Most arboviruses are Level 2 or 3 at the most. The fact was that he hadn't worked with the kind of lethal viruses that this monkey pathogen might represent. Nevertheless, he said he was going to Amsterdam to investigate and was taking a veterinarian with him. He left me with the impression that he considered this to be his exclusive turf, even if he had no experience with an Ebola infection. Things had changed from the days when we were the reference laboratory to help WHO. We were still that it name, but now politics were foremost.

There were more problems. When I got on the phone to health officials in Amsterdam, they told me that, of course, they appreciated the fact that we needed to know where the virus was coming from, but they intimated that CDC might not be welcome in the Netherlands. When we told Fred what had happened, he put me in touch with Professor Bruinsma, an eminent Dutch virologist and a professor at the Institute of Tropical Medicine in Rotterdam, who also happened to be an old colleague of his. As soon as I got Professor Bruinsma on the line, I could tell I had an ally—he was pleasant, positive, and knowledgeable.

"Don't take any notice of what these officials say," he told me. "Come on over to Amsterdam, I will meet you, and we will see what is going on. It's just the usual local politics."

He added that a meeting had been called by health officials in The Hague to discuss the outbreak. It was set for the next day. That would be a good time to come. For all of Professor Bruinsma's encouragement, I was still anxious about the situation. I felt I needed an official invitation. But when I told Joe and Fred that I had been specifically told not to come by the Dutch Ministry of Health, they expressed no doubt about what my next move should be.

"Go anyway," Fred said.

This was something we normally would do only under very unusual circumstances, for which, I supposed, this event qualified. We always had to receive an invitation from local authorities before we conducted any investigation outside of the CDC itself—whether we were dealing with an outbreak of hepatitis in Idaho or food poisoning in Nebraska. It was even more the case when an outbreak occurred on foreign soil. On the other hand, this was admittedly a unique situation. Besides, Fred was my boss, and he'd just given me a direct order. Technically, I did have an invitation, even though it was a private one from Professor Bruinsma. I reminded myself that the KLM representative in New York had assured me of the airline's full cooperation in Amsterdam's Schiphol Airport. As far as I could see, I had no choice but to go and take my chances.

I took a flight to Amsterdam that evening. The seat next to me was vacant, and the plane had just taken off. Great, I thought, I'll have more room to get comfortable. Who knows, maybe I'll even be able to get some sleep—and be fresh for what promised to be a very difficult morning ahead.

But soon after the seat-belt sign was turned off, a thickset, middle-aged man walked down the aisle and sat down next to me.

"I'm coming with you," he said. "I'm Chuck McCance, Division of Quarantine."

Quarantine worked out of a separate building at CDC. Our paths had never crossed. Now what?

In fact, Chuck turned out to be delightful company. He was also highly competent and very experienced. I was to benefit from his support and his advice; that he'd decided to accompany me to Amsterdam underscored just how seriously Quarantine took the situation. Chuck's division is unique at CDC because it enjoys statutory legal authority, and imported monkeys were very much a part of their business. Even so, I emphasized how difficult the Dutch officials had been. Chuck expressed surprise that I'd been given orders to proceed with the trip regardless. He'd assumed that I had had an invitation. It turned out that neither had he. There wasn't much we could do when we got to Amsterdam, except to wing it.

As promised, Professor Bruinsma was waiting for me when we landed. A short, dapper man, probably in his late fifties, he made us feel welcome at once, but he couldn't devote much time to formalities. The Hague was almost an hour's drive away, and we had to hurry. On the way, he told us that our predicament wasn't at all unusual. Europeans, he said, traditionally hate to be told what to do by ex-colonials.

"But the politics of Holland are less important than finding out if there is a real problem with Ebola passing through the airport," he said. He wanted to reassure us that our contribution was needed.

Once we arrived in The Hague, he led us into an old government building. An elevator took us up several floors. I couldn't tell whether what I was feeling was nerves or jet lag. Professor Bruinsma shepherded us into a room so small that the two tables placed in the middle of it left almost no space for anything else. There were several people already assembled around it. The only person I recognized was Jim Meegan from WHO. One thing became very clear, though: I was the only person in the room who had had any direct experience with Ebola. Now there was no question in my mind: I *was* nervous.

To my surprise, the atmosphere seemed rather friendly. I was even asked to say a few words about Ebola. When I finished, I had the impression that my contribution was welcomed. Chuck, observing intently, didn't say anything. After about an hour, the meeting began to wind down. By this time, it seemed to me, things had gone much better than I could have hoped. We'd even got permission to look into the monkey transport facilities. As the meeting broke up, however, one of the senior government officials walked around the table. He came up to me and said in a low voice, so that no one else could hear: "I told you we didn't need you to come."

"I'm sorry," I replied. "Normally, this is something I would never do. But I was ordered to come. I had no choice."

What else could I say?

The official nodded gravely.

"I didn't ask you to leave the meeting, because you are a lady," he said, "but tell your superiors never to do this again."

Chuck, who was sitting right behind me, couldn't help hearing this. Characteristically, he gave no sign of a reaction.

In spite of the sensitivities involved, not to mention the embarrassment, we had what we needed. The next day we paid a visit to the monkey-holding room in the animal hostel in the cargo area of Schiphol Airport. It was a large, airy, hangarlike building designed to accommodate all sort of animals, ranging from household pets to exotic beasts. There was even a special area for racehorses. Two small rooms held monkeys, birds, and other small exotic creatures. I was struck by how pleasant the facility was. It also seemed to be run with great efficiency. Actually, it compared favorably with the departure lounge in the main airport—and it was certainly less crowded.

We were told that birds and monkeys were sometimes quartered together. Virologically speaking, this made for a few interesting possibilities. Could it be a bird virus we were looking at? It wasn't such a farfetched notion. We occasionally had wondered whether Ebola might not be a plant virus, mainly because the only viruses that look remotely like Ebola are some plant viruses.

The manager of the facility assured us that the monkeys that ultimately ended up at Reston had only been here for a few hours. The records they gave us to examine indicated that the their stopover was no more than six hours. Was it possible that, even in that short span of time, the Reston monkeys had come into contact with any animals from Africa? The only possible African connection we could identify from the records were two primates: a baboon and a spot-nosed monkey from Ghana destined for a private zoo in Mexico City. Was there a link here? We had no way of knowing. The manager did mention to us that the African animals and Asian monkeys had shared the same water bottle. This disclosure hardly constituted convincing evidence that this was how one virus could have been transmitted to the monkeys, let alone two; nevertheless, it was not a good idea, and we told the manager that this practice ran the risk of spreading disease. Our view seemed to make an impression on him, because he hastened to assure us that he would take any necessary precautions in the future. He immediately gave instructions to his staff to obtain a fresh supply of bottles and some means of sterilization.

Late that night, I was awakened by the telephone. It was Joe calling from Atlanta. The news wasn't good.

"We have information that a shipment of monkeys bound for Texas is on its way to Amsterdam from Arusha in Tanzania," he said. "We've heard rumors that the facility in Arusha has been seeing a lot of monkey deaths."

He gave me the flight details. The animals were already in the air, on a KLM flight on its way to Schiphol. I looked at my watch. Two A.M. I set my alarm. At 6:30, I called the animal facility to warn them.

"We will check the animals very carefully when they arrive," they told me.

Whatever the truth behind the rumors, the monkeys were found to be healthy, and they were allowed to proceed on their way to Texas. But we weren't satisfied. We wanted additional assurance that the animals were safe. I called Joe and told him the animals were in the air, and we agreed to arrange a reception for them. Bobby Brown, chief of Animal Resources at our facility at CDC and a leading veterinarian in primate medicine, prepared to fly down to his home state to greet the monkeys.

Joe knew that it was of critical importance to discover if the monkey filovirus had come from Africa, or if it was from the same family but had come from Asia instead. My report that a few primates from Ghana destined for a private zoo in Mexico City had shared the same room as the first Reston shipment aroused Joe's curiosity. He managed to get the name of the intended recipient in Mexico City and gave him a call. Relying on his somewhat rusty Spanish, he ascertained that the animals had arrived in good condition at the zoo and were still healthy. But that wasn't enough for Joe; he needed a serum specimen to test for Ebola infection and contacted a friend at CDC named George Baer, a veterinarian who has spent a lifetime working on rabies. George knew everyone who counted in Mexico—or nearly everyone. It was through his auspices that Joe was able to obtain the serum specimens from the baboon and the spot-nose monkey. The primates were completely negative for antibodies to both the Reston and the African strains of the Ebola virus. As far as we were concerned, we had just ruled out the last possibility that an African virus could have been implicated. He was now convinced that what we were dealing with was a new Asian virus.

Although the animals in Mexico City checked out, we still didn't know for certain that the Texas shipment was safe, which was why Bobby Brown decided to return to his home state to take a close look at the monkeys from Arusha. His connections were working well; the veterinarian who was taking possession of the monkeys was an old friend. So he and Bobby collaborated. After carrying out careful physical examinations of all the animals, they could turn up no evidence of illness compatible with active African filovirus infection. They did discover that some of the animals had developed antibody, however, and even performed biopsies on these animals to ensure we had the best possible specimens to detect virus. Bobby even brought some of these antibody-positive monkeys back with him to Atlanta. We kept them for a long time at CDC to see if they would progress to a disease or show any evidence of persistent silent Ebola infection. Nothing happened. It was another false lead. But at least we had established that a healthy monkey, even with antibody to Ebola virus, poses no hazard to the health of other monkeys or to its handlers. This was critical information, because, after Reston, we were constantly getting calls from anxious vets with antibody-positive monkeys. Their animals were valuable and important to them, and they didn't want to have to sacrifice any of them unnecessarily. Everyone wanted to avoid a repeat of the Reston monkey massacre.

On my return to the States from Amsterdam, I went through New York. I was on my own. Chuck had returned to CDC directly. On my arrival, I was met by Steve Ostroff, who had come up from Reston to investigate the animal hostel at JFK. After we'd collected my bag, we went to the hostel together. It was run by the American Society for the Protection of Animals and managed by a large woman in her thirties, with long black hair. It was obvious that she had a big heart when it came to animals. In fact, the hostel was nothing less than an animal rights' haven. However, in terms of space, cleanliness and efficiency, it didn't begin to compare with its counterpart in Amsterdam. Several monkeys were crammed into small rooms. Some had to be kept in the hallways because there was no room for them elsewhere. Any monkey that didn't look well was taken from its cage and was hand-fed and nursed by the staff.

Steve and I freaked.

If there was any way to get Ebola infection from Africa into a U.S. citizen on American soil, this was about as good as it got. Our fears were further confirmed when we questioned the manager about any possible disease in the animals or the employees that might be construed as Ebola.

Not only did she know of such a case, she told us that, two years before, she herself had come down with an acute febrile illness that sounded suspiciously like a hemorrhagic fever. Though she'd recovered, we were anxious to test her blood. It turned out that she had a low level of antibody to Ebola, but we couldn't be sure what this meant exactly. A follow up specimen showed no change in antibody titer, which suggested that whatever she'd had, it wasn't associated with a recent Ebola infection. Maybe she had been infected, or maybe this was a false positive.

By the time I returned to Atlanta, I discovered that the Division of Quarantine had begun to take action to control the import of monkeys. Their main target were the "rogue dealers," who treated monkeys deplorably, transporting and holding them in cramped conditions, and selling them illegally as pets.

Eventually, in March 1990, the Division of Quarantine finally was able to secure a temporary ban on the importation of monkeys—from anywhere in the world, not just the Philippines. There was a tremendous uproar from outraged scientists. We had no idea how much researchers depended on these monkeys caught in the wild. To our amazement, we were told that more than 20,000 of them were imported into the United States each year, 16,000 of which are Cynomolgus—that is, the Asian crab-eating monkey, the same species that ended up in Reston. While the monkeys were mainly used for medical research, some were also used for product-safety evaluation. We were horrified at the scale of importation and the trivial use to which some of these animals were put. This was big business, and it also appeared to have its own "mafia." It was much bigger, and far more lucrative, than we had ever imagined. So we did something for the monkeys by limiting and better regulating trade.

But it took time for the ban to take effect, and monkeys were still being imported, some undoubtedly diseased. There was an urgent need to get information to veterinarians and researchers. Joe, Steve, and I had our hands full, putting together and publishing guidelines for handling imported monkeys with possible hemorrhagic fever viruses. We had some precedent to go by. After the Marburg epidemic in 1967, quarantine regulations were instituted that called for monkeys to be held for thirty days before release, but these regulations didn't go far enough. In quarantine, a few always die. It is possible that a spreading Ebola epidemic might be misdiagnosed or simply overlooked. As a result, some infected but asymptomatic monkeys might be released after the mandatory waiting period. We debated at length how to go about insuring that Ebola in a recently imported monkey wouldn't be missed.

There was worse to come. We learned of yet another outbreak of Ebola-like disease in recently imported monkeys. This time it had occurred among imported monkeys purchased by a different company in Pennsylvania. Like the monkeys at Reston, these animals, too, came from the Philippines, and this batch had crossed the Pacific. No African connection here. Now we knew the virus was new, that it was Asian, and that it was related to, but not the same as, Ebola. Yet it was becoming clearer and clearer that humans were not getting sick. What we didn't know was whether they could still become infected.

In January 1990, matters deteriorated further. Astonishingly, Hazelton Laboratories had managed to recover from its losses and had begun importing monkeys again—from the very same establishment in Manila that had provided the original animals. These monkeys started to die, too. They were also infected with the new filoviruses.

It was unbelievable.

Was it still lurking in their laboratories, or were they importing freshly infected monkeys each time? It seemed highly likely to be the latter.

As if this weren't enough, in early February we received news from Texas of a somewhat lethal disease, which was sweeping through an imported monkey population. I called the veterinarian of the Texas lab, Steve Pearson, who—no surprise—was another colleague of our own chief veterinarian, Dr. Bobby Brown at CDC.

In addition to being a skilled vet, Steve also had a cool head. The distress he felt over the loss of his valuable animals was communicated even over the phone. He really cared about them and was intent on saving as many of his monkeys as he could. We were now far enough along in our investigation to be reasonably certain that people were not getting disease. If it had been Ebola Zaire, we believe we would have had a number of infections and deaths in humans by this point. I suggested to Steve that what we had to do was establish how the virus was being transmitted. Otherwise, we could not hope to control it. He was quite willing to cooperate. He was taking all the risks himself, performing all the autopsies personally. We talked for a while. I described the specific risks he was facing and what precautions he needed to take to minimize them.

I realized that Steve could use a hand. To that end, we sent an epidemiologist named Peggy Tipple from the CDC to Texas to help the local state epidemiologist, Kate Hendricks. Peggy liked animals, was a horse owner, and was also obsessive when it came to gathering data. It was Kate, though, who produced most of the data in this case. She was young, somewhat excitable, full of unbridled enthusiasm, but a thorough researcher. Working closely with Steve Pearson, she was able to piece together the story of the dying monkeys in a way that gave us more insight into why these epidemics might be occurring.

The investigation was simplified to some degree because Kate and Steve were dealing with Asian Ebola virus in this instance and not a mixture of Ebola and simian hemorrhagic fever as in Reston. Most facilities, including the Texas one, administered a tuberculin skin test to monkeys as soon as they arrived. TB is a major problem in caged monkeys, particularly newly arrived animals. The test calls for a reagent to be injected in a tiny dose under the skin in the soft folds around the eye. If the animal is actively infected by TB, the injection produces a large bump. The reagent comes in vials containing several doses, while the narrow syringe itself can, in theory, accommodate eight doses; however, because of the "dead space" in the needle shaft, you can only draw up enough at one time for seven injections. A trivial detail, this turned out to be the key to solving the riddle of how this particular group of monkeys became infected.

Kate carefully examined the layout of the cages, paying special attention to the monkeys that were infected. Then she reviewed the procedures with Steve and his staff. Yes, they said, they always examined the monkeys in the same order. She asked them to show her how they went about their routine. Indeed, the handlers would proceed from one cage to the next in a specific order. The cages were arranged in two layers: a top row and a bottom row. Then Kate numbered the monkeys according to the order in which they were injected with the TB reagent. Once she had that information, she went back to the data compiled on the monkey deaths. She counted carefully. Every eighth monkey remained well, with no signs of illness. In fact, only monkeys, numbers two through seven, ever became sick. There was little doubt why. The lucky monkeys—numbers one and eight—were injected with clean needles.

This was the best evidence that any researcher had thus far collected to show that the Texan filovirus, like Marburg, Ebola, and even Lassa, could be spread by reused needles. Kate's findings provided the most plausible explanation for how the transmission of the virus had occurred. If the virus had been airborne, every first and eighth monkey would have been exposed the same way as the rest of their unfortunate companions. Kate later presented these data at a meeting of the American Society of Tropical Medicine and Hygiene in Boston.

CHAPTER THIRTY

The Infected Animal Handlers

Five animal handlers at Reston undertook most of the care of the recently imported monkeys. One of their jobs was to release the animals from their shipping crates. This was a pretty primitive process, involving crowbars and a lot of cheap wood being splintered in an ill-ventilated room. The poor, bedraggled, and terrified animals inside had to be grabbed—hopefully by someone wearing heavy gloves—and then consigned to a standard stainless-steel cage. It was a very messy business. The traveling crates were caked with the excrement of over a hundred scared animals confined for forty hours or more. The cages—they were like what we used at CDC—had pull bars that allowed the handler to control the monkey without putting himself in direct contact with the animals, at least until they were anesthetized. By pulling on two bars with both hands, it was possible to maneuver a false back in the cage. Unfortunately, once the monkey figured out how this mechanism worked—and it didn't take long—it would place its hands and feet against the front bars of the cage and brace its back. You pulled, it pushed. Often, the monkey would win. The only way to beat the monkey at its own game was to also use *your* feet. And, at CDC, we did this while wearing our space suits. There were times when I actually sat on the floor and assumed the same position as the monkey, with my oversized rubber boots propped up against the bars of the cage to get some leverage.

Because in the CDC Level-4 lab we always received everyone else's leftover monkeys, we became the hosts to some rather large and elderly animals. It wasn't unusual to have forty pounds of experienced simian determination on our hands. You have to treat these animals with courtesy, always avoiding eye contact, which most monkeys interpret as threatening. The animal is given a light anesthetic shot, so that it can be safely removed from its cage, examined, and necessary specimens taken. The routine we followed required us always to use a new needle for each monkey. It was also a rule that no procedure involving the monkeys could be carried out without two people in attendance; in fact, usually we worked in threes.

That was the way we did things at CDC. In many commercial holding facilities, however, handlers often preferred to take a macho approach and dispense with gloves. Many cages lacked pull bars, so the only way to get at the monkey was to open the cage door, and fight it out, primate against primate. In some circumstances, monkeys were put two to a cage, which made catching them an even more dangerous process. There were instances of animal handlers who were grabbed, scratched, or bitten by a monkey and who then contracted a disease known as Monkey B virus. This virus does not make the monkeys very sick; in fact, it is a herpes virus, which causes the monkey nothing more than blebs—cold sores. In human beings, however, it causes a rabieslike disease, which is usually fatal. It is so rare, that people had almost forgotten Monkey B could be a human disease, and most monkey handlers dismissed it as a thing of the past. In the mid-1980s, however, a monkey handler died from Monkey B in an establishment near Pensacola, Florida. Afterward, investigators went to his office and found a textbook on his desk, open to the page describing the symptoms of Monkey B virus infection in humans. He hadn't told anyone about his fears, even his wife.

One morning in January 1990, Steve Ostroff came to my office and told me that an animal handler at Reston had cut himself while dissecting the liver of a dead, infected monkey. Steve was surprisingly relaxed about it. Because Joe was attending a meeting on hemorrhagic fevers in Russia, I could not get in touch with him. I called Peter Jahrling and asked for his take on the situation.

Actually, it didn't sound good. Jahrling had already made an electron microscope prep and examined the liver. It was full of filovirus. Alarm bells rang. This was it. With this kind of accident, you did not escape Ebola.

I had no problem deciding that this incident needed to be taken seriously. I put in a call to Peg Tipple, who had only recently been dispatched to Reston to relieve Steve Ostroff and take over the surveillance of people who'd come in contact with the infected monkeys. She had already examined the animal handler. He was generally in good shape, she said, and not at all anxious. He was, however, middle-aged, overweight, and suffered from severe diabetes. Not a good combination.

I told her to check him every hour throughout each day but to do nothing to restrict his movements unnecessarily.

"Just don't leave his side until he is through the incubation period," I said. "That will be about a week from now."

After hanging up, I got in touch with Peter. We agreed that the best way to follow this man's progress was to obtain a daily blood sample.

The next day, Joe got back from Russia.

"Good job," he said after I told him what I'd done. "It was the right thing to do." Then he added: "This is it. If this is anything like the Ebola we know from Africa, the man has a good chance of becoming very ill and dying."

Peggy stayed as close to the man as she reasonably could over the next ten days, regularly examining him and taking his temperature. We watched and waited. Peter continued to process the specimens. Three days after the accident, I called him to find out if anything were different. Yes, something was.

"It's there," he said.

The latest blood sample was positive for filovirus antigen by the Elisa test. He'd grown the virus from a blood sample. There was no doubt about it now: the animal handler had been infected.

But nothing happened.

The man displayed no symptoms no fever, no sore throat, not so much as a minor headache. Even his diabetes remained under control.

We were also watching the other four animal handlers who had worked with the new monkey shipments. We were fortunate to have blood samples on hand, taken from the handlers when the first batch of monkeys became sick in November 1989. Three of them showed that they had seroconverted to the new virus, meaning that when we had first tested them, they had no antibodies to Ebola, but now their sera reacted strongly. What it came down to was that four out of five men, all of whom had had close contact with the monkeys, became infected with Reston virus.

Not one of them was sick. This was a new Ebola all right, but it was an Ebola that had let us off the hook. This time. Still, there was trouble ahead. Joe experienced it all firsthand. Let him tell it:

When the animal handler cut himself, but failed to become ill, it was clear to me that the virus had a very low pathogenicity for man. I thought that that was very good news. But, in fact, my assessment of the situation didn't please a number of people, including some of my own colleagues at CDC.

What I hadn't taken into account was how difficult it was for people to back down from their previous positions. My conclusion that this was not a human pathogen seemed difficult for some to accept, for whatever reason. I participated in some pretty unpleasant discussions at CDC about this issue.

In an incident that has since become well-publicized, C. J. Peters of USAMRIID phoned me and, in the most forcible terms, questioned my decision to put the first infected animal handler into the hospital instead of the "slammer"—as the military isolation facility was known. It wouldn't be right to claim that my decision didn't cause me some uneasiness, but experience and the published data had convinced me that isolation of patients with a hemorrhagic fever in such facilities is simply not warranted. In my view, it is based more on fear than fact. We'd already seen the treatment that poor Jenny Sanders had been subjected to by the British authorities. I had no compelling reason to put one of my fellow citizens through a similar ordeal. Indeed, nearly fifteen years of personal experience and the experience of others showed that solid basic barrier-nursing did not place hospital staff at undue risk, and provided the best care to the patient. The patient needs to be in a good hospital with an experienced tertiary care team. This was the only way to assure optimal care. Moreover, as I pointed out, the existing CDC guidelines for the management of patients with viral hemorrhagic fevers were quite clear on the issue. And since they were supported by extensive experience and scrupulously reviewed published data, I saw no reason to depart from them.

Nevertheless, in the highly volatile atmosphere prevailing at Reston, the pressure to put the infected handler in Level-4 patient isolation was enormous. Even if the patient had actually become sick, my position would have remained the same.

While it was true that we could take only so much comfort from the fact that the Reston virus was not pathogenic for humans—that it did not make people sick—there was always the threat of yet another strain turning up in monkeys imported from somewhere else, which might prove harmful to humans. I wanted to find out what I could about animal infections. It was obvious that we needed methodical data collection and epidemiological analysis, but the team responsible for the monkeys was primarily composed of veterinary pathologists, with no expertise in epidemiologic methodology. I suggested that Steve Ostroff might offer to assist the vets and bring his epidemiological expertise to bear on their outbreak. I further suggested his expert assistance to the military, which USAMRIID declined. One can only hope that data will eventually be published showing how the virus spread in the Reston facility. Without compelling evidence to the contrary, we have presently to assume that it was spread by contamination of needles and other instruments used on the monkeys, just as it was in Texas.

Our ignorance about the Reston strain of Ebola virus extended to the real cause of death among the Reston monkeys as well. There is some evidence that their deaths could have come from the coinfecting simian hemorrhagic fever virus. Alternatively, it's possible that the combination of SHFV and Ebola produced a more lethal illness in the monkeys than what either virus could have accomplished alone. Subsequent experiments in our laboratory by Sue and her team certainly suggest that the Ebola virus from Asia, even when given in large injected doses, was far less lethal than its African relatives. A more measured reaction by investigators might have led to a better understanding of the nature of the disease process. But Ebola virus, whatever its origin or strain, has never been famous for provoking measured reactions.

I said to Sue: "If this was anything remotely like the Zaire Ebola virus, we would have known by now. Asian filovirus is not a hazard for humans."

Far from reassuring our superiors, my assessment seemed to provoke disappointment, even indignation. I'll let Sue have the final word on Reston:

In March 1990—long after the animal handler had failed to become ill after infecting himself—Joe and Fred Murphy had a major falling out regarding several issues of the Reston investigation. Though scheduled to leave in late April, Joe left Special Pathogens for the Division of HIV/AIDS later that same month. For several years since he had started working with AIDS, he had been torn between hemorrhagic fevers, the study of which he loved and to which he'd dedicated so much of his life, and the need to contribute to the fight against AIDS, which claimed so many more lives, especially in Africa. In July 1989, Jonathan Mann, who was then head of the Global Program for AIDS, and Joe worked out an agreement for Joe to go to Geneva. They planned to establish a program to test HIV vaccines and drug therapies for AIDS and opportunistic infections. The sites they selected to conduct their studies were mainly located in developing countries, where Joe's experience would be especially valuable. WHO and the CDC Division of HIV/AIDS had initiated the complicated paperwork in the summer of 1989, long before the first Reston monkeys ever left the Philippines.

So suggestions in the popular press that Joe quit CDC because of any controversy over the Reston outbreak were wholly unfounded. But Joe need not have left Special Pathogens so precipitously or so acrimoniously. That the CDC would allow his expertise in the field of hemorrhagic fevers to be lost is a serious indictment of the agency's management. With the departure of Karl Johnson several years earlier, the loss was even more profound.

For me it was a personal loss as well. Through all the years that I had worked with hemorrhagic fevers, Joe had been my mentor and my guide. With him gone, I felt very much left on my own.

CHAPTER THIRTY-ONE

The Monkey Expedition

It was late March 1990. All of the paperwork was finished, Joe had transferred to the Division of HIV/AIDS, under whose auspices he was due to move to Geneva in May to join Jonathan Mann. I continued to keep in touch with him, seeking his advice. There was no one else around CDC who had any real understanding of the field of hemorrhagic fevers. Although we had a competent technical staff, the director of the Division of Viral Diseases was new and was working under a handicap: he had no medical or epidemiologic background. His solution was to put in a caretaker as branch chief, who knew little about hemorrhagic fevers and had never set foot in Level 4.

There was a lot of work to do, so I got on with it. The Reston outbreak had produced an avalanche of specimens, mostly monkey sera. It seemed that anyone who owned a monkey or was conducting research on one was worried about Ebola. We had no idea that there were so many monkeys around. We ended up performing serosurveys on monkeys in holding establishments all over the United States. A serosurvey is usually conducted to assess the prevalence of a particular infection, which indicates how much of an infection there is in a population, in this case, a population of monkeys. We were using this method to try to determine how many monkeys had antibody to Ebola viruses, a marker of how many animals had ever been infected. From this, we'd be able to determine whether the viruses were common.

We faced a major technical hurdle, however. We used an antibody test, which had been developed for Ebola at the time of the 1976 outbreak. It worked well in an outbreak situation, when the test was performed on individuals with recent infections, because antibody levels are usually high in the immediate aftermath of an illness. Unfortunately, the same test produces ambiguous results when specimens are taken from a large population (of people or monkeys) in which no recent infections had occurred and where there was no clear history of Ebola disease. Karl Johnson had tested San Blas Indians from Central America when he was trying to evaluate the original test, and he'd found about 2 percent positive for Ebola antibodies. Other researchers later tested sera from Native Americans in Alaska and found a similar percentage positive. But the meaning of these results was so uncertain that no one really knew what to make of them.

This lack of specificity is a peculiarity of Ebola virus. (Marburg virus doesn't appear to present the same problem to researchers.) The closest that anyone ever came to a solution was Tom Kzaisek at USAMRIID. He developed a fairly complex test based on the Elisa system, which was an improvement over the IFA. In April 1990, I suggested—unsuccessfully—to the caretaker branch chief, who had taken responsibility for the serological testing at that point, that we introduce Tom's test to CDC.

We experimented with other test systems, like Western blot, which is used to confirm AIDS, but we had no better luck. As a result, we found ourselves suddenly inundated with monkey sera, without having a well-defined and evaluated system to test them with. Moreover, the process of carrying out these tests is incredibly repetitive and tedious. A great deal of expertise and patience is required to produce reliable results even within the limitations of the test itself. Cuca Perez, of our monkey team, put in countless hours, performing many of these time-consuming tests. Of the several hundred monkey specimens we'd received, about 10 percent showed some reaction to Ebola virus antigen, though it was usually low-level.

In an effort to understand the significance of these low-level reactions, we decided to look more closely at the origins of the monkeys from which these sera had come. Most of them were the crab-eating monkeys (Cynomolgus), the same variety as the Reston monkeys, most of which came from the Philippines or Indonesia. However, crab-eating monkeys are found all over Asia. While they are common in the wild, they often turn up at tourist sites, where they boldly solicit visitors for food. Charming at times, they can also be evil-tempered and maliciously destructive. In many parts of the continent, they are considered pests, not the least because they breed rapidly, travel in large troupes, and have a tendency to pillage crops.

The mysterious new filovirus (not just the antibody to it) appeared only in monkeys that came from a single holding and shipping facility in Manila. So why, then, were so many apparently healthy monkeys from other areas exhibiting antibody to the virus? That was the puzzle.

Carefully examining the data, I looked to see where most of the infected monkeys were coming from. The result was a surprise. It turned out that the highest percentage of antibody positives weren't in monkeys from the Philippines at all. They were from Indonesia.

I got into contact with a senior virologist in Indonesia whom I knew from working with dengue hemorrhagic fever in Thailand. I also discussed the Indonesian connection with officials at the U.S. Embassy in Djakarta. In addition, I had the benefit of the collaboration of a U.S. Navy virology unit, based in the Indonesian capital. Because the export of wild monkeys was a major source of hard currency for Indonesia, the government was understandably concerned. We were formally invited to come to Indonesia to see if we could pinpoint a source of filovirus.

In May 1990, I left for Indonesia with Steve Ostroff. By this point, Steve, just in his thirties, had acquired a distinguished track record at CDC. He'd done much of his work with the diarrheal diseases branch in the Division of Bacterial Diseases. He had reservations about viruses, though. He said that he didn't trust them because he couldn't see them. Steve was good company, and very smart. He quickly picked up enough Indonesian to be able to identify useful epidemiological markers; for instance, he was able to distinguish monkey meat restaurants by the simple expedient of translating the sign over the door. In Asia, just as in Africa, monkey meat is considered a delicacy.

We were received in Djakarta by Jerry Jennings, a virologist at the U.S. Navy unit there. A tall, delightful man, Jerry was working in Java on the epidemiology of dengue virus infections. The navy had arranged accommodations for us at a beautiful old hotel—though we had little time to enjoy our rooms. Djakarta has traffic problems that make good old American gridlock seem insignificant. We had to leave the hotel by six in the morning and be sure to be back before four in the afternoon, if we expected to be able to move in traffic at all. Laboratory hours were set accordingly.

In tandem with Jerry, we began our investigation at the four monkey holding facilities in Djakarta. Most of the monkeys, we were told, came from the island of Sumatra, which is to the northwest of the main island of Java, where Djakarta is located. To all appearances, the monkeys we inspected looked healthy and well cared for. But we did make one disconcerting discovery. We were told that when the handlers found a sick monkey, it was placed in a gang cage with other sick monkeys. Of course, all these monkeys might have very different diseases. Because these gang cages could hold anywhere from twenty to thirty monkeys, we didn't think that this was a very good idea. All it would take was one infected monkey to pass virus on to all the others, especially since they would already be weakened from whatever was making them sick in the first place. (Was this how the Reston monkeys simultaneously acquired SHFV and Ebola? We wondered.) But monkeys were too valuable to sacrifice; so, if any survived the gang cage, they were placed in the shipments along with the healthy monkeys. This gave the survivors, who might still be contagious, a chance to infect an entire shipment.

What we still didn't know was whether there were any monkeys with an Ebola-like disease in Indonesia. We assumed that such an illness might resemble the one we'd seen in Reston, but since the animals in Reston were also infected with SHFV, we couldn't be absolutely sure what we might be looking for. All we knew now was that there were monkeys who showed positive antibodies to Ebola virus, indicating they might have been exposed to a filovirus. But in Djakarta, we had found only healthy monkeys. If we were to find the answer, we would have to travel to the monkeys' natural habitat. That meant a trip to Sumatra.

After a week checking holding facilities and testing sera in the navy lab in Djakarta, we were ready to launch our expedition. We boarded the ferry for Sumatra late that night, and passed the remnants of Krakatoa in the dark. From the port, we drove to a city called Lampung, where we checked into a hotel that offered a superb view of the bay. It was, in fact, unbelievably beautiful—but, again, there was no time to sit and enjoy. We quickly had breakfast and went in search of the local health authorities so we could find out about Lampung's monkey holding facilities and, ultimately, find the places from which the monkeys came. We had two critical questions: Were monkeys in the forest dying of Ebola-like disease, and were the people who were catching them getting infected?

The Lampung health officials lent us a guide to show us the way into monkey territory. After a brief rest, we set out for the jungle along a narrow road choked with terrifying traffic. The passing landscape was mainly jungle and dominated by palm trees. All along the way, we would stop and ask people whether they had heard anything of particular interest about monkeys. People in Djakarta had told us about the monkey catchers. Apparently, they all belonged to a single tribe from Java. We were unlikely to learn much from them, our informants said, because the tribe was very secretive. They were reputed to have special magic powers, which allowed them to communicate with the monkeys. In the evening, it was alleged, they would visit the trees in which the monkeys slept and talk to them. Talking in magic language, which only the monkeys understood, they would tell them to abandon their refuge in the treetops. Then the catchers would spread a net below the tree and leave. In the morning, the monkeys, presumably acting under the power of suggestion, would clamber down from their perch, only to find themselves trapped in a net. It sounded interesting, but we thought it more likely that the catchers just tempted the monkeys down with food.

The journey into the jungle was the longest and most grueling I can recall. Without having got any sleep the night before, we had to travel for another twenty-four hours over terrible roads before we succeeded in reaching our first destination, a monkey camp. The catchers slept in a shelter, built out of logs, bamboo, and banana leaves, and raised off the ground on stilts. They were happy to show us their latest catches, which huddled in bamboo cages. Among the captured monkeys was a mother and her baby. The tenderness the mother displayed toward the small creature nestled in her arms touched us. We later learned that the baby died on the road to Lampung. Only the strongest monkeys could survive the journey to Djakarta and on to the cities across the Pacific.

This whole business upset me. I hated seeing monkeys in such distress. When I caught the eyes of one of these animals, I felt I knew what the slave trade must have been like.

While Steve interviewed the monkey catchers with the help of an interpreter, Jerry and I took blood samples from them. They told us, though, that if we really were interested in finding the main monkey-catching sites, we would have to travel much farther north. This involved yet another long journey over dirt roads that took us through vast sugar-cane plantations.

After about four hours, we arrived at a camp close to the major monkey-catching site on the island. As we approached the camp, the monkey catchers suddenly appeared beside us on the road. There weren't many of them. It was dark, and we could only make out their faces sharply illuminated as they came into the path of our headlights. It was a surrealistic vision, a scene painted by Caravaggio. Curious and surprised, they ultimately proved very cooperative, despite what we had been told about their secretive nature.

We went through the same procedure as we had with the first group of catchers, working as best we could, since the only illumination we had was from the headlights of our truck. Again we asked about monkeys. Did they know of monkeys that had got sick or died? Had they discovered any dead monkeys? Did they know of any people who had died of a high fever with bleeding? Time and again, the answer was the same: no. When we were finished with our questions, they vanished into the darkness as suddenly as they had appeared. I wondered whether I'd been dreaming.

In the time available to us there was little more we could learn. The only thing to do now was to drive back to our hotel. We got back about four in the morning and caught a few hours' sleep. Awaking, I adjusted to the daylight and saw a favorite batik dress that I'd worn the day before. I had bought it long ago in Thailand and had worn it for years in my travels throughout Africa. Twenty-four hours in the jungles of Sumatra had battered it beyond restoration. I threw it in the trash.

The following day, we returned to Djakarta and began to test the sera taken from the magic monkey hunters. All of them turned out negative. There wasn't even a single questionable low-level positive. Given the urgency of the situation, our very limited resources, and our small sample, we'd accomplished what we could: we found no significant threat of hemorrhagic fever to humans in Indonesia that might have come from the trade in wild monkeys. If wild monkeys were infected with anything like an Ebola virus, it was not being seen in their captors and transporters, who were the humans who'd come most directly into contact with them. Most important of all, we had found no filovirus-like disease in monkeys.

Our findings made the Indonesian government quite happy. It was our conclusion that the presence of a low-level antibody reaction to Ebola virus in a monkey did not represent any risk to humans or to other monkeys in Indonesia.

Back in Atlanta, I still had two serious scientific questions to answer. The first was to discover whether the new Asian filovirus had any real pathogenic potential; the second was to establish once and for all whether monkeys with the antibody to Ebola were safe to work with. Could monkeys get rid of the virus once they'd recovered? Was the virus capable of establishing a persistent infection? Obviously, these were major concerns to veterinarians, animal handlers, and researchers of the new filoviruses in monkeys. The second question arose because so many veterinarians were asking us for our guidance when they discovered that their monkeys had antibody to Ebola. The usual reaction, based completely on fear, was to destroy the monkeys. Again and again, I found myself talking to a vet on the line who said, "Do I really have to kill my monkeys? They're great animals, and we're in the middle of a very important and very expensive medical experiment." We always advised them to leave the animals alone, continue with the experiment, and not worry unless they became sick. The presence of antibody was unlikely to represent any current infection, nor any threat of infection, if the monkey was otherwise healthy.

But we needed to come up with definitive, publishable data. I got together with our monkey team and Bobby Brown, a large-framed man who sported a string tie and tooled leather boots. He looked like a casting agent's idea of the quintessential Texan. We had a lot of trouble finding a space suit large enough for him. I devised an experiment using crab-eating monkeys from Asia and some green monkeys (*Cercopithecus aethiops*) from Africa. Like the crab-eating monkeys, African green monkeys are plentiful and considered pests, but they are much nicer animals to know. We tested thirty-two monkeys—sixteen African and sixteen Asian—with two types of filoviruses, African and Asian. The Asian virus we were going to use came from Reston; it had been purified by Peter Jahrling, so that we were all now confident it was free of contamination with SHFV. (We did our own tests to confirm that there was no evidence of it.) The results of our experiments were just as we had predicted. The African viruses were lethal for nearly all the monkeys. On the other hand, the Asian viruses, while they certainly made the monkeys sick, killed far fewer of them. In addition, the Asian disease was slower and much milder. Most of the monkeys infected with the Asian strain eventually recovered after about a month. Furthermore, the African animals were much more resistant to the Asian virus than the Asian monkeys. Nearly all of our friendly green monkeys survived.

It now seemed reasonable to consider that SFHV was probably more responsible for deaths in the Reston monkey than the filovirus alone. It was also very possible that the combination of two infections had caused more severe disease and more deaths. We could now state with certainty that the Asian filovirus was much milder in monkeys than its African counterpart, and that the viruses did not persist in animals surviving acute infection. We were also able to reassure the public that the Asian virus did not cause disease in humans. We held on to the surviving monkeys for nearly two years, conducting exhaustive tests to see if we could find any trace of virus in any of these animals. We found none. Although they continued to have high levels of the antibody, they posed no risk to other animals or their human handlers.

There was one final issue that had to be resolved. Would it be possible, we wondered, to make a vaccine for Ebola?

If we were to produce a safe Ebola vaccine, we needed to know whether an infection with one live filovirus could protect against an infection with a second. If we couldn't protect an individual using this method, then all the work that we'd have to invest in producing a genetically engineered vaccine would be useless. Ebola virus did not appear to produce neutralizing antibodies in survivors—antibodies able to prevent the virus from infecting new cells.

It occurred to me that the Reston filovirus might be useful as an experimental live vaccine for Ebola in monkeys. In theory, it looked like a good possibility. At the very least, it would demonstrate that it was possible to give protection, but in practice, based on our limited knowledge, it was simply too risky to try to create a vaccine in this manner. As an experiment, I went ahead and tested two monkeys that had survived the Reston filovirus to see whether their exposure would also protect them against the deadly Zaire Ebola strain. I gave both of them very large doses, larger than might be expected in a natural infection.

The results were mixed. One monkey was completely protected and never even had a fever, while the second died of Ebola. In retrospect, the experiment did achieve something unprecedented: This was the first time that even a single monkey had been protected against a lethal dose of injected Zaire Ebola virus. After that, though, I gave up working with monkeys. I had got to like them too much.

As we completed this work, another virus was waiting for me. It was one we had pursued before and found it to be, if anything, even more elusive than Ebola. Now it was to reveal itself more fully. We were about to get a closer acquaintance with Crimean Congo Hemorrhagic Fever.

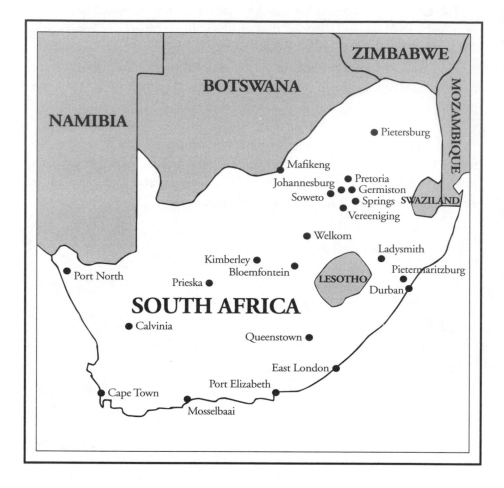

Desert Fevers

When several cases of Crimean Congo hemorrhagic fever (CCHF) occurred in Saudi Arabia, in the holy city of Mecca and in nearby Jidda, the Saudi government was alarmed. No one in the country knew much about the disease or its treatment. Bob Fontaine, a CDC epidemiologist who directed the Saudi Field Epidemiology Training Program (FETP), proposed to the Saudis that they invite me to lend advice.

It was just before the hajj, the annual pilgrimage that Muslims from all over the world make to Mecca. As part of the tradition of the hajj, the pilgrims are supposed to sacrifice an animal and distribute the meat to the poor. With two million pilgrims visiting Mecca each year, that is a lot of slaughtered animals. Bob had already traced the source of the virus to the slaughter houses; each instance of illness involved someone who had handled fresh sheep meat. Most of the victims so far were slaughterhouse workers. The Saudis themselves would never do this kind of work, so it was left to migrant laborers from poor neighboring countries. These rough-and-ready butchers had a regrettable habit of placing their bloody knife between their teeth while they worked with their hands. We were faced with the alarming prospect of a major outbreak associated with the hajj. Bob and his team instituted control measures.

Invited to give a talk about CCHF, I arrived in Mecca uncertain of the kind of reception I would receive. I needn't have worried. My audience was remarkably attentive. They knew about the cases in Mecca and Jidda and were anxious to find out as much as possible about this devastating virus. I began by telling them about an outbreak of CCHF at a hospital in Dubai, an oil-rich state on the Persian Gulf (part of the United Arab Emirates) and a neighbor of Saudi

Arabia. A man came into the emergency room bleeding and in shock. A surgical team struggled to save him, at one point even resorting to mouth to mouth resuscitation. In spite of their heroic efforts, the patient died. Within the next few days, the emergency team started to fall sick with shock and bleeding. And then they, too, died.

I looked out at the sea of faces in front of me. I went on to recount a similar episode, only, this time, one that had occurred in Pakistan.

"In 1976, in Rawalpindi," I began, "a shepherd was brought into the hospital vomiting blood. The surgeon, who had no idea what was wrong with him, operated but was unable to save him. A few days later, the surgeon himself died. . . ."

I detected a movement from the back of the hall. A man stood up and said, "Yes, I knew him. It was terrible. We were all very frightened."

There was a hush. I waited to see if he wanted to say anything more. He did not.

I continued with my story—The virus was first described by the Russians after World War II. What happened was this: Russian soldiers were dispatched to the Crimea to help with the harvest in the war-devastated countryside. Within a short time, many of them were being struck down by mysterious infections characterized by bleeding and shock. The mortality rate was alarmingly high. The virus spread across the Black Sea to Bulgaria. In fact, the disease became such a serious problem in Bulgaria that they manufactured their own vaccine, out of mouse brain infected with live virus, which was then inactivated with formalin. I was shown this vaccine in a hotel room in Thessaloniki, which lies just south of the Bulgarian border, in Greece. I was among a small circle of international experts in hemorrhagic fevers who were gathered around a senior Bulgarian scientist. He was holding up an ampoule.

"This is the vaccine!" he proudly declared, as if the sight of the ampoule was to convince us of its efficacy.

He claimed that all their soldiers who were sent to the frontier received this vaccine and that none of them had ever come down with the disease, but gave us no further details. His was a strict communist regime. We paid attention but were generally skeptical.

The disease was then found again, this time halfway across the world in Xianjiang, a remote region in China, north of the Himalayas. For this reason, CCHF is known to the Chinese as Xianjiang fever. This region contains both the Yaklamakand and Gobi deserts and boasts one of the most inhospitable climates in the world. Hidden in these desolate wastes are the ruins of ancient trading cities along the Silk Road. Modern Chinese have also used the deserts to conceal their top-secret nuclear facilities. Tom Monath, the yellow fever expert who was also from CDC, related an experience he'd had in China that recalled my encounter in Thessaloniki. When he inquired about the disease, an official showed him an ampoule of similar formalin-inactivated vaccine and expounded on its effectiveness—all without ever explaining how it worked or what it was made of. Despite reservations, we must remember that CCHF is an acute disease that might very well be prevented by a simple vaccine of this kind. Maybe both they and the Bulgarians were on to something. In fact, China and the former Soviet Union have always had a keen interest in hemorrhagic fevers. We know that the Soviet military had established a major experimental program to investigate these diseases. But their interest in the viruses was not necessarily altruistic. More ominously, they may have been motivated by the prospect of developing CCHF and other diseases for use in biological warfare.

In 1956, the virus was finally isolated from a dying boy, but in the African Congo, which was how the virus came to acquire the Congo part of its name. The virus was shown to belong to the family known as bunyaviruses; carried by ticks, these viruses can be transmitted by many types of animals, especially sheep and goats. CCHF is spread mostly by hard ticks, which flourish in hot dry areas. Humans may get infected by tick bites, by contact with fresh blood from infected animals, or by contact with blood or tissues from infected humans.

Once I had related the history of CCHF to the Saudi audience, I went on to tell them about the CCHF investigations we'd earlier conducted in Africa. In South Africa in 1986, Joe and I had looked into a series of cases of suspected hemorrhagic fever that had broken out in a semidesert area north of the Kalahari. Many of these cases, we learned, were linked to remote sheep farms. CCHF also took us to Senegal in 1989. Joe and I became involved in this investigation because of a researcher named Mark Wilson, who was studying animals and insects in a remote Sahel region of the country called the Ferlo. In the course of his investigation he found that a very high percentage of animals, mainly sheep, had been infected with CCHF virus. Was there a similarly high rate for humans? It was very difficult to know. The Sahel was another desert region where hardly anything grows. It is a land without hospitals, doctors, clinics, stores, or transport. There is nothing here at all, save sand, thorns, and the occasional water hole, around which the nomads' animals congregate, lowing and bleating. There is plenty of dust, too, swirling, getting into your clothes, into your eyes, your nose, your mouth.

The inhabitants of this inhospitable region are mainly nomadic tribesmen who drive their animals many hundreds of miles across the desert along the southern rim of the Sahara to get to rivers and grazing land. They are a charming people. Predominantly Muslim, they lead a rigorous, ascetic life, but while the men tend to the animals, it is the women who do most of the work. Tall and lissome, they are the ones who carry the water and pound the millet—their staple diet—swinging six-foot wooden pestles way over their heads in a marvelous rhythm.

This is a place where a man may have four wives, but where it is not permitted for you to count his animals. That would bring misfortune, because only Allah should know how many animals you own, lest He take them away. This belief caused Mark some trouble, I explained to my audience. To assemble the data for his research, he needed to have a fairly good idea about the size of the animal population. Since he couldn't ask anyone how many animals they owned without running afoul of local custom, he devised an alternative method: the turd count. Each morning he went to the thorn bush compound in which the animals had spent the night and counted fresh turds. While it wasn't an ideal method of taking a census, it was the best he could do under the circumstances.

When our team got to the Sahel, we had to camp with the nomads. With their customary generosity, they had built two grass huts for Mark, which they connected with a grass pergola. But we preferred sleeping on camp beds outside. The only problem with this arrangement was that it put us in close proximity to the cocks. When a cock is crowing at a distance of about three feet from where you're sleeping, you become acutely aware of its presence. They were impervious to all manner of abuse. The only way I could imagine it would be possible to shut one up would be to process it into coq-au-vin or chicken soup.

So what happens, we wondered, when people got sick in this region? What did they do for medical care? It was difficult to find out. We spent weeks questioning the nomads intensively. People were very friendly and cooperative, but we could never discover any evidence of CCHF disease or deaths.

We found plenty of CCHF virus in animals, but no obvious human disease. "Perhaps"—I concluded to my Saudi audience—"the virus was different in this region, affecting animals without harming the human population. Or maybe the nomads had developed immunity to it. Or maybe everyone who was infected died, and we never got to hear the true stories. These tribes guard many secrets. We were never able to find out. So, nearly half a century after it was first identified, Crimean Congo hemorrhagic fever is still very much a mystery."

Before driving to Mecca, I went first to Riyadh and then to Jidda. It was in Jidda that I ran into trouble. I was traveling alone, something I didn't think about until I tried to check in at the four-star French hotel where I was booked. The clerk looked up at me with apprehension.

"Excuse me, Madam," he asked, "but where is the man who is accompanying you?"

"What man?" I asked. "Why do I need a man?"

The clerk looked embarrassed. It was only then that I discovered that I wasn't even supposed to travel around this country without a male companion, let alone put up at a hotel without one. I asked to see the manager.

When the manager appeared, he recognized me. He had been the manager of the Mammy Yoko Hotel in Freetown, Sierra Leone, also run by the Sofitel chain. Apologetically, he explained that he couldn't afford to risk the chance of checking in a single woman. He told me the ubiquitous religious police would close him down.

"What am I supposed to do, then?" I asked. "I am a guest of the Saudi government! What will they do if I have to sleep in the street?"

We resolved the problem by a complex series of sleights of hand that shifted around a lot of paperwork. So, at least, I had a roof over my head. Assured of that much, I went to the coffee shop for a late lunch. The waiter came over and politely invited me to the "family room." I pointed out, equally politely, that I was not with a family, and that I was quite comfortable, thank you, right where I was. The waiter nervously served me where I was, but it was clear he had instructions to move all women, even western ones, out of sight. Any ideas I might have had later of cooling off in the inviting swimming pool were also quashed. Men only. I was effectively confined to my room.

The next day, I met my Saudi hosts, doctors from the Ministry of Health traditionally dressed in long flowing white robes and checkered kaffiyehs. They were courteously concerned about my well-being. Was I all right? Was my hotel in order?

I laughed.

"Yes, everything is fine, no problem at all. But it was an experience."

"What do you mean, it was an experience?" they asked. "So there were problems?"

"No, not at all," I assured them. "It was just an experience, that's all," I repeated. I let a moment pass, then added, "I guess I know what it's like to be black in South Africa."

Gradually, they began to laugh. They knew what had happened. After that bit of awkwardness, the rest of my stay in the country went smoothly enough, but I swore nothing would ever persuade me to come back.

Bob Fontaine and his team completed the investigation. They determined that CCHF was probably going on all the time in Saudi Arabia but at a very low level. It was only with the hajj that the virus found its chance to spread. Animals were imported from all over the globe to satisfy the demands of the ritual sacrifice. The countries from which many of the animals came included the Sudan, Iraq, Yemen, Iran—all of them good places for finding hard ticks and thus, presumably, CCHF virus. True, there were also nice, clean young lambs bred in New Zealand. The problem was that the slaughterhouse workers crowded them all together in large holding pens for a couple of weeks before slaughter. So the infection could easily spread from the Iraqi lambs to their New Zealand cousins. By the time the lamb was slaughtered, it would just be becoming viremic.

With Bob's guidance, the Saudis managed to deal effectively with the problem by instituting obligatory testing for all imported animals. The result is that CCHF is no longer a problem before the hajj the way it used to be. Later, when I worked in Pakistan, I would tell men whom I'd meet—Muslims, of course—that I had given a lecture in Mecca. This was a show stopper. A white woman in Mecca! Impossible!

The hunt for CCHF virus took Joe and me to South Africa in 1986 when we were asked by Horst Kustner, head of epidemiology at the Department of Health and Population Development in Pretoria, to investigate outbreaks of the disease. The department performed much the same role as CDC in the United States, albeit without the advantage of CDC's resources. Horst, who had worked with CDC, was in his fifties and occupied a rather lonely position as head of the only epidemiology unit in South Africa. When we met him at his office in Pretoria he looked anxious. Crimean Congo hemorrhagic fever had health authorities worried. Until the risk factors for the disease were properly understood, there would be no way to set up an intervention program. After Horst briefed us, we decided to set up a case control study.

Crimean Congo hemorrhagic fever is a very difficult virus to study because cases are always widely dispersed, both in terms of geographic area and the timing of their occurrence. Experimental work on the virus is hampered by the fact that it is impossible to make any animal sick with the virus—except, unfortunately, for human beings.

Our next step was to call on Bob Swanepoel, an old friend who was very active in the field of hemorrhagic fevers and operated a Level-4 lab in Johannesburg, which, after the one in the States, is probably the best in the world. Not only did we need Bob's help, but we knew that it was he who was making the laboratory diagnosis of the cases. A short, energetic veterinary virologist, Bob had left Zimbabwe (then Rhodesia) for South Africa several years earlier. He became famous for going into Kruger National Park game reserve and catching and bleeding a startling array of birds and animals, many of them exotic. To his surprise, every single one had antibody to CCHF—with one exception: an eland. It was quite a remarkable feat. The information he'd acquired from this study was of more than passing interest to us, since it showed just how widespread this virus is in the wild. Bob had also investigated an outbreak in Tygerberg Hospital in 1984, which had involved seven cases and two deaths. He told us that all the secondary cases had come into direct contact with patients and their blood. Up until that point, all South Africa infections had been associated with either the slaughter of animals for meat or with being bitten by ticks and, surprisingly, ostriches., farmed for their feathers.

We didn't think there would be any problem getting together with Bob, but when we telephoned him he seemed strangely reluctant to see us. We insisted, and a time was set up for a meeting. Horst sent along a statistician who worked for his office named Annamik Middlekoop. As we waited for Bob in his office, Annamik fidgeted uncomfortably. When Bob came in, he was amazingly hostile. At length, he told us what was troubling him: he and Horst weren't on speaking terms, and the battle, which seemed to me baroque in its complexity, had something to do with turf.

Our problem was that while we needed Bob's help for our investigation—because of his laboratory facilities and expertise—we also needed Horst's help in providing us with epidemiologists. We had been invited by the epidemiologists to assist them with a problem, and we found tension between them and the laboratory. This was not new to us, since it often seems to happen. This time, however, it was particularly unfortunate, because we counted both groups as friends.

We decided to sidestep this hurdle and concentrate on the problem at hand. We lent an ear to Bob, and once he got things off his chest, he seemed to relax into the person we knew.

Bob told us that he'd been cataloging cases of human CCHF in South Africa since 1981, all of which he'd diagnosed in his laboratory. The first patient whose blood he'd seen had come from a twelve-year-old schoolboy who had gone with his friends on a Pioneer Camp outing. The boys hiked into the bush and then camped out overnight under a tree in the veldt. All South African kids went on these expeditions, which were designed to teach them about their country. No sooner had the boy returned home, than he fell ill. An examination revealed a suspicious bump on the top of his head: a tick bite.

The boy's condition became desperate. He bled. He died.

After isolating the virus from his blood, Bob drove out to the site of the camp and dragged the ground under the tree in question with a blanket. This is the classical entomologist's method for trapping ticks. The blanket came out full of them.

We hammered out an agreement with Bob. We would proceed to Kimberley with Annamik and some of Horst's young recruits. Whatever specimens we found, we would send to Bob for testing. He also agreed to allow his contacts in Kimberly to provide us with access to information we might need for our study. Meanwhile, Horst would remain in Pretoria. That way, the two wouldn't have to meet.

There were a great many specimens to collect. We bled six hundred humans on sheep farms and in hospitals. To these were added sera from two thousand animals, specimens collected by an energetic veterinarian.

Although Kimberley is known as a diamond mining hub, most of the CCHF patients were farmers from remote sheep farms north of the Kalahari Desert. In the last year, Kimberley Hospital had seen at least nine cases. One of the patients, a farmer, had died. That a nurse had died of the disease the year before suggested laxity in barrier-nursing techniques. We found the hospital to be a modern, well-run facility. A special ward had been set aside for CCHF patients. They received good care there, and many of them survived, which was unusual, given the high mortality associated with the disease.

Assured of the hospital's cooperation, we assembled a team made up of two groups, one of which would conduct the hospital investigation while the other surveyed the prevalence of CCHF on some thirty-six farms. These were vast spreads set in a limitless landscape. Some are so arid that there is no more than one sheep per acre. There was both grandeur and monotony in the desolation. It takes a certain type of individual to survive here. I found the Afrikaaner farmers highly hospitable and remarkably rugged, but sometimes difficult to comprehend. While we visited one farm, we became aware that we were being watched. Two pairs of eyes followed us from behind the door of a farmhouse. When we approached, we saw that they were two men, sons of the farmer who owned the place. It was obvious from their expressions and mannerisms that they were severely retarded. This prompted us to speculate on the amount of inbreeding there must be on these farms. Usually, no more than one white family and a number of hired hands, all of them "coloreds," lived on a farm. With such enormous distances separating the farmers, it was no wonder that intermarriage was common. Yet who could question the resilience and stoicism of these people? One farmer brought out his rainfall book to show me. It indicated precipitation for the last twenty years. The last entry had been made in 1977. This was 1987. Not a drop for ten years!

Once you entered the houses, you walked into a whole different world. The desert outside might just as well not exist. Living rooms were filled with plushly covered easy chairs sporting antimacassars. There were always neat little coffee tables, and wherever you looked you'd see brass ornaments and knickknacks—souvenirs brought back from visits abroad: Dutch clogs decorated with blue windmills, miniature Eiffel Towers. Having a stranger to talk to was a rare treat. We were usually offered a big lunch of stewed sheep. In the Falkland Islands, mutton is called "365" because it is served every day of the year. The same situation applied here.

The only way to communicate with the outside world was by using an antiquated phone system. You had to crank the handle and hold an earpiece up to your ear while you shouted into a mouthpiece. Joe said it reminded him of growing up in Georgetown, Indiana, population five hundred, in the 1950s.

One farmer we met described to us how he had got CCHF. He recalled crushing a tick with his hand. As it happened, he had an open cut right across his palm. Three days later, he was hit by a headache and had muscle pains and chills that were so bad he had to lie down. He told us that it was rare that he ever had to take to bed like this, because he was never sick. But he wasn't going to allow even CCHF infection to keep him down. After a couple of hours, he got up and went out to help his "boys" round up sheep. As he was doing this job, his nose started to bleed—uncontrollably. His platelet count must have been approaching zero at this point; the blood wouldn't clot. He was taken to the hospital in Kimberly and treated. He was one of the lucky ones. He survived.

When we published the results of our investigation, we were able to show that the Crimean Congo infection in humans was associated with the raising of sheep, specifically in the way that lambs were handled by the farmers. That explained why it was the farmers, and not their employees, who were usually infected. We were also able to show that the death from CCHF at Kimberly Hospital was an aberration and did not indicate an absence of proper sterile techniques. It was true that, in addition to the nurse who had died a year earlier, a laboratory technician was found to have antibody to CCHF. Her positive status was peculiar for two reasons. For one thing, she'd always taken great care when she worked with specimens, had worn gloves, and hadn't had an accident. For another, she never fell ill. It was difficult to explain. It was even possible that it had nothing to do with her employment at the hospital, and that she'd been unknowingly bitten by a tick. We concluded that in spite of high levels of virus transmission in livestock and the constant exposure of human beings to ticks and blood from domestic animals, human infection was actually infrequent.

But when the virus did infect humans, it could prove fatal. It seemed to us that the most important thing we could do was to see whether we could treat this disease with an antiviral. After Joe's success treating Lassa fever with ribavirin, we wanted to make CCHF our next target. While we discovered that the CCHF virus is highly sensitive to ribavirin in the laboratory, we had no idea whether it would work in humans. We couldn't use animals to test it on because animals never show any symptoms of the disease. Joe and I drew up a protocol for conducting experimental tests on humans who had become infected. The last we heard, ribavirin enjoyed great success among the patients who received it. Bob presented data on thirty such patients at a subsequent conference in Berlin; only one of the patients given ribavirin died, and, even in that case, he'd succumbed to complications not directly related to the acute viral infection. Mortality in CCHF is at least 30 percent and, in some outbreaks much more, so these results represented a dramatic improvement. It will be good to see these data published and a treatment finally established for this terrible disease.

CHAPTER THIRTY-THREE

Cathedrals

I was divorced—Sue had been down that road herself—in the late 1980s, and I left the Special Pathogens Branch in March 1990. Sue and I began to see more of each other socially, and we discovered many interests in common, particularly the outdoors and music. We went skiing with my three children and backpacking in the wilderness of Wyoming's Wind River Mountains. Both of us were settling into life that went beyond the narrow scientific field of viral hemorrhagic fevers. I was working on HIV, and later became involved with HIV and malaria projects in Western Kenya and elsewhere, while Sue was making her way in the division of Bacterial Diseases, where I'd started out at CDC twenty years before. But the atmosphere at the division had changed. Ironically, we ended up seeing more of each other after I left the branch than we did when we were working together at CDC.

Two years later, in March 1992, I scheduled a ski trip to Vail, Colorado. I'd been there with my children a few years before, and it was such a wonderful experience that we all decided to go back. It was then that I had an inspiration. I proposed to Sue, suggesting that we get married on top of Vail Mountain.

It would be a cathedral in the air, she said.

I made some calls to find out if people ever did such a thing, and discovered that, while it wasn't exactly a commonplace occurrence, several marriages did in fact take place on the mountain each year. The judge who performed these ceremonies had the memorable name of Buck Allen. It turned out that he lived at the base of the lifts. Not surprisingly, he was a superb skier. We made the trip from Atlanta in record time. My kids, Kit, Peter, and Anne, together with a couple of friends, took turns at the wheel, driving nonstop all the way. We arrived in Vail twenty-four hours later. My seventy-three-year-old mother flew out for the occasion. In order to reach the summit for the ceremony, she had to take the gondola up the mountain and then make the rest of the journey up a 200-foot hill called the Eagle's Nest, helped by ski poles and encouraged by Kit and Peter.

It was a gray St. Patrick's Day. Buck Allen arrived on telemark skis and positioned us so that we were standing against the backdrop of the spectacular Gore Range. The ceremony, brief but immensely moving, was followed by a walk down the slope to sign the documents and celebrate with champagne at the ski lodge. We didn't linger; after signing the marriage certificate, we skied off into the powder in the back bowls, and certainly distinguished ourselves as the only people on the slopes that day who were wearing fresh flowers pinned to our jackets.

Marriage wasn't the only major change in our lives. Soon, we would find ourselves in a different part of the world altogether, in a place where we'd never planned to be. More to the point, it was a place that we had never even cared to visit. Let Sue tell this part of the story.

In September 1991, Joe received an unexpected phone call from Dr. Jim Bartlett, dean of the Aga Khan University Medical School in Karachi, Pakistan. Before he could get out more than a few words, the line went dead. He called back a few moments later. "Don't worry," he said, "this is normal. We always get cut off."

Jim went on to say that David Fraser, one of Joe's CDC colleagues who had worked on Lassa fever, had suggested that Joe might be interested in a job in Karachi. Before Joe had a chance to learn much more, they were cut off again. But as soon as I heard what Jim had called about, I said, "Karachi? Forget it."

I knew enough about Asia to be aware that Karachi had a reputation as a large, dirty, polluted city with few attractions. Besides, Pakistan was a Muslim country that took its religion very seriously. For women, that can be a grave impediment. My experience in Saudi Arabia had done nothing to endear me to that culture. Joe was no more disposed to relocate to Karachi than I was. We both laughed and put Karachi out of our minds.

But the idea of moving overseas did appeal to us. We both wanted to get back to where things were really happening. Optimally, we hoped to have the opportunity to work together again. We even contemplated setting up a program of our own if we could find the right circumstances to do so.

Then David Fraser called from Paris. He'd recently left his post as president of Swarthmore College to take up a position as advisor to His Highness, the Aga Khan, leader of the Ismailis, a Muslim sect whose several million adherents can be found throughout the world. David's position made him responsible for health, housing, and welfare in many parts of the Muslim world. After he had the chance to explain the purpose of the organization, he asked Joe to reconsider. *He* didn't get cut off.

As universities go, the Aga Khan Medical School was very new, having opened in 1983. The Aga Khan had established the school with the objective of educating young Pakistanis as doctors, using western teaching methods and standards. As a result, the school offered a level of training far surpassing that of any other institution in the area. The position that David and Jim wanted Joe to consider was chair of the Community Health Sciences (CHS) Department. Whoever assumed the job would be expected to introduce medical students to the field of epidemiology. The CHS department offered opportunities extending to several countries in Asia and East Africa, where the Ismailis had established an extensive network of hospitals, clinics, schools, and rural development programs. Although these countries have significant Ismaili populations, it isn't necessary to be Ismaili to receive these services. Anyone in the community is eligible. It became clear to us that the Aga Khan's program was both enlightened and well organized.

By this point we'd become sufficiently intrigued to at least visit Karachi. We found that it lived up to its reputation. It was hot, dusty, violent, and chaotic. On the other hand, we also found it vibrant and bustling. The streets are filled with nearly every form of transportation known to man for the last ten thousand years: donkeys, camels, bicycles, push carts, three-wheel took-tooks belching out noxious exhaust, colorfully painted lorries, motorbikes, imported limousines, and four-wheel drives. With so little space, all these animals and vehicles are often trapped for hours at a time, just trying to inch ahead. Although the poverty is pervasive, appearances can be deceptive. While it's true that you will occasionally come on beggars—often veiled women holding "rent-a-babies"—homelessness is not the problem that it is in the cities of the West. The family is the safety net of last resort. If someone needs a roof over his head, there is usually a cousin or an uncle somewhere who will be able to take him in. And as the commercial hub of Pakistan, Karachi is also a city with relatively immense wealth, albeit distributed inequitably. The rich and the abject poor are side by side.

If we had serious reservations about Karachi, Aga Khan University was another story. It was clearly an opportunity for us. We could introduce some ideas to a few faculty and students that might bring about some changes that would have a favorable impact on this part of the world. Karachi is one of the new breed of megacities that are rife with public health problems. There is little in the way of an organized public-health effort and almost no public-health training. The idea of a high-quality private university building a public health program was intriguing, particularly in a country that apparently did not know what public health was.

But the potential is there. Many of the young people we met at Aga Khan impressed us with the quality of their intellect and their keen desire for achievement. That the CHS was also developing programs to target infectious diseases was also of great interest to us. If there was one thing that Karachi had in abundance, it was infectious disease. We were told that if Joe decided to take the offer, there would also be a job for me as head of the clinical microbiology laboratory—and at the same time, I'd be taking charge of establishing a molecular diagnostics laboratory, which was intended to support studies of infectious diseases in the field.

Almost in spite of ourselves, after long deliberation, we agreed to give it a try. On June 1, 1993, Joe left for Karachi, and I joined him two months later. For my part, I had been increasingly dissatisfied with the politics of CDC. I was happy to give this a go.

The university and medical school turned out, in fact, to be a cathedral of sorts. Constructed of pink marble, it stands in striking contrast to the rest of Karachi. We found a lovely house in a quiet, residential district, which became our refuge from the disorder that prevails in many parts of the city. But there was much to do. And no time to lose. Although we were well aware of the threats that cholera and typhoid posed to the health of Pakistan, we were soon to discover that another disease was rampaging through the country, a disease equally pervasive and even more dangerous. There was an epidemic, only it was silent. To bring it into the open, we undertook a trip to the interior of the Punjab, to a city called Hafizabad.

With an eager gleam in his eye, a tall slender man with blond hair—looking for all the world like Ichabod Crane, Washington Irving's schoolmaster pursued by the Headless Horseman—was negotiating his way down a narrow alley in an agricultural town in the Punjab, trailed by his acolytes. His entourage was a motley crew, among them three or four recent graduates from the Aga Khan University. Tagging behind them was an unusually colorful collection of children, goats, chickens, and assorted young men in *shalwar kamiz*, the Pakistani national dress, which on men consists of a very large and very long shirt, with tails, and on women is a robe decorated with embroidery and mirrorwork. The crowd had turned out for what promised to be at least an entertaining, if not downright comical, break in the routine of their lives.

We were tracking down viruses again—this time hepatitis—and Ichabod (he was really Steve Luby) was in charge. Steve had recently been recruited by Joe to head up a new epidemiology program in Aga Khan's Community Health Sciences Department. The practice of epidemiology was almost unknown in Pakistan. So the offer to come to Pakistan held appeal for Steve, even though he'd never laid eyes on the country. For a recent graduate of the CDC Epidemic Intelligence Service and Preventive Medicine Residency, it represented an exciting, albeit formidable, challenge. He'd arrived in Karachi, sight unseen, in mid-September 1993, with his energetic and capable wife Jenny, who is a teacher, along with four very lively small children and an elderly cat. Now just two month's later, he found himself exploring the Punjab in search of viruses with his own team of enthusiastic raw recruits.

On this occasion, Steve acted as guide as well as leader. To find his way, he was using an oversized hand-drawn map covered with annotations. As we went along, he kept peering down alleyways where the walls were caked with cow-dung patties (each bearing the imprint of the molder's hand) that would later provide fuel for cooking. The map was the only one that existed of the city we were surveying. Hafizabad, with a population of 120,000, lies in the hub of an agricultural area, about three hours' drive from Lahore, the ancient city of the Mogul emperors. We'd come to Hafizabad because people were falling ill with jaundice, a sign that they were suffering from some form of hepatitis. But we weren't just looking for one virus. Hepatitis can come in various flavors, as it were, designated by letters of the alphabet. Basically, hepatitis is an inflammation of the liver, and one of its effects is jaundice: the victim's skin and the whites of his eyes turn yellow. Other symptoms include nausea and weakness. Most of those who are stricken are infected with one of two viruses—hepatitis A and hepatitis E—which are usually spread in conditions where sanitation is poor. It was no surprise to find that hepatitis A and E thrive in a country like Pakistan. The only sewage system in the entire city of Hafizabad, which is hardly atypical, consisted of open drains. Waste ran directly from each house into these, and you could see some very nasty things floating slowly downstream. Piles of rotting garbage added to the stench.

But the absence of sanitation alone didn't necessarily account for the outbreak or explain why there were so many victims. There was only one way to find out what was going on. We would have to undertake a survey, collect samples of blood, and test them. However, several obstacles stood in our path. Because no census of any kind has been conducted in Pakistan since 1981, most of our information was out of date. We would have to come up with our own method of conducting a reliable epidemiological study. To do this, Steve planned to choose one house in each of twenty-seven districts. If you wish to draw conclusions about a particular population or location, you need to devise a survey that relies on random sampling. After all, you can't possibly interview and take blood from every single individual who lives in a city where, say, a viral epidemic has broken out. It would take far too much time and money. Instead, you rely on a mathematically based sampling of people that accurately reflects conditions affecting the entire population of the area under investigation. The same principle applies to taking polls intended to gauge an electorate's opinions of presidential candidates or viewers' preferences for TV shows.

In theory, it should have been easy to identify houses for the survey. In reality, things were far messier. Once you actually got into the tangled alleyways of Hafizabad, it was far more difficult to determine the precise point Steve had designated on his map. All that we could say for sure was that we were probably standing at a particular intersection defined by the grid that represented our starting point. Since this was a random survey, Steve chose the "first" house in the district by a simple method, which relied on one of the most indispensable tools of the epidemiologist: an empty Coke bottle.

In all parts of the world, you make use of what you can find that does the job. And there's hardly any object that is more common than an empty soda bottle. Since we were all from Atlanta, the home of Coca-Cola, the choice was obvious. A resident physician named Karim Qamruddin was given the job of carrying it, an assignment he seemed to consider a privilege. He placed the Coke bottle on its side at the point indicated by the grid, and then spun it. We couldn't have been any more riveted than if it had been a roulette wheel. When it stopped we looked to see where it was pointing. We had our first house.

The time had come for Steve and me to fall back and allow the young Pakistani trainees to take over. In any case, the people who lived in this district spoke only Urdu or Punjabi, and Steve and I spoke neither. The secret to the success of any investigation is to get yourself invited into the house. Once an investigator is inside a house, he is in a much better position to explain his purpose and to conduct an interview, as well as to obtain consent to draw a sample of blood. Achieving these objectives sometimes requires charm, empathy, and a silver tongue. Fortunately, our young Pakistani colleagues had these attributes in abundance. But the occupants of the houses we'd selected also had good reason to cooperate. They realized that there had been a number of unexplained cases of illness marked by jaundice, and they were obviously concerned. Indeed, it was some of these people who had actually asked us to come up and help. As soon as we explained that we were trying to find out the cause of the infection and prevent more cases from occurring, they were more than happy to open their doors for us. Sometimes it was difficult to escape the hospitality and get on with the job, so we worked long hours. Because there was no hotel in the town, the team members were compelled to commute an hour each way to the next large town, Gujranwala, where their hotel was located. The route took them along a narrow, congested road that cut through a succession of rice fields. Given the potential for colliding with a car or truck, the commute was probably the most dangerous part of the investigation.

After three weeks of knocking on doors and talking their way into houses, our Pakistani team had interviewed three hundred and twenty "random people" and obtained samples of blood from most of them. On the basis of what we found, we tentatively concluded that the virus responsible for the majority of cases of jaundice was likely to be hepatitis E, one of the viruses passed by fecal-oral transmission, often contaminating water, and known to be common in the area. We wouldn't know for certain, of course, until we had the results of the blood tests.

Whatever the results, though, there were some obvious precautions that the population could take to avoid infection. We could advise people about the importance of clean drinking water and eating only properly prepared food. But we had to spread the word. Gratifyingly, we found that people were willing to listen. They were so interested in what we had to say that they invited us into their homes. But they didn't want us to think that they were inhospitable and so they insisted on offering us elaborately prepared, excellent dishes like rice pilaf with fresh-baked chapatis, and freshly peeled fruit accompanied by heavy, milky, sugared tea. No one was more acutely aware of the dangers of food contamination than we were. Up until this point, we'd avoided eating until we were back at our hotel. But on the other hand, we didn't want to offend our hosts. So we ate and we drank, and waited to see who would get jaundice. We survived, and so did our livers.

The real surprise didn't come until we sat down at our computers and analyzed the data we'd collected from our study, as well as the laboratory results from the serum specimens. Out of every hundred people we'd bled—all them apparently healthy—seven were infected, but not with hepatitis E. Someone afflicted with hepatitis E can recover without too much difficulty. This was another hepatitis virus, one which is spread, like HIV, in the blood. It's a silent virus, which, like HIV, infects you without exhibiting any obvious signs, and then lurks in your body for the rest of your life, slowly destroying essential cells. AIDS ultimately sabotages the immune system; this virus, however, has a different target. Many years after the initial infection—the incubation period can even extend into decades—this virus destroys the liver, slowly and painfully. Bellies swell up, people vomit copious amounts of blood. When the virus's work is done, its victims die of liver failure, sometimes liver cancer. The virus is hepatitis C.

We hardly knew what to make of this finding. In the U.S., far fewer than one person in a thousand is infected with hepatitis C. Here it was seven out of every hundred. Put another way, if you had a hundred blood donations, seven would have this deadly virus in them.

What was going on? We had several clues. Like HIV, hepatitis C can be spread by shared needles, which is why the disease is commonly found among drug addicts. But when we searched for drug addicts in Hafizabad, there were none to be found.

While it was already clear that hepatitis C is transmitted much less by sex with an infected partner than HIV, we still couldn't rule sex out as the means of transmission. But this was a traditional agricultural town where there were strong social structures, little or no prostitution, and no widespread homosexuality. Besides, some of those infected were Muslim housewives and children. In our attempt to determine what was going on, we had to ask intimate and probing questions. Some of our Pakistani friends told us that we had better forget it. Posing intimate questions was just not done in this society. But there was no other way we were going to find out what was going on. So Steve persisted.

"Let's try," he said. "What have we got to lose?"

So we returned to Hafizabad, armed with a more detailed questionnaire, including some pretty penetrating questions. The worst that could happen was that people would slam the door in our faces. But this didn't happen at all. It was amazing and gratifying. In spite of the potential for embarrassment and anger, most of the people we queried proved to be very frank and helpful. They understood the importance of what we were trying to do, and they wanted to cooperate.

The first clue to unraveling the mystery had come when we took our blood samples to the local hospital laboratory, where they could be separated by centrifuge and packed for shipment back to Karachi. The lab was clean and orderly compared with many I've seen in the developing world. Nevertheless, it was equipped with little more than an old centrifuge, an ancient microscope, some chipped slides, and a pathetic supply of old, chipped test tubes in a scarred wooden rack, sitting on a cracked tile bench. Alongside these were three needles and three syringes, the entire supply of the lab. Outside in the hallway, a line of at least six patients was waiting to be bled.

We continued to investigate. Although the inhabitants were mostly poor, the town was still very well supplied with general practitioners, a few of whom were actually qualified. The others just put up a sign stating that they were doctors, and the trusting patients came. The common wisdom was that all someone needed was a piece of wood and a tin of paint to become a general practitioner. However, the general practitioners didn't have the field to themselves; they had to compete with the hakims, traditional healers who dispensed charms and potions. Patients went to hakims only if their complaints were minor. When it came to more serious symptoms, they preferred a doctor, especially if they were feeling weak. They were convinced that if you wanted to feel strong again, you had to get an injection—never mind of what; that was up to the doctor to decide. And if the doctor believed that an IV drip was warranted, so much the better. Most of these injections, it turned out, were merely vitamins. Some contained nothing more therapeutic than water. The doctors had every incentive to promote injections, regardless of whether there was any need for them or not: injections are a money spinner.

It wasn't difficult to find needles or syringes in clinics and doctors' offices, but what was so striking was just how few of them there were in comparison to the sheer number of injections the doctors were administering. We looked for sterilizing equipment, but the needles and syringes we found were the plastic disposable type. These are difficult to sterilize; indeed, they are designed to be discarded, not reused. If you boil plastic syringes, the markings will come off, making it impossible for the nurse to read how much of the drug to draw up. Mostly, they ended up being rinsed out in tepid water.

In any case, our search turned up little in the way of sterilizing fluids, nor could we find any apparatus for boiling instruments. In one surgery, we did discover an electric sterilizer, but the building had no electricity. In short, it was clear that disposable needles and syringes were being reused on a massive scale, without sterilization. Hepatitis C is transmitted with remarkable efficiency under such conditions.

We continued to analyze our data. It soon became evident that there was a direct correlation: individuals who harbored hepatitis C in their blood regularly made visits to the doctor to obtain injections. But indiscriminate injections weren't the only means of spreading hepatitis C in this population. The infection is also spread by blood transfusions, and no one was testing blood for C, even though the test had been available since 1992. The problem was that a blood transfusion in Pakistan costs the equivalent of twenty dollars; the test kit for hepatitis C costs fifteen—much more expensive than the test for HIV. No one in a town like Hafizabad could afford it. Actually, the test isn't intrinsically expensive, but a single American company holds the patent on the reagents needed for the test and can set the price at whatever level it wants. When we contacted a representative of the manufacturer, he sounded apologetic and gave us a couple of free kits. But that was it. This is big business.

What, then, was our responsibility? What could we do to check this insidious epidemic? We could set out to determine the extent of the outbreak and the nature of its transmission. We could also let the world know what we'd found out by publishing our findings in medical journals. What complicated the situation, though, was that this was at a time when AIDS was beginning to show up in the population as well. The problem involving the transmission of hepatitis C was overshadowed.

Was Hafizabad exceptional? Or was the same hidden epidemic happening all over the country? To get a sense of just how widespread C was, Steve led a team of Aga Khan medical students to Dur Mohammed Goth, a village about twenty miles outside of Karachi. While in many ways the village is typical of thousands scattered throughout Pakistan, Dur Mohammed Goth boasts a clinic and a family-planning center as well as two schools, one public and one private, which offer instruction in English, Urdu, and Sindhi—and this in an area where most of the children are brought up speaking yet a fourth language, Baluchi. Community workers proudly offer visitors glossy booklets describing the improvements they've been able to make in the lives of the inhabitants. So it wasn't as if the village was being completely neglected, or that its population was not being educated.

Positioning themselves outside of the clinic, Steve's team interviewed the patients as they emerged. After inquiring as to whether they had received an injection, the investigators then would ask if they could take a blood sample. Most of the patients had gone to see the doctor, who made regular weekly visits to the village, for minor complaints—fever, backache, cramps, or diarrhea, conditions that, for the most part, did not warrant an injection. The result of the sample poll was staggering: 82 percent of the patients had received an injection on that particular visit.

Later that same night, Joe and I were watching CNN when the phone rang. It was Aamir Javed Khan. He was one of our best medical students, and he'd been in on the Dur Mohammed Goth survey. He sounded upset.

"I'm with Shaper," he said, referring to Shaper Mirza, the very able young technician doing much of the laboratory work. "We've got a problem."

"What is it?" I asked.

"Shaper has just run the hepatitis C tests on the patients we bled leaving the clinic. Something is wrong. Over 60 percent of them are positive."

Over 60 percent? This was unbelievable. More than six out of every ten patients? I asked to be put on the phone with Shaper. I wanted to be sure that there wasn't some mistake.

"Did you wash the beads adequately? Were your controls all correct?"

"Yes," she replied. "Every one had a control. I even put in extra ones to be sure. We even put in Amir's blood as an extra control, and he is negative."

"What were the readings?" I asked.

"Very high," she said. "All of them were very high."

We slept on it. The next day we decided to redo the tests to check Aamir and Shaper's work. They had made no mistake. The results were the same: over 60 percent positive. We were now forced to accept the fact that the majority of a random sampling of patients in a small Pakistani village were infected with hepatitis C, and the only conceivable source of the disease had to be the general practitioner. We estimated that these doctors were using, on the average, about one needle for every three patients they injected, though some needles were likely being used for many more patients without sterilization. It was a nightmare.

When we began to compare our notes with those of other epidemiologists and doctors who were doing similar work, we realized that Dur Mohammed Goth was no more an exception than Hafizabad was. We needed only to visit the medical wards of the public hospitals to find scores of patients suffering from terminal liver disease, much of it caused by hepatitis C. We'd always been aware of the problem. Now we knew how it had happened. The situation was not unlike what had happened when Ebola or Lassa had broken out in Africa. Sometimes well meaning, sometimes mercenary, doctors were spreading lethal viruses by using modern medical treatment in poor conditions without giving any thought to basic hygiene. Unlike the case with Ebola and Lassa, though, this virus would lie low, like HIV, waiting for years before working its deadly biochemistry on the unsuspecting host. Few, if any, of the practitioners we interviewed seemed to be aware of what they were doing or its implications. Some, apparently, did not believe they were doing any harm or simply didn't care. The bottom line remains ruler of the world.

CHAPTER THIRTY-FOUR

The Stricken Surgeons

We were seated around a table late one evening in December 1995 in the elegant lobby of the Serena Hotel in Quetta. There were four of us: Leslie Horvitz and myself, and two young surgeons, Dr. Jamil Khan and Dr. Shafiq Rehman. Joe was in bed in his hotel room, nursing a high fever—fortunately, nothing more than flu. Both Dr. Jamil and Dr. Shafiq lived and worked in Quetta, which is the principal city of Baluchistan, a rugged and sparsely populated province in northern Pakistan near the Afghan and Iranian borders. We'd come to talk about Crimean Congo hemorrhagic fever. The two surgeons knew it intimately. It had nearly killed them, as I well knew, since I had treated them.

We were listening to Dr. Jamil, a man in his thirties, with a kindly round face and intelligent eyes. He spoke quietly and methodically. His English was good, measured out in the lilting cadence peculiar to the subcontinent.

"Actually it was December fifth last year. I was sitting in my room when I got an emergency call from a hospital near Radio Pakistan. They told me that 'we have admitted a patient for you, and this patient has got abdominal pain and he is bleeding. He has vomited blood, so please come and see this patient.' When I went there I saw the patient, and I called the gastroenterologist to discuss about that patient. The gastroenterologist told me that it was not possible to say anything at all about the reason for the bleeding in this patient and that he must do a gastroscopy."

We were silent. He went on.

"The next day, in the afternoon, my colleague did a gastroscopy to that patient because in the morning we are busy in the hospital for our normal duty. In the evening, when I was sitting in my clinic, they asked me go ahead and do surgery, because most probably, as we had thought, he has got ulcer in the stomach from where he is bleeding.

"I think it was eleven o'clock in the night. We started surgery with five to six units of reserved blood, because we thought definitely he would bleed during the surgery. Dr. Shafiq was with me, assisting me, along with another assistant operating theater technician."

Dr. Shafiq, sitting beside him, shifted uneasily in his seat. He was a few years younger than Dr. Jamil, taller, and strikingly handsome. Dr. Jamil continued in his low voice: "When I opened the abdomen, the whole surface of the gut was oozing blood. It was not possible to control the bleeding, though I tried both with diathermy [the application of heat to body tissues using an electrical device] and stitches. It was just not possible to control the bleeding. So we thought maybe this patient had taken some analgesics, which have produced acute erosions of the stomach. There was just bleeding and oozing from the surface, no clinical ulcer."

I listened without surprise. Here we go again. The same story that we had heard from surgeons in Sierra Leone, in Rawalpindi, in South Africa, Dubai, and China. More recently, surgeons had become infected during an operation in Kikwit, Zaire. Again uncontrollable bleeding. And this time it was Ebola. But what Jamil was describing wasn't Ebola, although it was something very much like it. He continued: "We decided to take out the whole of the stomach to control the bleeding. I performed a total gastrectomy and connected the esophagus to the duodenum. The spleen was very fragile and friable. When I tried to mobilize the stomach, the spleen was torn and I had to take it out. It was enlarged. So was the liver, which was dull red in color. Not bright and shiny like a healthy liver. The whole operation took us almost two and a half hours. Dr. Shafiq pricked his finger with a needle covered with blood from the patient. I tore my gloves several times. It was a long struggle. Eventually, about two o'clock in the morning, we finally shifted the patient to the recovery room, and we left the hospital. The major thing was to stop the bleeding.

"The next morning, I went there and saw the patient, who had low blood pressure. He was in his senses and was talking, but his blood pressure would not come up. Around three o'clock in the afternoon, when we went to visit the patient again, I met his brother, who told me he was dead. The man was just about forty-five or forty-seven years. He was from Sibi.

"Then I thought more about this operation. One thing which alarmed us during surgery was that when the anesthetist was passing the tube into the stomach through the nostrils, there was blood oozing through the nostrils. The anesthetist was unable to control the bleeding, so he packed the nostril. He said he was also scared that there is something wrong with this patient because he was bleeding too much from the nostril. Also the patient was febrile."

This sent a shiver down my spine. It all fitted. Short, severe illness in a healthy man from an area in Baluchistan where a viral hemorrhagic fever is known to occur. Fever, uncontrollable bleeding, low blood pressure, enlarged fragile liver and spleen, vomiting blood, sometimes abdominal pain. A patient brought into hospital with such symptoms might often be treated as an abdominal emergency. There would be blood all over the place, and it would be full of virus.

At this point Dr. Jamil turned and indicated Dr. Shafiq.

"On the fifth day after doing the surgery—I think that was a Friday, in the morning—Dr. Shafiq's wife telephoned me at my home, saying that Dr. Shafiq was calling me to come to his residence. She said, 'Please come, because he is having high-grade temperature, and he is complaining more of headaches and body aches.' So I went at once to visit him. I just joked with him, saying, 'What happened to you? Probably you are suffering from malaria or something like that.' I stayed with him for two to three hours. Two or three of his cousins were also there. He was just crying with pain, all over the body.

"He was telling me all this, and I was laughing. But he said that this is not the pain of temperature. This is the pain of death, and I am going to die. He is very fond of a sweetmeat, a delicacy called ras malai. He said, 'I am going to die, and my last wish is to eat ras malai.' He asked his brother to go to the bazaar to fetch some as his last wish."

Both surgeons laughed for a moment at the memory.

"Dr. Shafiq asked for one of his professors of medicine to visit him from the medical school in Quetta. I went to the hospital to do my rounds, and while I was there the professor visited Dr. Shafiq. The next day he told me Dr. Shafiq had a temperature throughout the night and still had terrible body aches. The professor said he thought he [Dr. Shafiq] was probably suffering from enteric fever and gave him Amoxil. Then Dr. Shafiq started to have diarrhea, and they had to give him a drip to replace his fluids.

"I got hold of the physician who had endoscoped the patient, and I told him that I had operated on the patient and there was no ulcer in the stomach, just the oozing of blood from stomach. I told him, 'My friend is very ill, and I am very much scared of this.' So he came along with me and visited Dr. Shafiq, and he said he thought that we should reinvestigate him for malaria."

At this point I stopped him.

"Were you already connecting Dr. Shafiq's illness with the patient you'd operated on?" I asked.

Dr. Jamil shook his head with determination.

"No, we were not connecting it with the patient. We were not thinking that this is all due to that at all."

I let him go on. But I wondered. It seemed to me that they must have been in some sort of state of denial. They knew what CCHF was, and they knew that it had killed surgeons in Pakistan. Not once, but twice. And the second surgeon had been a close friend of Dr. Jamil's.

Dr. Jamil continued. "In the afternoon that Saturday, the day after Dr. Shafiq fell sick, I was in my outpatient clinic again seeing the patients. It was now five days since the operation. Quite suddenly I think that I am feeling some body aches. After completing my clinic, I went to my ward and asked my ward sister to please bring the thermometer, I think I am running temperature. My temperature was 102°F. That afternoon I had two private patients scheduled, so I contacted my anesthetist and told him that I will not be able to operate on both of them. So we will operate one in the afternoon and postpone the other for night.

"As I was operating on the first patient, I was having rigors, shivering and shaking with fever. When I had finished, I went to my residence and told my brother that I am not feeling well, and I am going to have sleep, and I am going to wake up and go to clinic late in the evening. During the evening, I went late to my clinic. At that time, after taking some analgesics, I felt a little better, but I phoned one of my colleagues and asked him to come around to the hospital. I said, 'I have to perform a cholecystectomy, and I am not a hundred percent fit for that.' He arrived as I was scrubbing for the operation. I started the surgery myself, and he was just assisting me. But then I felt such terrible body pains, I felt I could not even stand in the operation theater. I just left the theater and asked them to carry on the surgery. I went to the surgeons' room and just lay there crying in pain. I was really crying."

I interrupted him again. "Didn't it hit you then that you and Dr. Shafiq must have the same thing?"

"No. At that time it was not in my mind, though both of us were scared that something is wrong, but we were not putting these things together. I didn't stay too long after the surgery was complete. But it was not possible for me to drive home in the night. So I called my younger brother to please come over and take me home with you because I am not able to drive. There was so much pain in my body. Before I left the hospital I called one of my physician colleagues. He came to see me and asked me 'Where are you feeling pain?' I said the back muscles. He just started pressing my back and told me it would just be fine, don't worry about it. Just go home and rest.

"The next day I was no better. I got a colleague to take some blood, and sent it off for blood tests. My platelet count was very low. We sent off blood from Dr. Shafiq as well, and his platelets were also very low. Then we thought this is it. This is CCHF. "

"We knew we were going to die. I had had a close friend, one year senior to me at medical school. He died of the same disease after operating on a patient at the same hospital in Quetta. That was at 1987, and this was 1994. I remembered his story, and I told Dr. Shafiq I am very much scared of his death. The whole story came back to me.

"I was working in Karachi. I think just three or four days before his death he became engaged to marry a doctor from Quetta. I had also become engaged to be married about three months before operating on our patient. I told Dr. Shafiq that this is the same story which is repeating itself.

"What happened to my friend was in the newspapers later, but I was involved very early. The day just before his engagement, my friend came to my room in Karachi. You know people from Quetta very much stick together. He told me he was running a temperature. He said, 'Take my pulse; it is one hundred twenty per minute.' I took his pulse; it was one hundred twenty per minute. He said, 'I have a fever, and I operated on a patient in Quetta a few days ago, and the patient died next day.' We said no more. Then he said, 'Tomorrow, you must come to my residence in Karachi for my engagement ceremony. I am going to marry a doctor who has graduated from Fatima Jinnah Medical College in Lahore.' I went to the ceremony. My friend was there, looking very smart and bright, but he was running a temperature. We took him to the emergency room of the Aga Khan University Hospital, where he was given a prescription, which was later printed in the newspaper and was also in the medical news. It was for Septran tablets and Panadol. They sent him for a chest X-ray, and as he stood there for it to be done, he collapsed. Nobody was recognizing the fact that he might be suffering from such a severe disease."

I listened to Jamil's description and shook my head. Taking a good history is essential to any clinical examination. We had seen it in Chicago, in the Middle East, in Pakistan, across Africa, and in so many other places. It was surprising that the local medical profession was not more aware of what had happened to the surgeons who were earlier stricken with CCHF in Islamabad. After all, the incident had been well published.

Dr. Jamil continued: "The next day, and I think I exactly remember the date because it was April first, I was again sitting in my room preparing for my surgical fellowship examinations on April fourth. My registrar came into my room and said, 'Your friend has died.' I replied that he was making an April fool of me. He was just here two days back and was okay. And then we went to his house and all the way there I did not believe the story. When we arrived, the people there told me that they had already taken the dead body to the grave. He had died of uncontrollable hemorrhage from the gut. He had CCHF. We went to the graveyard after prayers at the house.

"This terrible story ran through my head the whole time *I* was ill. Like Dr. Shafiq, I knew I was going to die. I talked to my professor and told him that we must have CCHF. Our blood count is very low; we have a temperature. I told Dr. Shafiq that I am planning to get us to Karachi tomorrow, to the Aga Khan Hospital, because I think there we will be able to get some platelet transfusions. There was nothing like that in Quetta. Also, maybe there will at least be somebody who knows something about this disease. I went back to my residence and talked to my father, who is a retired deputy director of a school. I told him that this has happened to me and that I am leaving tomorrow for Karachi. Tonight I will not stay at home. I will stay in the hospital because something might happen to me during the night. So Dr. Shafiq and I both went to the hospital. I phoned my close friend Dr. Shahid Pervez at Aga Khan Hospital, and I told him to arrange our immediate admission there, so we should not have to spend much time in the emergency department. He said he would make arrangements, so that we would be immediately admitted to a room and the treatment started.

"That night was a very difficult night. We were very scared, and we could not sleep. We were both very frightened and depressed. At one stage, I dozed off, and then suddenly woke up with difficulty breathing. I asked my brother, who was sitting up watching us, to take my blood pressure, telling him I was not feeling well. It was ninety over sixty, very low. My brother immediately rushed into the lab to get cross-matched blood, and ran to find my professor, saying please come along with a physician quickly because Dr. Jamil's blood pressure is dropping. The physician and my professor, both of them, came at four o'clock in the morning. They started a drip, and my blood pressure slowly came back up.

"The next morning after this terrible night, the news that two surgeons were admitted to the hospital with a viral infection, or something, was in the newspapers. Reporters often visit hospitals in Quetta, looking for stories. The story went, the surgeons are ill in room number such and such and such. I think from seven o'clock in the morning doctors and nurses and friends started visiting us, and I think five or six hundred people visited us. I talked with Dr. Shafiq, discussing our very low white-cell count. We thought we might get some infection from all these people, so I arranged for us to have masks to protect us from people till the afternoon."

Oh, I thought, so that was why they were both wearing surgical masks when I found them in their room at the Aga Khan Hospital. I hadn't known about all their visitors in Quetta. At Aga Khan, we were strictly controlling people from going into the room, except for essential medical and nursing staff.

Dr. Jamil turned to me and said, "Dr. Fisher-Hoch knows very well the rest of the story."

But I pressed him to go on. I only knew my side. He continued: "We caught the regular afternoon PIA [Pakistan International Airlines] flight from Quetta to Karachi. There were two or three ambulances waiting there, including the one organized by Dr. Shahid. Dr. Shahid was also there with my brother-in-law. Another ambulance had been organized by Dr. Shafiq's brother-in-law, who was a brigadier in the army. He was also there. So we rushed to the hospital and were admitted about seven o'clock in the evening. We were seen by the routine duty doctors, but no consultant saw us.

"The next morning the consultant came there, and I told him the story, but I think he was not impressed. He didn't believe me that I was suffering from such and such disease. He told me that we will be doing urine culture, throat culture, and blood culture, and he was suspecting some bacterial infection or some viral infection, but not a serious disease. I called Dr. Shahid to me again, and I told him that this man does not realize the situation; please do something, otherwise we are going to die. So he went to talk to his professor, Professor Khurshid, and told him that two of my friends are here and they say that they are infected with CCHF virus. Dr. Khurshid recognized that this was a serious story, and he and Dr. Shahid went at once and found Dr. Fisher-Hoch."

This was a day that I will never forget. I was sitting in my office, writing on the computer, when Professor Khurshid and Dr. Shahid rushed into the room. I was surprised to see them. They then told me about Jamil and Shafiq. After they'd described the operation, adding that both patients had very low platelet counts, I put my head in my hands and said, "This is CCHF!"

We hurried to their room. As I was taking their history, Joe walked in. We both recognized how serious the situation was. The patients were febrile, and their blood picture was bad. And they also had one or two ecchymotic spots—purple discolorations—typical of CCHF. Given what we knew about the disease, the prognosis was very poor. I thought they would probably die.

And why was their room so crowded? None of these people should have been there. Joe and I immediately emptied the room, and placed Dr. Jamil's brother in charge of keeping everyone out, with the exception of authorized personnel. We then set about trying to explain to the staff what we really meant when we talked about "barrier nursing." To drive our point home, we had to be very blunt. Otherwise, we realized, we wouldn't have been taken seriously enough. What made the whole exercise more frustrating was that this was unquestionably the best hospital in the whole country. Already we could picture the headlines: CCHF KILLS SIX AT MAJOR KARACHI HOSPITAL. It would be a great sensation, but for the hospital a terrible disaster.

The sad fact is that the need for safety does not seem to be understood, or it is simply not regarded as all that important in this culture. Either you can't get people to take reasonable precautions, or else they finally do get the message, panic, and refuse to look after the patients. In this case, though, the nursing staff at Aga Khan complied with our instructions without panic or protest; and in this respect, they did better than the physicians—who couldn't be told *anything*.

One reason why Joe and I were so concerned was that the two surgeons had been sick for four or five days before we saw them. While we knew from the South African experience that there was a good chance that ribavirin would be an effective treatment, it was also true that the best results were achieved early in the disease. In the case of Shafiq and Jamil, we thought it was better that the drug be given intravenously.

Dr. Jamil went on with the story: "Dr. Fisher-Hoch and Dr. McCormick visited us along with the medical director of the hospital, Dr. Mirza. I narrated my whole story to Dr. Fisher-Hoch. I remember at the moment she said that, 'Yes, a hundred percent you must consider yourself infected with CCHF.' When she said this, I think we relaxed a little bit, that at least someone had diagnosed our problem. Then she said we must have ribavirin, and we started running after a supply. Injections of ribavirin were not available in Pakistan, and my brother and my friend were worried. There were capsules, however, and she told us to send someone to get hold of them at once, and start taking them immediately."

I had told them that I didn't care what else they did, as long as they got hold of those capsules and took them every six hours on the dot. If they couldn't swallow them, we would find some other way of getting the drug into them. In the meantime, we left no stone unturned trying to find intravenous ribavirin. We made calls all over the country, and when that yielded nothing, we tried Singapore and Europe as well. There had to be a supply of intravenous ribavirin somewhere that was accessible, we reasoned. We reasoned wrong.

"Dr. Fisher-Hoch said that if the injections are available, yes, there is chance that they will survive, and with capsules I don't know. They may or they may not, but it must be tried. She said to my younger brother: 'Start these capsules immediately, and if we can get the injections we will change over at once to intravenous.' My brother-in-law contacted the manufacturer of ribavirin, ICN

Pharmaceuticals in the United States, and persuaded them to send us the injections. But it took four or five days for it to reach us, by which time we were getting better on oral ribavirin. When I asked my younger brother what Dr. Fisher-Hoch said, he did not tell me the whole truth. He said that she will save you, and you will be all right, and there is nothing to worry about. When we recovered, then he told me that Dr. Fisher-Hoch actually said she did not know whether we would die or not."

Even treated with the oral ribavirin, our two patients did very well. But we had other worries. We were told that there was a sweeper—the local term for janitor—at the hospital in Quetta, who'd had the task of washing out the blood-soaked cloths after the same operation from which Shafiq and Jamil had become infected. He was now known to be home, sick. I immediately was reminded of what had happened at the hospital in Aba, Nigeria. There a student nurse had washed blood-stained operating drapes after surgery. And there, too, both surgeons had died and the nurse had fallen ill.

We made urgent calls to Quetta in an attempt to locate the sweeper. The director of the hospital personally went to look for him. As soon as he found him, he insisted the sweeper accompany him on a flight to Karachi. When the sweeper was brought into Aga Khan, he was placed in a room opposite the one the two surgeons occupied. When I went to examine him, I was appalled. I couldn't believe that the admitting doctor had put him in with another patient. It was a situation I made sure was rectified at once.

I didn't need to take a history to figure out that the sweeper was in very bad shape. He was bleeding from his rectum and had a large ecchymosis—the purplish bruise that results from bleeding below the surface of the skin. He certainly had CCHF. We started ribavirin.

Luckily, we got to him in time. He, too, did very well on the medicine. As soon as he was better, he left Quetta, declaring that it was no longer a safe place for him to be. But the last I heard of him, he was back in Quetta and working at his old job.

Dr. Jamil went on with his story: "We were very sick. We had evidence of bleeding, with small skin lesions, spots on the abdomen and arms. We were scared of going to the toilet, thinking that the bleeding will start, and we will die of hemorrhage. Shaving was forbidden. Brushing teeth was also banned, in case it made us bleed. We became very drowsy. For two or three days I don't remember what happened to us. My brother told me later that he used periodically to make some sound to make sure we would still wake up from such a deep sleep. Later one day, when I saw my pulse, it was around fifty or sixty per minute. I told my brother to get a physician and get our ECG done, because I think this virus has infected our heart muscles as well. Then we saw our urine become yellow and thought we were getting jaundice, but Dr. Sue said she thought it was just concentrated because we were not drinking enough. Another day I felt pain in my upper abdomen, and I remembered the liver of the patient who died. I told Dr. Shafiq that our livers are just like the liver of that patient. It will be swollen and friable.

"After about seven or eight days, the pains stopped, and Dr. Fisher-Hoch came into the room in normal dress, without a gown and mask, and shook our hands and said, 'You are all right, and you can leave the hospital because you are no longer a danger to other people.' The course of ribavirin was not finished, so we took it seven days in the hospital and three days at our residence. We were advised to take about six weeks' rest. In the recovery phase, we used to get tired even sitting with friends for an hour. We were fully recovered, I think, after six weeks, and I started my work again in the hospital. When I went home my mother was awake and said, 'Where are you, why are you doing so much work in the hospital? Take some rest.' And than everybody was after me not to go to the clinics and not to work in the private hospital. But I was okay. We had both lost many pounds in weight, and we ate and ate after going home. Sometimes four, five, six meals a day.

"After two and half months I got married, but I was very much worried about this, that I should marry during this period, because I was scared that I may transmit this disease to Saima. Dr. Fisher-Hoch told me when I asked her that this was not a problem.

"The worst part was seven days in the AKU, remembering the story of my friend who died."

All this time Dr. Shafiq had been quiet. We turned to him, and he began to talk. "I did not have any other experiences beyond those that Dr. Jamil told you about, but I was much more depressed. My wife was so depressed. Dr. Jamil was unmarried. I am married, I have three kids, so I was worried about them. What would happen to them after my death? Who will care about them? I was very much conscious of this all the time, even from the very beginning, even when the physicians came, Dr. Jamil came, and my professor came."

We were all quiet for a moment. It was after midnight. We gathered ourselves together, said good night, and then went on our separate ways to sleep.

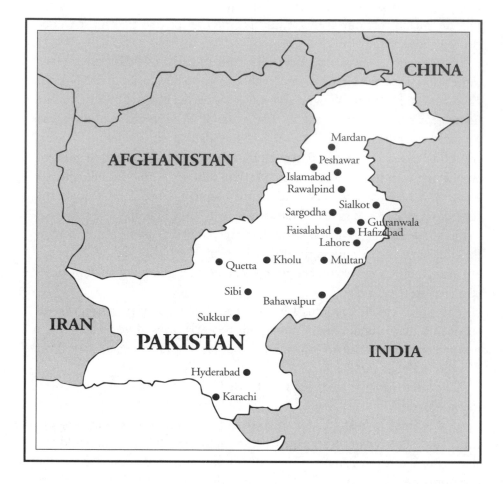

Between Two Worlds

Physicians such as the two surgeons Sue helped save from CCHF are essential assets to an underdeveloped country. They reflect the hope of their community and the pride of a respected profession. Intelligent, hard-working, and dedicated to serving their own people, the progress they create is tangible. Our young recruits to the epidemiology program at AKU are recent medical graduates and young faculty, and are likewise bright, enthusiastic, and tireless. They are also eager to do something meaningful and are not averse to adventure. Best of all, they are interested in working in their country. The Aga Khan University had inaugurated a good medical school to train high-quality physicians, but its graduates, along with many able graduates of other medical schools in Pakistan, were inclined to leave the country to seek further training in the West. Many never return. We tried to impress on our students that they could find no greater excitement or satisfaction than pioneering work in their own country, where public health is so desperately needed. In an internship or residency in the West, they would see one patient at a time, and even if they returned to Pakistan as tertiary care physicians, their impact on health would be marginal. If we could offer inspiration and create a critical mass of epidemiologists, then, I thought, there was a chance that the entire health system could—sooner than later—be shifted from a dominantly clinical approach to a preventive approach, more in keeping with the resources available and the needs of the people.

Sue set about building a virology department at AKU from scratch. In Pakistan, the study and practice of virology didn't really exist. Like me, Sue works with young people in her lab and in our public health research program. Our programs range from studies of diarrhea, pneumonia, and tuberculosis (among the biggest killers in the country) to injuries, burns, and violence, to studies of health care delivery among the urban and rural poor. There were six epidemiology recruits in the first year. As soon as we thought they were ready, we challenged a few of them to hunt down viruses. Aamir Javed Khan we dispatched to one of the most remote regions on the face of the planet, the mountain deserts of Baluchistan, to work among nomads and see what evidence he could find of Crimean Congo Hemorrhagic Fever. It was a grueling assignment, and he couldn't have been happier.

Sue and I read Aamir's report: "In the last week of August 1995, Taj Mohammad, a Baluchi shepherd from the Khetrani nomad community of Narkot in the Barkhan District of Loralai Division, Baluchistan, took his family's herd of sheep and goats up the surrounding hills, his choice of pasture over the preceding month. On this day, however, he came back to tell his family that he was feeling weak and feverish."

Unquestionably, Aamir Javed Khan is one of the best. He readily rejected the easy option of instant emigration, even though he had passed his exam to qualify for an internship in the States. A thickset, handsome young man, with an intermittent growth on his chin, depending on the state of his studies (more book time, more beard), he has an insatiable enthusiasm for food. His Central Asian Uzbek ancestry shows in the roundness of his face and in the brightness shining out of his dark eyes. Aamir had already played a major role in several surveys, including one for AIDS among prisoners in the Sind, as well as our hepatitis C study.

The idea for the survey of nomads originated as I lay in bed with the flu in a room at the Serena Hotel in Quetta. Jamil, Shafiq, Aamir, and Sue had collected in the room for an impromptu meeting, and were talking about various ways we could identify the risk of CCHF. We agreed that a survey was needed. But who should be surveyed? Suddenly, I sat up.

"Nomads," I said.

We already knew that the nomads were in close contact with sheep and ticks. One evening previously we met a Baluchi tribesman in a tranquil courtyard at AKU. He was wearing the traditional tribal garb, a white robe and a large white cloth wound loosely around his head to form a broad flamboyant turban. He told us that there had been many deaths among his people.

Aamir headed off into the mountains in pursuit of cases we had heard about from the Baluchi tribesman at Aga Khan. He finally reached a valley called Kholu.

"The road is terrible," he wrote in his report. "It took about twelve hours to get there from Quetta. You have to go through passes that are so narrow you can almost touch both walls. There are no hotels, but we were lucky; we were put up in a government hostel. The valley itself is desolate but beautiful. It's a dry mountain desert. At night it's very cold—below freezing; by day, it's filled with bright sun. There is little vegetation, yet the valley floor is studded with the tents of nomads and their herds of camels, goats, and sheep. The Baluchis range over large areas of Afghanistan, Iran, and Pakistan. These are a people of wonderful carpet weaving and embroidery of incredible fineness.

"There is no electricity. Water is scarce. There is no gas and little cooking fuel. This is ironic, because less than two hundred kilometers from the valley is Sui, which is Pakistan's largest natural gas field. But these people receive no benefits, no compensation, not even any gas. The war in Afghanistan just over the border has prevented them from making their traditional summer migration to the higher mountain pastures in that country. The best that they can hope for is to scratch out a living in the dusty landscape. A worldwide symbol of basic influence from the outside world is the presence of a Coke advertisement. There are no Coke advertisements in Kohlu."

Aamir was clearly moved both by the stark lunar beauty of the place and the plight of the people. Perhaps his empathy was natural: the Uzbeks come from northern Afghanistan and Russia, which made them neighbors of the Baluchi, and many of their tribes are suffering a similar fate.

His account continued: "Over the next two to four days, Taj Mohammed became progressively weaker and developed severe backache. He at first opted to rest at home, but on the 28th of August his gums started to bleed, and the family decided to take him to the government civil hospital in Barkhan town, a half-hour jeep ride from Narkot. The doctor at the hospital examined him and informed the attendants that the patient required snake antivenom, not available in Barkhan. He advised the family to take the patient to Kohlu, a fifty-minute jeep ride away, where the antivenom could be procured. Around this time, the patient also developed a nose bleed.

"On August 30th at about two in the afternoon, Taj Mohammed presented himself at Dr. Usman's two-room clinic in the small, remote market town of Kohlu, a twelve-hour jeep ride from Quetta city. Dr. Usman examined the patient on the same day and, along with his compounder, also cleaned the patient's gums. [A compounder is the man who "compounds"—grinds up— herbs and indigenous medicines with a pestle and mortar.]

"The patient began to vomit fresh blood on the second day of admission. Dr. Usman called the two other practicing doctors in Kohlu to examine the patient—Dr. Aurangzeb and Dr. Khair Mohammed. The trio decided that the patient had been bitten by a small snake endemic in the area, 'whose bite is not always noticed by nomads.' The patient was given snake antivenom, and the attendants prepared on the advice of the doctors to take him to the city of Multan, where there was a large government hospital. But Taj Mohammed asked to be taken back home to Narkot. There he became unconscious and died. Khan Mohammed, his brother, recalls bluish-black bruises on the patient's body and seeing fine red dots on his abdomen while he was ill.

"Khan Mohammed was with Taj Mohammed throughout his illness and remembers wiping his blood. Khan Mohammed became febrile three days after Taj Mohammed died. He developed abdominal pain on the first or second day of illness and recalls distinctly the presence of fine red petechia-like spots on his abdomen. He also had lower backache and knee-joint aches. He was ill for about three days, but recovered.

"Dr. Usman developed a fever on the 2nd of September. Over the next two days, he continued to work, though he complained of weakness and body aches. On the morning of the 5th, Dr. Usman's uncle noticed that the doctor's gums were bleeding. At this point, Dr. Usman told his family and friends that he had contracted some virus from the index case, Taj Mohammed. He agreed to go to Dera Ghazi Khan, a town six hours by jeep from Kohlu. There Dr. Usman was seen by Dr. Yousuf in a private clinic. Dr. Yousuf tested Dr. Usman for his platelet levels and reported a count of 16,000 [normal is more than 200,000]. He advised the patient to seek medical attention at the Nishtar Medical College Hospital in Multan. Dr. Usman finally reached Multan, but had great difficulty in getting a doctor to actually see him. It was past midnight on the 6th, when he was admitted to the family medicine ward at this government hospital. He was treated for typhoid and malaria, but to no avail. He continued his fever and illness, and his platelet count remained very low. Dr. Usman remained conscious and was eating soft foods, but on the morning of the 7th, he started passing fresh blood per rectum, which continued throughout the day. He was given over five units of whole blood, and in the afternoon, the treating physician advised evacuation to the Aga Khan Hospital in Karachi. He started vomiting fresh blood, and before he could be moved, at around 8:30 P.M., Dr. Usman died in Nishtar Hospital.

"Just two hours prior to his death, a large caravan of forty or so of his friends and family, all males, had reached the bedside of Dr. Usman in Multan. [Sickness and death in Baluchistan, as in Africa, are not private affairs.] There was a lot of blood around, and many of these people were exposed to large quantities of infected blood. The body was brought back to Kohlu and buried. Five days later, one of the relatives, Jawand Shah, who reached the bedside just prior to death and had contact with Dr. Usman's blood, developed an acute febrile illness. He was seen by Dr. Aurangzeb, who had seen Taj Mohammed, and he diagnosed snake bite. Malaria tests were reported positive, but the next day, the patient developed gum bleeding. Jawand Shah was also taken to the DG Khan Civil Hospital, where he was seen by Dr. Manzoor Ahmed Qamar. His platelet count was very low, and, on the morning of September 19th, he had a nose bleed and vomited copious fresh blood. Jawand Shah died at noon that day, and the body was taken back to Kohlu for burial the next day. The body was washed by his uncle Azim Khan."

At this point in Aamir's report, we stopped reading and began counting the casualties—about ten in all, as we figured it, of whom at least four had died. Aamir had reviewed tick exposure and other related issues. With ticks thriving on the hides of sheep, goats, and camels, the potential for infection was obviously enormous. It would not take an unreasonable jump to imagine a chain of infections, leading from a city like Multan. It is one scant hour by plane to Karachi. From Karachi to Europe, eight hours; to the United States, just thirty hours. Geography is no protection, and a fat bank account no guarantee of escape from emerging diseases. We ignore the afflictions of the underdeveloped countries at our peril. Karachi, the city where we now live, is a vital link in the chain.

To get to Karachi's Civil Hospital it is necessary to make one's way—slowly—through streets pulsating with cars, buses, lorries, rickshaws (motorized three-wheeled vehicles), and carts drawn by mules and camels. The old Bundah Road leads to the hospital and the port, past a desolate landscape of broken sidewalks, crumbling buildings, and flowing sewage.

Dr. S. N. Bazmi Inam, a pediatrician specializing in public health at AKU, is a slender, intense man in his late thirties. He has a certain air of world-weariness about him, perhaps acquired from years of struggle against poverty and intransigent bureaucrats. Although he has been working at AKU almost since its inception, he regards one of the two government hospitals, the public Civil Hospital in downtown Karachi, as his "parent." There he received his basic medical training. From time to time, Bazmi returns to his "parent" to see how it is getting along. As he drives, he glances out the window at the traffic and grimaces.

"I always drive with my windows closed."

The air is thick with noxious fumes spewing from myriad cars and trucks. The pollution, Bazmi says, is only getting worse with the population explosion. A recent study, he says, found that, of all the major cities in the world, Karachi has by far the highest levels of lead in the air, despite the sea breezes that freshen its atmosphere. Pakistan, it turns out, adds more lead to each gallon of gasoline than any other country in the world.

The Civil Hospital has the desolate institutional look of a skeleton with a few rags hung on it. The facade is disintegrating, the gloomy entryway needs paint. Bazmi says that the hospital has deteriorated markedly over the last several years (as have most of the public hospitals in Pakistan). He attributes the decline to politics in a system that allows doctors and administrators to reach the most prestigious posts, not because of their abilities, but because of favoritism and connections. The government has failed to operate an effective medical system. Medical schools assign little importance to prevention and family medicine. They prefer clinical training, particularly in specialties. Moreover, less than one percent of the GNP goes to health care and education combined, whereas about 35 percent of the GNP goes to the military, and 30 percent to debt servicing.

"We are suffering from the problems of both the underdeveloped and developed world," Bazmi says, "and we are dealing with both of them poorly."

The Civil Hospital is the medical facility of last resort for thousands of Karachi residents, as is the Mama Yemo Hospital for those in Kinshasa, the Connaught Hospital in Freetown, the Treichville Hospital in Abidjan, and the public hospital in Kikwit, site of the recent Ebola virus epidemic. Civil Hospital is no different from the countless, but destitute public hospitals in cities and towns in developing countries around the world. The emergency room entrance resonates with the wails of frightened children and the moans of the injured. Some of the women are bundled up in *dupattas* and colorful *shalwar kamiz*, others are all in black, their faces heavily veiled. The men sport an admixture of styles, ranging from Western garb to more somber *shalwar kamiz* and turbans. Yet whatever their differences, they are united by their need for urgent medical care—and by their poverty. The two seem inseparable. Patients are not charged, but the sad fact is that patients will be required to purchase most of the medication and supplies they need.

This 1,700-bed hospital is the largest in the country. Typhoid and cholera account for a large number of patients, cardinal signs that clean food and water are not provided in the city. While the hospital does provide food to patients, families are allowed to bring their own.

"It's better that they do," Bazmi notes dryly.

Drugs and blood for transfusion are difficult for patients to obtain, given the scarcity of funds. The reuse of needles and syringes and transfusion with unscreened blood are common. Students at the Karachi Medical School recently formed an association to raise money for the general care of poor patients. Their motto, "We Feel, We Care," hangs prominently above the blood bank.

Karachi's growth, like that of megacities throughout Africa and Asia, has been meteoric. When Pakistan became independent in 1947, Karachi was a sleepy port city with a population of about 300,000. Now its population of close to twelve million is increasing at the rate of about 6 percent a year. No census has been taken since the early 1980s. The allotment of government jobs, money, and political representation in the provincial and national assemblies is distributed according to the population of each demographic group. Because of ethnic and provincial conflicts, the government has been unable or unwilling to take a new count. The political and ethnic tensions that have accompanied Karachi's growth are reflected in each night's tally of the number of detained, wounded, or simply slain in what are commonly referred to as "encounters" with police. Bazmi calls the situation a "low-grade civil war." Countless variations of this may be found in other megacities worldwide. Violence is the last resort of the world's poor and oppressed.

It's the same story that we have seen in Africa and in South America, as farmers and peasants abandon the countryside for the city in expectation of finding work. Karachi's reputation as the commercial hub of the country has also attracted people from Central and Southeast Asia, and as far away as Russia. There are thousands of squatters who construct their own makeshift quarters, called "encroachments," which are regularly torn down by police and just as regularly resurrected by the squatters.

As more people are born or immigrate, the infrastructure, already strained, breaks down a little more. A single whiff of Karachi night air returns the stench of raw sewage. There are many parts of the city where no sewer lines exist at all and where waste runs through open ditches. Not only is the city's tap water unfit to drink, in many cases, it has to be delivered by truck, either because there aren't enough pipelines or because the lines have been "shut down for repairs." Without sanitation, fecal material runs into the water and food. People get typhoid and other salmonella infections, shigella, cholera, E. coli, and many other bacterial—as well as viral—gastrointestinal infections. They are treated with broad-spectrum antibiotics, often inappropriately or inadequately, which frequently has the effect of selecting bacteria resistant to many antibiotics. An endless cycle floods bacteria through millions of people treated with antibiotics, which select out resistant strains that are fed back into the water. Over the past ten years in Karachi, *Salmonella typhi*, the cause of typhoid fever and the most common bacterium isolated in blood culture in Karachi hospitals and clinics, has gone from being sensitive to the commonly used cheap antibiotics to being utterly resistant to three or more antibiotics. Similarly, a new strain of cholera, called 0139, has emerged in South Asia, incuding Karachi, and it is resistant to commonly used sulfonamides.

So what can a nomad do? Remain in a desert valley, where they are restrained from practicing their traditional lives because of war, and starve or, perhaps, die of myriad infectious diseases? Or is it better to take a chance in a city like Karachi?

It would be difficult to find a more dramatic contrast between the Civil Hospital in downtown Karachi and the hospital at Aga Khan University, located in a fashionable residential district about twenty minutes' drive away. Aga Khan University, which is possibly one of the best facilities of its kind on the Asian subcontinent, was built of marble, tile, and stucco ten years ago at what for Pakistan was the stupendous cost of $300 million. While there are many beds here reserved for charity patients, for the most part, the hospital is available to the relatively more wealthy; there is no medical insurance in the country to speak of. The university encompasses a medical school, with a department of Community Health Sciences, a nursing school, and an Institute of Educational Development in addition to a private hospital, all constructed around a series of courtyards that successfully evoke the golden age of Sindi architecture. The red sandstone walls of the buildings use a traditional desert design, which creates a constant shadow and mitigates the impact of the sun in the middle of the day. Patients and their families regularly gather in the peaceful gardens to chat. Nevertheless, like many high-quality institutions throughout the world, AKU is struggling with how to provide excellent health services to all, including the poor, and pay for it. It is a social-economic schizophrenia, for which there is no easy cure.

Dr. Saeed Hamid is a genial, smooth-spoken gastroenterologist who received his postgraduate training in London. His low-key manner masks a certain impatience. Since hepatitis is a major public health problem, he has come to understand that the illnesses he treats are economic and social problems more than medical conditions. Four or five mornings a week, he leads about half a dozen enthusiastic residents on rounds.

The first patient we meet is a dark-skinned sixty-year-old man with a gray beard, who is suffering from severe hepatitis B. Like hepatitis C, this is a virus spread in blood, which slowly destroys the liver, but, unlike hepatitis C, there is an effective vaccine. Trouble is, countries like Pakistan that really need the vaccine do not give it to their children. Like many patients, this man has been in and out of the hospital over the last several years. The only thing that can be done for him now is to treat the symptoms, drain the fluid from his abdomen, and give him a shot of albumin (cost $100) to elevate his serum protein. These are temporary measures, which hardly constitute a cure. Saeed says that for the amount of money it costs to treat him this way, fifty or more people could be vaccinated. Saeed's practice is overloaded with such patients.

It is also end-stage liver disease that afflicts Saeed's second patient of the morning. A woman of fifty-two, she, too, has constantly been in and out of the hospital. Seldom does she see the same doctor twice. This is another big problem, Saeed says. It is unusual for people to have any regular primary health provider; as a result, there is no one available to catch the disease in its early stages, when patients could be treated most effectively. The third patient is not only suffering from liver failure, but from kidney failure as well. This is no coincidence. Patients who undergo a great many transfusions and regular dialysis are at greater risk of being injected with non-sterile needles and acquiring hepatitis viruses.

"We keep telling the kidney centers to dedicate different dialysis machines for hepatitis B and C patients to minimize the possibility of infection of others who share the machines," Saeed says, "but it's expensive. The centers have only managed to dedicate machines for hepatitis B, not for hepatitis C."

It is yet another illustration of the poverty-borne improper use of Western technology: There is no hepatitis B vaccination, but there is a renal dialysis program at the public hospital.

Saeed escorts his residents down one floor and into the emergency room. He wants to take a look at a man, also afflicted with end-stage liver disease, who was admitted two nights before, vomiting blood. The day after his admission, Saeed recounts, the man checked himself out because he felt better. He went home and immediately started throwing up blood again. So he came back.

"It would have been cheaper if he'd stayed in the first place," Saeed says. But he can understand why the man left. "The meter is ticking."

A drive of only a few miles from the center of downtown Karachi, with its office towers, banks, and boutiques, will bring you to a district called Essa Nagri—a slum, which, in Urdu, is called a *katchi abadi*. The word means poor or temporary dwelling. Essa Nagri is one of five slum sites in which the Community Health Science Department at Aga Khan University operates. Here, AKU is represented by Shaista John, a pretty and engaging woman of twenty-nine who is a community supervisor (social worker), and Safia Dhouri, a personable woman in her early fifties who is a lady health visitor. In this country, only women can visit houses as health workers; males would not be allowed in. The community center is reached by one of seven dirt lanes that thread through the district. They are jammed with children, carts, and donkeys, mingled with drug addicts and water buffalo. When the center opened in 1987, these lanes were so mired in sewage that it was necessary to walk on planks to reach your destination. Not that the sewage has entirely disappeared now—the brook that runs through the middle of the district still exudes a rank odor—but at least the planks are gone and you can walk on something that resembles solid ground. The center itself is nothing more than a spartan cement structure of two floors, equipped with a few wooden tables, some folding chairs, a couple of old metal file cabinets, and a hot plate to maintain a plentiful supply of tea for visitors.

At first, the center's role was to promote health awareness and prevent disease. Only gradually did it expand its function to encompass social outreach and community development. People have been made more aware of the need for hygiene. Shaista is especially proud of a program she has developed called Children to Children. Because so many mothers are at work, older siblings are obliged to look after their younger brothers and sisters and make sure that they stay out of trouble. An aspiring playwright, Shaista uses playlets and skits to encourage youngsters to assume responsibility for their households in their parents' absence.

If the children of Karachi have a future to look forward to, it will be on account of people like Shaista. Regrettably, there just aren't enough of them.

All over the world, December 1 is commemorated as AIDS Day. But, until recently, Pakistan had ignored the event, discounting the threat of the disease to its people. For years, Sue and I and many others in the public health field have said to anyone who would listen—and, probably, a greater number who would not—that Asia was at particular risk for AIDS. It isn't that anyone wanted to assume the role of a Cassandra, but the evidence of the threat was little short of overwhelming. I'd seen what happened in Africa; I didn't want to have to watch the same thing repeat itself on another continent. The sooner that Asian nations took action, I argued, the more chance there would be of mitigating the impact of the epidemic. Indeed, I gave such a talk in 1985 in San Antonio, Texas, at an international meeting, only to receive an irate letter from a Thai physician accusing me of slandering his country, because I had pointed out the rise of prostitution and drug addiction in Bangkok. Only when the HIV rates skyrocketed in 1989 did the Thais realize the sad validity of what I and others had been saying.

Today there is no longer any question that AIDS is spreading throughout the continent, which is why the inaction of Pakistan is so disturbing. It is always the same response: "It won't happen to us. We are different." Yet WHO recently issued a statement declaring that the center of gravity of the infection is beginning to shift from Africa to Asia; with 60 percent of the world's population, Asia is expected to dominate the HIV picture in terms of total numbers infected. India is expected to have the largest number of infected individuals of any country in the world by early next millennium.

Countries in the Asian region particularly prone to social disruption, primarily because of the extreme poverty associated with violence and chaos, including Cambodia, Vietnam, and India, are at especially grave risk for AIDS. By the end of 1995, Pakistan had eight hundred confirmed cases of the disease, but there's no question in anyone's mind that the real number is much higher. Pakistan is a country where 70 percent of the population is illiterate and where victims of AIDS are stigmatized. Public health workers may quite literally take their lives into their hands when they attempt to question people with AIDS in order to learn more about the spread of the disease. On one occasion, a man with AIDS warned an investigator that he would "blow his head off" if he ever dared to come back to his house. He was holding a Kalashnikov—the assault weapon of choice in Pakistan following the Afghan war—and he left absolutely no doubt that he was prepared to use it.

But even in the face of such problems, we and our colleagues have refused to give up. Ever since we arrived at AKU, we have been trying to galvanize people into taking action against AIDS before it overwhelms the country. We initiated a small, inadequately funded project spearheaded by a young physician, Shehla Baqui, who had just finished four years working in the Bronx. It was slow to get started, because the idea was foreign to so many. Now we have the collaboration and involvement of Dr. Sharaf Ali Shah, manager of the AIDS program of the province of Sind. The official National AIDS Program has been very poor and has only succeeded in consuming precious resources, producing little. After ten years, there is little (if any) accurate data on AIDS in Pakistan. Our collaboration with the Provincial Ministry of Health (with support from the University of Alabama at Birmingham and the Fogarty program of the U.S. government) has allowed us to expand a working group of professionals who had an interest in seeing that more was done about the problem. We are now in a position to be able to train more personnel and set up special clinics in Karachi.

The organizer of the AIDS Walk, Dr. Sharaf Ali Shah, intelligent and highly motivated, has proven an able diplomat, insuring that the needs of the program are met while at the same time making certain that political and religious interests continue to be satisfied. To drive home the point that AIDS is an issue that affects everyone, he recruited as many schoolchildren as possible. Education about AIDS has to begin with the youngest members of society; unlike many of their elders, they are still open to new ideas.

Joe and I left Karachi at the end of 1996. Although the city was tumultuous and overall a difficult place to live and work, we had nevertheless achieved a number of our goals. Most important, we had created programs training a substantial number of young people in public health, family medicine, epidemiology, and microbiology. Our intent was to provide a nucleus of young people with identifiable skills who could ultimately make the changes needed for their country to make more rapid progress in improving its people's public health. Karachi was and remains a very difficult place to work in public health despite the desperate need.

Further deterioration of the social structure has been brought on by more assassinations in downtown Karachi. These assassinations are carried out with automatic rifles. Targets include religious sects and political elements, and Americans working on nothing more controversial than developing local oil projects for the Pakistani government. The summer of 1998 brought the nuclear arms race with India. Danger for expatriates is now compounded by terrorists over the border in Taleban-controlled Afghanistan—a loveless society, without pity or charity, particularly for women. The uncomfortable, eerie feeling of Karachi increased in August 1998 after the embassy bombings in Kenya and Tanzania. We knew at least one of the victims in these bombings. The subsequent missile attack on Afghanistan was just over the border from the area where we had searched for Crimean Congo hemorrhagic fever.

Steve Luby was still there in 1998. He stayed on after we left, nurturing his enthusiastic young crew of epidemiologists and community health residents and expanding an excellent program of field studies. He was the bedrock of the master's degree course Joe had established. By June 1998, following the Indian and Pakistani nuclear tests, Steve told us he felt he had to get his young family back to the U.S. By the end of August 1998, even the American school where his children had studied and where we used to jog regularly was closed. Many of our young Pakistani colleagues and friends are currently in the U.S. obtaining higher education. We remain in frequent contact. We hope things will improve and that they can return to their homeland as they intend, with new skills and knowledge in public health, epidemiology, microbiology, and other disciplines. We are sure that most of them will do that, but they are presently anxious about and depressed by social and political developments.

CHAPTER THIRTY-SIX

Lassa Fever Vaccine on the Fast Track

When in December 1996, Joe and I returned to the U.S., we were considering taking up positions at a university in the South. However, Joe had been approached by Maxime Schwartz, the Director of the Institut Pasteur in Paris, and we had visited Paris in late November 1996 to see him and discuss possibilities. Late in the day we dropped in on a friend of Joe's at the Institut Pasteur, Marc Girard. We chatted awhile, then left him at about 5 P.M. to return to our nearby hotel to finish packing and leave to catch our flight back to Karachi. Just as we were loading our bags into the trunk of the taxi on the busy cobbled street of Avenue Suffren, the receptionist ran out calling after us.

"Monsieur McCormick, there is a telephone call for you."

Joe went back in and picked up the call. It was Marc.

"I have just spoken with Charles Mérieux. He wants to see you at once."

We had heard of Charles Mérieux. He is the grand old man of vaccines, the creator of the Mérieux group of companies that later became the Pasteur Mérieux Connaught vaccine company. PMC is the largest producer of polio and meningococcal meningitis vaccines. Charles Mérieux was born in 1907, the son of Marcel Mérieux. Marcel Mérieux had worked with the legendary Louis Pasteur in Paris, then returned to his native city, Lyon, in the south of France, to set up the Institut Mérieux in 1897. The stuff of which history is made, and this famous man wanted to see us—at once. But we were in Paris, had a flight in two hours to Karachi, and many commitments back there.

Joe replied, "But Marc, we can't do that now, we are just about to catch the plane back to Karachi. Can I call him from there?"

"OK," said Marc, "but I warn you, his French is most particular, and even

I have difficulty understanding him over the telephone. The other thing to re-member is that he is ninety years old, and he is in a hurry!"

He was indeed in a hurry! This was the beginning of an intense exchange of phone calls (which were very difficult to understand), and multiple faxes, which are the Docteur's preferred means of communication. He worked then, as he does now, at ninety-three years of age, every day in the office. He has read all the newspapers before 8 A.M. and is still at his desk at 6 P.M. Now re-tired from the companies he founded, he runs a charitable foundation with a wide variety of excellent public health projects. He is constantly traveling, at-tending meetings, starting new projects, and in general driving his vision of public health as far as he can. His faxes tend to be fired off first thing in the morning (French time), which was fine in Karachi, but the exchange contin-ued when we returned to Atlanta just before Christmas. There faxes rolled in at 3 A.M. on a regular basis, including Christmas Day. The phone would ring five times. We buried our heads, and left reading the fax until we made morn-ing coffee several hours later. We can now joke with the Docteur that the only way to get a full night's sleep had been to come to France.

"Il faut venir, au plus tôt possible"—You have to come as soon as possible— was the recurring theme.

The 3 A.M. Christmas morning fax decided us. We booked a return flight to Geneva for December 27, coming back just after the New Year.

We were received like royalty and with great kindness. Charles Mérieux, known as "Le Docteur," despite his wealth and fame is a totally unpretentious and immensely generous man. We were taken to Annecy, the lovely alpine country home of the Docteur, which now doubles as a conference and train-ing center. Les Pensières, at Veyrier-du-Lac, is several hundred years old and sits on the shore of the lake. There is an ancient water gate where the abbots of Annecy used to step ashore from their boats, having been rowed across the lake from the great abbey of Annecy for their summer retreat at the old house. As we arrived it was snowing lightly on the wide lawns and the great old cy-press trees, and the mountains were shrouded in mist. We showered and changed our clothes. Pretty soon came a telephone call from Lyon.

"Tout est changé"—Everything is changed. "They must come to Lyon at once," he told our hosts, Masha and Micha Musset, who ran the conference center attached to the house at Veyrier-du-Lac.

So we drove to Lyon and were shown into his office. We found a magnifi-cent old man with a wicked gleam in his eye and a full head of thick white hair, but with a modesty and charm we found utterly remarkable in such a fa-mous figure. Behind him was a spectacular view of the gray walls and towers of the thirteenth-century abbey of Ainay, sharply outlined in the still-falling snow. A younger man was presented to us.

"Jean-Marc Tourret, architect," he said, offering us his hand.

On top of a fractious array of papers and pictures on the table was a sketch for a Biosafety Level-4 suit laboratory. We were already familiar with the project, as we had seen the sketch, and knew this was why the Docteur was after us. The Docteur wanted to build a high-containment laboratory.

"Il faut faire un P4"—We must build a P4—he said. P4 is used in France for high containment, where it has some meaning, the P being associated with the French word *protection*. BSL stands for Biosafety Level, which means nothing in French. Charles Mérieux had been wanting to build one for twenty years, he said, and produced books and papers to prove it. In his youth, his sixties I guess, he had been responsible for one of the first Biosafety Level-3 laboratories ever. It was for cows with foot-and-mouth disease, one of his great passions. He has been to a major degree the driving force behind the vaccine for this disease.

The new laboratory was to be built in Lyon and was to commemorate his son, Jean, who had died in an automobile accident four years earlier, leaving a legacy that would fund the construction.

Any reader who has read this far will appreciate that this was a project after our own hearts, and in a beautiful city in southern France. We knew that Europe and the world needed such a center. CDC could not cope with everything: the capacity to supply reagents, establish collaborations, underpin field studies. Other countries also needed an opportunity to contribute to this field of research, currently denied them because they lacked good facilities. This was an opportunity difficult to turn down. Jean-Marc described the project as it had progressed thus far.

"The only way we could make enough space for the laboratory on the site the Docteur had provided is on the roof of the Laboratoire Marcel Mérieux," he explained. Then he showed us the architect's drawings. They were very elegant. The Laboratoire Marcel Mérieux is an S-shaped building on a busy highway. We suppressed our consternation.

"Is it all made of glass?" we finally asked.

"Yes," he replied. "But there is a metal frame from which it is suspended. It is a box within a box, and we have three floors, two in the glass enclosure, and the third on the roof of the building below." This part was correct design. The upper floor must be for air handling, the middle floor for the internal box, which would be the laboratory, and the lower floor for heavy equipment and fluid effluent decontamination. The exterior structure was beautiful, but there were problems with the interior design. Lots of problems. Obviously these plans were not drawn up by anyone who had ever been in a suit laboratory or a BSL4 laboratory of any kind.

"How did you come up with this design?" we asked. "This is beautiful, but a bit of a surprise." All other BSL4 laboratories so far have been concrete fortresses. We were trying not to sound too shocked.

"I couldn't find a way to do it for a long time," he replied. "It needed more space than I could find inside the building. But then one night I dreamt of a cigarette lighter floating horizontally on the roof." He was referring to the "Briquette" type of cigarette lighter commonly used in France. This is a shiny, metal rectangular box, with a circular device on the side that you roll with your thumb to light the gas flame. The design he was showing us was indeed a glass rectangular box, on its side, attached to which was a semicircular construction that had space for offices and a small ordinary laboratory. This he called a "half Camembert," after the famous round cheese.

We did not know at that point that Jean-Marc, like us, had been hijacked that day by the Docteur. Jean-Marc had been at his chalet in the Alps with his family. They had not left to ski as planned because the weather was bad—since they were locals they did not ski if the weather wasn't good. The phone rang, and the Docteur's voice came briskly over the line.

"Jean-Marc, come at once. They are here." Click—the phone was down. The Docteur is always to the point. Jean-Marc, who knew him well, did not hesitate.

Back in the office in Lyon, late in December 1996, overlooking the abbey of Ainey, we discussed our future assignment. There were problems with the design of the laboratory itself. The interior had to be completely rethought. He knew what he wanted and told us. He wanted it well done, and immediately. He explained to us that at his age he only took on five-year projects, and he planned to see them through.

"How soon can you start?" was the chief question. The only other condition was doing it right.

All this required thought, but nevertheless led us precipitously into an unexpected French adventure.

The Institut Pasteur in Paris was seriously interested in getting Joe to set up an epidemiology program. Maxime Schwartz, the Director, had read Laurie Garrett's influential book *The Coming Plague.* Laurie had extensively quoted and described Joe in this book. One day in mid-1997 Maxime had rolled up in New York and had sought out Laurie at Newsday.

"Where can I find Joe McCormick?" he asked her. "I want to hire him."

Laurie laughed, and told him we were in Karachi.

She must have been surprised. Maxime's appearance and demeanor is not at all in keeping with the directness of his question or his desire to find someone who has been described as a "Disease Cowboy." Maxime is very formal in both his impeccable dress and courteous manner. He understands English well, but seems shy in speaking it, something he shares with the Docteur. Maxime is an excellent scientist, a man with some vision, but with a customary reserve. He is a man struggling with a very well entrenched system, resistant to change. He knew that the traditional "purity" of the Pasteur science was not keeping up with modern developments and that he had to inject new energy into the system. In addition, his father was the grand old man of epidemiology in France, and epidemiology was therefore a path to broaden the horizons of his institute. Indeed, his father had been urging him for years to do such a thing. This would not be an easy task in a country where the word *epidemiology* is used to describe data analysis using mathematics and biostatitics. France had little or nothing in the tradition of shoe-leather epidemiology, such as is practiced by CDC. Indeed, its only program was derived from a course in field epidemiology, given initially by CDC in France and now given by those graduates of the course who work with the nascent French epidemiology program. That original CDC course in France was started by none other than Charles Mérieux.

So we had to think this out. We both had excellent job offers in the States, and nothing had been signed with the Institut Pasteur. We decided to go to Paris for the day and discuss our options. If Joe was to be in Paris, I could work on the laboratory in Lyon. The Docteur was clearly unwilling to "steal" Joe from the Institut Pasteur. This was a question of French politics. In Paris, his BSL4 laboratory project was described as *"fumée"*—smoke. Complicated. But at least the commute would be minimized by the operation between Paris and Lyon of a high-speed train, the TGV (Train Grande Vitesse), which goes nonstop, each hour, at 175 miles per hour between Paris and Lyon, covering 475 kilometers in exactly two hours. It glides swiftly through the beautiful French countryside, an elegant blue snake, with two tiers of passengers sitting in comfortable reclining seats. During the whole journey you do not pass through a town.

We passed New Year's Eve 1996 with the Docteur and some friends in an excellent small restaurant. (Lyon is the gastronomic capital of France, and it is very difficult, if not impossible, to eat badly.) On January 2 we purchased our TGV tickets for Paris. It was cold, the coldest winter Europe has had for many years, with the canals of Holland thick with skaters and snow right down to the Côte d'Azur on the Mediterranean. We set out at 6 A.M. In the dark, we could just make out the snow on the embankments as we slid past at amazing speed. We reached Paris and conducted our business. Prospects for Joe looked good.

At 6 P.M. we returned to the Gare de Lyon to find our return train. We stood under the departure announcement board. No platform, just a general announcement: *"Au delà de Valence, rien ne circule plus"*—beyond Valence, nothing is moving. This was unsettling. Valence is the gateway to Provence, about eighty miles south of Lyon, and the route to the Côte d'Azur. You expect things to be warm down there. We stood below the board for two hours before we found out (1) that there was no hope of any train that night and (2) that the Rhône valley south of Lyon had been hit by the most vicious ice storm driven by the fierce icy mistral wind. Everything down there was frozen in place. Trains had frozen to their tracks. The next day we discovered that 3,000 passengers had been stranded all night in unheated trains.

We were lucky not to be on those trains. At 8 P.M. we still stood in a crowd of undecided people milling around the Gare de Lyon in Paris, not sure what to do. We had no hotel, and we needed absolutely to get back to Lyon to conclude business with the Docteur.

"There is a car rental place upstairs," I said to Joe. And we made for it.

The car rental people were very nice and happily rented us their last car, a tiny Opel Corsa, two-door, midget car—about the size of a large rowboat. The good thing was that we could drop it off at Lyon at no extra cost. We took it.

We set out, and were amazed that in Paris the temperature was minus five degrees centigrade. This is very, very cold for Europe. We quickly reached the Périphérique, the Paris ring road. With the cold and the Christmas/New Year vacation there was hardly any traffic. We set out on the A6 Autoroute for Lyon.

There was no one on the road except the great trucks and us. For a while all went well. Normally, this is a three-to-four-hour drive, so we were not concerned. We expected to be in by midnight. It was very cold, but dry.

As the signs came up for the exits to the cathedral city of Auxerre the road rose to pass over the Morvan, the northernmost part of the Massif Central, the great mountainous area of central France. We stopped for strong black coffee and then set out again. The snow started. Our tiny car hardly held the road at all. Each time a truck passed us, the car swung all over the place. We slowed down and kept going by running in the tracks gouged by the trucks.

We reached Lyon after eight hours of driving. It was now 4 A.M., and there was several inches of trackless snow along the banks of the Rhône. We pulled into our hotel, where we collapsed.

At 8 A.M. the phone rang.

"Vous êtes là?"—You are there? It was the Docteur. Philippe Sansonetti, whom we had met at the Institut Pasteur, also called us from Paris, wondering what had happened. He had heard the news about frozen trains on the morning news and was afraid we were still sitting in an unheated train, somewhere in deepest France.

Our trip became an apocryphal tale. We had felt that with all the travel difficulties in Africa and elsewhere, it was a matter of principle not to be beaten by a little cold and snow in France! People in the Dakotas are probably laughing at this tale, but they also probably might not dream of venturing out in something as small as a Corsa, even in good weather.

My first reaction when I knew I was to be responsible for a new BSL4 laboratory was to track down Lee Alderman. With Joe's help I found him at Emory University, where he had a position in the Department of Biosafety after his retirement from CDC. I called him.

"Help, Lee!" I said. "Would you like to come to France and work with us to build a suit laboratory?"

"No problem. Be delighted," Lee replied with his usual calm, direct manner.

I needed Lee. Though Joe and I knew most everything there was to know about the use and operation of a Biosafety Level-4 laboratory, Lee is far more knowledgeable about the engineering and biosafety intricacies essential to its function. He rose rapidly to the occasion, picked up some French tapes, and got himself ready for long working hours, rewarded by samples of great French cuisine.

It has been a wonderful experience. Lee and Jean-Marc Tourret got on immediately despite their lack of a mutual, comfortable language. Details of laboratory design bridged that easily. Lee stood a whole day before Jean-Marc's team and detailed the performance requirements and the tricks we had learned over many years of experience to achieve these. There are only minimal guidelines for BSL4 laboratories and no French norms at all. The only way you can build a BSL4 laboratory is by drawing from experience. Because the design was a glass box, balanced on six pillars, we could not use the traditional massive fortress-like concrete walls of the previous laboratories. We had to do some careful thinking and a lot of improvisation. We also had to work with French authorities. They were initially quite unsettled to find they had no rules or regulations covering BSL4. In France, bureaucracy is highly developed, and everything must be done by the book. In the process of working with them, therefore, we created their book for them. This must now be the most highly documented laboratory in existence, and we are all happy.

Lee has now made ten visits, the latest to test the operation of the systems he designed with us and with the architects. Throughout this experience his disciple was Jacques Grange, who has now developed into the French "Lee Alderman." Jacques is an experienced scientist, but above all a superb technician in all respects. He has taken on the laboratory with passion and dedication, and its realization and effective function will in great part be due to him.

The laboratory is now very nearly complete and has received official authorization. In March 1999, President Jacques Chirac officially inaugurated the laboratory.

Jean-Marc's structure is superb, breathtaking in its daring design and extraordinary use of modern technology. Jean-Marc's team drew from their own extensive experience in building pharmaceutical laboratories and then from the nuclear industry, which is well developed in France. They have made stunning innovations in design that will mark laboratories of the next century. The light structure of the laboratory protects it from stress, and ensures only limited places where we have to deal with air leaks. All the weight hangs on the external framework or on the building beneath, leaving the laboratory itself quite independent. The walls are made of special panels, clamped together, and sealed with epoxy and silicone. It is more like an airplane than a fortress. High-technology air handling and electronic control and information work well. We have newly designed French space suits that are lighter and much more comfortable than the old blue suits and which provide additional protection in several ways. Also, being French, they are clearly much more chic. Even the Docteur at the age of ninety-three wore a suit for a photographic session, standing up for a full hour.

The laboratory is therefore not *fumée* (smoke, to use the French expression). In fact, the only smoke we see in Lyon is in the hands of Lee Alderman and Jacques Grange, as they patiently track potential air leaks with their smoke pencils.

With Sue establishing the laboratory in Lyon, I could now turn back to one of my most important goals. The reader will easily understand that this is to get a Lassa fever vaccine into the arms of those people who so desperately need it in West Africa. This project is also in the tradition of the Docteur. The story about him (of which there are many) which most amazes us was his intervention in the meningococcus epidemic in Brazil. I actually met him during the time I was a young EIS officer working in Brazil on that epidemic, described in Chapter 3. Docteur Mérieux also arrived in Brazil at that time and saw the ravages of the disease. He returned to his factory at Marcy l'Etoile outside Lyon and told everyone there that they were to stop all other activities at once, drastically increase output, and make, in the shortest time possible, 80 million doses of meningococcus vaccine. Everyone thought he was mad. They thought it could not be done and that he would be ruined, since he had no contract in his hand.

They were wrong. He produced the vaccine successfully used to control the epidemic in Brazil. This was the first time such a vaccine had been used to control a meningococcus epidemic. He was also eventually paid for his vaccine, but to his great credit, that was not his primary concern.

For this and for many other similar achievements, his stature in Lyon and in France is unquestioned. He is one of the great public health men of the century.

The BSL4 laboratory is the brainchild, the "folly," to use his own words, of Charles Mérieux. We are trying to develop it as an important center for collaboration in public health. There is clearly a great need among European and African scientists for a unifying force. The laboratory facilities make a natural focus. We have collaborators in Europe, the U.S., and hopefully, Canada. In Africa we are working well with centers in Gabon, in Central Africa, and in Senegal, in West Africa.

It is worth reflecting a little on the need for such initiatives and on the measures we can use to prevent diseases. In Karachi we saw the worst of both worlds. Uncontrolled development and population explosion give it polluted air, with some of the highest lead levels in the world. Toxic discharge and waste from industry flood the rivers and the bays. Death from injury in the home from kerosene stoves, on the sidewalk from fallen high-voltage lines, in the road from the battered, killer buses from which the passengers hang like ripe fruit. In one careful study, our students measured that only 3 percent of the traffic in Karachi consists of buses, but buses cause nearly 50 percent of the road deaths. Diets heavy in animal fat and widespread smoking lead to early, preventable death on the one hand, but there is malnourishment in many women and children on the other. Underdevelopment means that each and every newborn child, rich and poor alike, has to battle as soon as it is born with nearly every microbe known to man. For children under five, mortality is above one hundred per one thousand in most areas of the city, and even higher in the rural areas.

Daily in Pakistan we had thrust before our eyes the toll of preventable infectious diseases. Typhoid is a major player. Sue's laboratory reported that nearly 40 percent of the organisms isolated from cultures of blood from patients with fevers grew *Salmonella typhi,* the bacterium that causes this deadly fever. Most of these patients were under the age of two years. Worse, two-thirds of the patients had a multiply resistant typhoid bacterium. A multiply resistant bacterium is an organism that has managed to acquire a way of resisting the commonly used antibiotics. In the case of typhoid, there are only two antibiotics that still work, and they are both very expensive. Only one of these can be used in children. Since all the specimens in the Aga Khan Laboratory came, of necessity, from children with parents of means, it is not difficult to imagine the enormous toll this bacterium must be taking among the poor children who never get diagnosed. Based on these observations, Sue believes that much of the infant mortality blamed on acute respiratory infections may be simply *Salmonella typhi* blood infection. In a small child this will manifest as fever and rapid respiratory rate.

We have more or less eliminated typhoid in the developed world by food hygiene and clean water. We have a vaccine, but we do not need to use it unless we travel to developing countries. Then it is mandatory. This is one of many diseases for which there are reasonable if not excellent vaccines developed or under development, but the vaccines remain unavailable to the majority of the world's population. Failure to distribute the vaccines is due to poverty or lack of political will, usually both.

This dilemma is well illustrated by the current treatment for AIDS. Many have the idea that AIDS is vanquished now that it can be treated. It would be a poor joke indeed to tell an African or Asian with AIDS he or she must take fifteen to twenty pills a day and that this will cost a thousand dollars a month or more. A monthly salary may, for someone very lucky, healthy, and hardworking, be at the most one hundred dollars. Since the vast majority of persons with AIDS fall into that category, the clear, immediate, and desperate need for vaccine, even an imperfect vaccine, is obvious. Their plight apart, we neglect the needs of the poor of the world at our peril, since this uncared-for, infected population is a massive reservoir of viruses.

So we are entering the era when we must fight for the so-called orphan vaccines. By this we mean vaccines for which the economic base for commercial development is insufficient to get them into people's arms or mouths. Among the preventable diseases by vaccination we list typhoid, shigella (the bacterium that causes bloody dysentery), rotavirus (which kills large numbers of infants), our old foe the Lassa fever virus, the scourge of malaria, and so on and so on. The pitiful list of suffering and early death is unending. Why are we investing so little in these problems in the West?

There are several reasons, all of them difficult to overcome. The most obvious is the cost of development. We depend on commercial companies to develop vaccines, and this they will only do, obviously, if they see return on investment. Would you invest in a company that did not take this line? However, even they might be persuaded by the size of the potential market if the cost of development were not so prohibitively high. This cost is directly related to safety requirements, and safety is clearly paramount. However, we have reached a stage in our litigious society at which we expect perfect safety or someone must pay. This drives the cost out of reach of most of the world's neediest populations. Many of the vaccines we now use, such as yellow fever, polio, and some drugs, like aspirin and penicillin, could not be developed today because of their side effects. Because we have used them for so long, we accept side effects that would never be allowed in a new product. So, you may wonder why public bodies, funding, and international agencies are not supporting more vaccine development. Vaccines need research, and agencies funding research do not traditionally give money to develop a vaccine. Development agencies say this is not their problem—they are concerned only with development. Public bodies serve voters with enough money to buy vaccines and avoid disease, so it is not their problem either. So the cycle goes on.

We may now be entering an age when prevention of disease where most disease exists, among the very poor, will have greater priority. Witness the polio eradication campaign, which, if successful, can rid the globe of a massive scourge. It was difficult to drive by a single major crossroads in Karachi, as in many cities, and not see someone begging who was crippled by polio. It is miraculous to consider that children in poor countries can grow up in the future without major disabling paralyses. We, in our richer countries, will no longer have to worry about imported poliomyelitis.

Another problem is that the vaccines needed in developing countries are often not useful in richer communities, either because the disease has been controlled or because the age at which patients need the vaccine is different, usually older ages. In developing countries, because of the massive assault of the microbial environment, it is often necessary to vaccinate at birth. Neonates are notoriously difficult to vaccinate effectively because their immune systems are immature, and they need specially designed and tested vaccines. We are very encouraged by the recent initiative by the Gates foundation to address some of these issues by promotion of childhood vaccines in developing countries.

In conclusion, we think the world and the pharmaceutical industry are only now beginning to appreciate the number of people who need to be vaccinated, and thus the vast size of the market, given a product at the right price. It has always been our experience, in many countries, time and time again, that people of all races, creeds, and cultures will make big financial sacrifices for the health of their children. They will make the same efforts for their own health, including paying for vaccines, if only they understand the benefit, and provided the product is not, for them, absurdly priced.

Thus our interest in developing a Lassa vaccine, born several years ago when we were at CDC, nurtured by our constant experience of Lassa fever in West African villages. We developed a vaccine at CDC, very effective at preventing illness in monkeys. Despite our estimates that it might be given to as many as 50 to 100 million people, at that time there seemed to be no ability to translate that eventuality into a financially viable program to develop a human use vaccine. Our challenge now is to overcome this roadblock by finding funding to finish the development of the vaccine and to pay for field trials.

Sierra Leone called to us. This lovely country had been distant while we were in Pakistan, but it returned to our attention with some force when we came back to Europe and saw television pictures of the appalling conditions and ravages the civil war had brought to the Eastern Province. Women and children with severed limbs. Segbwema and Panguma were and are right in the middle of the horror of the civil war over the past six to seven years.

Sue was reading some reports from a relief agency that stated, as though it were a new discovery, that Lassa fever was endemic in the Segbwema area, that rodent control was the solution, and that the disease could be treated with ribavirin. She found a phone number, for Richard Allen, Merlin Relief Agency. She dialed, and Richard answered.

"Hi," she said. "I have just been reading your reports. I wondered if you were aware that there is a lot of published information on Lassa fever. We might be able to help you."

The conversation was a long one. It was clear that the Lassa fever problem was much worse with the insurrections, that Merlin was trying to do what it could, and was finding many cases on the agency's wards in Kenema.

"But we never saw much Lassa fever in Kenema. That is why Joe worked mainly in Panguma and Segbwema," Sue said.

Richard replied, "Problem is that there is no medical care outside Kenema now. The hospitals in Segbwema, Panguma, and Tongo Fields have been destroyed, and there is nothing out there. It is usually not safe to go there either.

Those people we see have been brought miles in terrible conditions." This explained why the mortality was so high. Patients must have been coming in late, much sicker than those in the past.

Sue tells me she at once saw the Segbwema hospital vividly in her mind's eye and wondered whether the Lassa fever ward and laboratory had been destroyed. Probably so. The rebels would not have known what to do with them, and the staff had fled. It turned out that Richard had found some of our old staff in Kenema, and they remembered some of the techniques we used. They had no diagnostic reagents, however, and he had great difficulty getting ribavirin.

This is another tragedy. Despite its clear effect, the FDA had never licensed ribavirin for Lassa fever. As can be imagined, it is an "orphan" drug—too expensive to develop because there is no market among people who can pay.

Sue decided quickly and invited Richard to Lyon. He accepted, and we all met at the Fondation Mérieux, in the Docteur's offices.

Sue also sought out Alf Lindberg. Alf is the head of Research and Development of Pasteur Mérieux Connaught, and though he is a very busy man, we were lucky enough to find him in town, and he was interested. Alf is a brilliant Swede, very experienced in vaccinology, with a very clear mind and great humanity. We sat all afternoon in a gradually darkening office listening to Richard's tales of the tragedies in Sierra Leone, the difficulties and the suffering from Lassa fever. They were trying to control Lassa fever by controlling rodents, but rodents are pretty uncontrollable in Africa in times of peace. In war it is impossible.

"Just think," said Sue. "In comes a relief agency with a pile of rice, and has to put it down before distribution. On the ground, I guess, covered with tarps, since few buildings will be in good shape. First to get to it will be rodents, happily urinating Lassa virus all over the grains." Probably not far from the truth.

Alf was vehement. "I don't want to hear about another mouse or another monkey," he said. "I want a vaccine in a human arm."

We all applauded, particularly Richard.

It is wrong and not even fair to assume that all corporate individuals are only interested in the bottom line. It is true that without a profit the company cannot exist, but that does not preclude concern about human beings everywhere. Alf decided that he would help us develop a vaccine, but that we would have to find the development costs somewhere else. Commercial backing was out of the question. Working with Richard and Merlin, we thought we ought to be able to make a case for funding, so we agreed to start. PMC has now made the first Lassa fever vaccine for human use. We are delighted to have as our first project one so clearly in the tradition of Dr. Mérieux.

We believe that the cost of actual testing and production of the vaccine will be low, because it is not high technology, and the delivery systems are already in place in many parts of Africa where the vaccine will be used. Our primary target is the refugees. Two million people are now displaced, wandering homeless in the Lassa fever epicenter between eastern Sierra Leone and western Liberia.

We are also mindful that yellow fever vaccine, around a long time and cheap, is not effectively used in West Africa, where recurrent epidemics continue to devastate rural populations. There will not be an overnight eradication of Lassa fever, but for a disease that kills an estimated several thousand people a year (accurate information is not available), an intervention with the long-term possibility of reducing this carnage is better than cynicism or resignation. Thus, we hope to launch our first tests of safety and immune responses in people within the next year or so. Exactly where we will turn for the million or so dollars required for developing and testing the vaccine, we do not yet know. One thing we have learned: it is no longer sufficient for the scientist to simply make the vaccine in his or her laboratory and assume that someone else will carry on from there, especially if there is little if any financial reward to be had for the effort.

The lesson that we should have learned from AIDS, and from the problems presented by an overpopulated world, is our interdependence. Whatever infectious diseases can affect the poorest in that population can also affect the richest, and we cannot therefore ignore any microbe. Isolationism is surely ultimately self-destruction. Sharing knowledge, skills, and resources with others and ensuring that people at risk are educated and that they use that education to make wise and informed choices, individually and collectively, are the duty of us all. Charles Mérieux's new laboratory in Lyon allows us to move to a new level of activity with these goals. Here we will be able to test the Lassa vaccine prepared for human use, and with our friends and colleagues coordinate the efforts to get it into the human arms that need it. This is yet another living legacy of Charles Mérieux. Sue and I are back working on our unfinished project, and he got us there—prevention of Lassa fever.

Loss and Hope

Joe received a great blow recently. Since 1997, two very important friends shared our regular March skiing week with us in Vail, the town where we were married and which is now our chosen home. They were Jonathan Mann and MaryLou Clements-Mann, who like us came to the remarkable fulfillment of a relationship formed in later years. Like us, they knew the joys of working together. This couple found, like us, that relationships need hard work and commitment, but give you a rare happiness and inner strength. Would that we had all been wise enough or lucky enough to understand this as young people. Joe and Jonathan had been close friends since Joe had recruited Jon for the AIDS project in Zaire, a story already told in Chapter 19.

One Thursday morning, recently in Lyon, Joe went online to check e-mails and read the headlines (we had no television). I heard him cry out:

"Sue, Swissair has lost a plane."

We always fly Swissair whenever we can. Their meticulous, friendly service puts them in a class of their own. Swissair was our lifeline in Karachi. Twice a week the white plane with a bold white cross on the red tail materialized magically out of the polluted dust over the end of the Karachi airport runway exactly thirty seconds before expected time of touchdown—the only miracle, recurring or otherwise, I knew of in Pakistan.

In September 1998, there had been e-mails from Jonathan confirming that he and MaryLou would be in Geneva and that we would spend the coming weekend together in Annecy, with the Docteur. Joe knew about the possibility of their being on that flight—I was slower in making the connection—and tried to believe they would not fly out of JFK since they lived in Baltimore.

* * *

As Sue relates we have lost not just two of our closest friends, but two individuals whose contribution to the health, welfare, and human rights of the poorest was not just extraordinary but, worst of all, unfinished.

Jonathan Mann, who features so much in this book and our lives, had told us by e-mail that he wanted to discuss a meeting on AIDS the following year and the Docteur was delighted. This would be in the Les Pensières conference center that had been converted out of old farm buildings a few hundred yards up the lawns behind his house. This center has become an important meeting place for many scientists and public health workers, and it hosted an important meeting on AIDS and human rights in 1987 while Jonathan was still at WHO. Annecy is less than an hour by road from Geneva. The four of us were also going to plan our annual ski week together for 1999. That weekend, as we discussed our plans, we had intended to hike in the mountains above the lake. We wanted to give our friends a special treat, so Sue had booked a table at Père Bise, a Michelin two-star restaurant on the lakeshore.

When I woke up on the morning of Thursday, September 3, I checked the news on the Internet after reading last-minute e-mails from Jonathan about arrangements for Saturday. We had agreed we would pick them up on Saturday morning in Geneva from their hotel, the Hotel Cornevin, right beside the rail station. We would then drive to Annecy and have lunch with Dr. Mérieux. I was immediately struck by panic. I read that Wednesday night's Swissair flight from JFK to Geneva was lost off Canada. There were no details. All it said was that this was the worst disaster ever for Swissair. Although not certain which flight they were on, I told myself Jon and MaryLou must not, could not, have been on that flight. Surely they must have taken a flight from Dulles in Washington.

I called the 800 number to get information, but it was the early hours of the morning in the U.S. We are six hours ahead in France, so really it had only just happened. Though only immediate family members were being given information, Swissair was already making files on friends. I called Sue in tears, telling her my worst fears. She had one of her staff members call the Cornavin hotel in Geneva, and they were told that Mr. Mann had checked in. Sue called to tell me the good news. I tried then to call the Cornavin myself, and they said yes that a Mr. R. Mann had checked in. But Dr. J. Mann had not. I called Sue again in a flood of tears. I was certain now that they were on that flight. Toward noon our friend Robin Ryder called from Yale to confirm the terrible truth: Jonathan and MaryLou were dead. They died together, our dearest friends, in the cold waters off Nova Scotia. We could not imagine what horrors they suffered in their last minutes, but we knew they would have been calm and holding on to each other. It was a nightmare from which we should surely soon wake up, but there has been no awakening from this one.

Many are blessed with multiple friends, but few of us are blessed with friendships that grow over twenty years or more based on both strong professional as well as deep personal ties. Such was my relationship with Jonathan. The loss of Jonathan to us all is irreplaceable. He was an articulate and impassioned spokesman for health and human rights and thereby for the world's poor and oppressed. After leaving the Global Program on AIDS that he created at WHO, Jonathan had gone on to found a Health and Human Rights Institute at the Harvard School of Public Health. This institute is a forum for teaching and propagating the rights of the poorest and to pursue Jonathan's view that denial of human rights is denial of health.

Jonathan's wife, MaryLou, was a soft-spoken, impressively competent, and patient physician and scientist. She worked initially on vaccines for children, and later AIDS. She was deeply respected by all who knew and worked with her. Her quiet perseverance in pursuit of her visions gave her an inestimable value to us all. Her loss to those she served cannot be calculated. She and Jonathan were about to embark on a joint chapter in their lives, much as Sue and I had. They had gone through much soul-searching in trying to determine what this should be, and they were nearing a resolution. They wanted to work together in some way to combine their skills and strengths in the service of the poor and diseased of the world.

Jonathan and MaryLou's loss to their families can only be felt fully by their loved ones, particularly Jonathan's three children and MaryLou's mother. I do not have a brother, but if I lost one, I cannot imagine being more grief stricken than I have been at losing my friend.

When Jonathan left WHO, an African friend told me that it was a devastating day for Africa, because Jonathan was the voice of Africa's agony and need. Through his Center for Health and Human Rights at Harvard, Jonathan continued to be a voice for change and particularly for improving the rights of women and children to reduce their risk of contracting AIDS. He was outspoken about the failure of the WHO leadership to understand and support the WHO programs on AIDS. He left a legacy that can and should continue to deeply influence policies of health and politics throughout the world to improve the rights of the weakest among us.

With these thoughts, and with our new vaccine project, I am preparing to go once again to Sierra Leone. I do not know what I will find, how horrifying it will be to see so many innocent people and children with traumatized lives, the most egregious deprivation of basic human rights. Sue is fearful for me, more so because of the loss of the Manns, but also urging me to go, because there is work to be done.

A Viral Pedigree

Note: What follows is restricted to the viruses discussed in this book.

1. FILOVIRUS FAMILY

A unique family of single negative-strand RNA viruses, consisting at present of Marburg and Ebola viruses. They are related, but serologically distinct, viruses of unknown ecology and natural reservoir. Bats have been suspected of being the vectors, but nothing has been proved. For the most part, cases of disease originate in Africa within five degrees of the Equator. The viruses are long, thin, and snakelike when viewed under the electron microscope, often taking the form of strange twists and loops. Some even look branched. This appearance is quite unique among human viruses. They are called filoviruses because of this threadlike appearance (*filo* is the Latin word for thread). All are Level 4. Molecular analysis currently suggests some distant relationship with respiratory syncytial virus in humans.

The incubation period for filovirus-caused disease is three to ten days, and the diseases are essentially similar. Onset is acute, with severe headache and muscle and body pains accompanied by high fever. Symptoms include a very painful sore throat, vomiting, and diarrhea. Blood pressure drops precipitously, and victims go into a state of surgical shock, which is the cause of death. Bleeding is common, but is not always severe, though usually there is some bleeding from the gums and from the gut, with small hemorrhages in the whites of the eyes. Patients may vomit blood.

Contrary to recent popular nonfiction and fiction, the guts do not "dissolve." In fact, at autopsy, most organs appear normal, and there is little to see except some blood-stained fluid in the body cavities and scattered patches of tissue destruction in some organs, particularly the liver.

Mortality is high in primary cases but seems to be reduced with human-to-human passage of the viruses, so that secondary, tertiary, and later cases tend to survive. There are currently no vaccines and no treatment except good intensive medical care, which may improve chances of survival considerably. Transmission stops as soon as barrier precautions are introduced.

Marburg virus (Green Monkey Disease)

The first filovirus ever seen. It emerged in 1967 in monkey facilities in laboratories in Marburg and in Frankfurt, Germany, and in Belgrade, Yugoslavia. There were thirty-one cases and seven deaths that year. All primary cases had direct contact with blood or tissues from recently imported African green monkeys from Uganda, hence the popular name green monkey disease. All three laboratories affected received monkeys from the same source: an island on Lake Victoria, near the border with Kenya, where they were held before being shipped, via London's Heathrow Airport, to Germany. There were no deaths in six secondary patients, all of whom had had close contact with primary cases when they were in the hospital. A woman married to a veterinarian who was stricken with Marburg became infected herself; it is believed that the disease must have been transmitted sexually, since Marburg virus was isolated from the veterinarian's semen several months after he recovered.

Since 1967, there have been sporadic cases of Marburg disease in Kenya, and one instance where a man apparently became infected in Zimbabwe and later died in South Africa. He was responsible for causing several secondary cases in the latter country. All the Kenyan patients were infected in the Mount Elgon area of Western Kenya, near the Ugandan border. Two of these patients were known to have visited the same place: Kitum Cave, which was later investigated as a source of the virus, without success. To date, there is only one strain of Marburg virus.

Ebola viruses

First seen in 1976 in simultaneous, but independent, outbreaks in northern Zaire (Yambuku) and southern Sudan (Nzara and Maridi). The two outbreaks were caused by different viral strains, known as Ebola Zaire and Ebola Sudan. In Zaire, with 318 cases, mortality reached 88 percent, and in the Sudan, with 151 cases, mortality was 53 percent. Transmission in both epidemics was caused by close contact with blood, particularly through contaminated needles, which were reused, and improper nursing care. A second Ebola outbreak occurred in Sudan in 1979. In both Sudan outbreaks, the index case worked in the same cotton factory in Nzara. Ebola did not reemerge until 1994, first in an isolated instance in the Ivory Coast, and then on a much larger scale the next year in Kikwit, Zaire. The outbreak in Kikwit was primarily due to poor hygiene and medical practice. From the hospital, the disease then spread into the community. Sporadic cases continue to occur in West Africa; the disease most recently has appeared in war-torn Liberia, and in 1996 in Gabon. The latter outbreak is linked to the handling of a dead chimpanzee. Presumably, the chimpanzee itself died of Ebola. It is likely that what appears to be a marked increase in the number of Ebola cases is a result of heightened awareness on the part of the medical and scientific community, rather than an actual increase in the spread of the virus to the human population, but that possibility cannot be ruled out. The virus derives its name from the Ebola River, a small body of water in northern Zaire.

Reston viruses

These viruses are most closely related serologically to Ebola. They were found in monkeys imported to the United States from the Philippines in 1989. The virus infects man but does not cause disease, so it is essentially a veterinary problem in spite of its Level-4 rating. It causes disease in monkeys similar to Ebola, but it progresses more slowly and is less severe. It is named after Reston, Virginia, where the first monkeys died in a commercial laboratory.

2. ARENAVIRUS FAMILY

A family of negative-strand RNA viruses that are parasites of rodents. These viruses infect the rodents as neonates, and then persist in them for life, causing little or no ill effects. The rodent excretes the virus in urine, often in very large quantities. Rodent urine is probably the main source of primary human infection, most likely through scratches and abrasions, though some instances of infection by inhalation have been reported. The most common arenavirus is the lymphocytic choriomeningitis. This is a virus of mice, which is worldwide in distribution, and which can infect man and cause inflammation of the membranes surrounding the brain (meningitis). This is a painful illness, but it is not lethal. The virus is ranked as Level 3.

The remaining arenaviruses are Level 4. African and South American arenaviruses cause severe hemorrhagic fever, sometimes with high mortality. The incubation period is seven days to three weeks. Onset is slow, with gradually increasing fever, headache, and body pains. These diseases are difficult to distinguish from common treatable diseases like malaria and typhoid. Many patients then recover, but others develop vomiting, diarrhea, bleeding from the gums, and small hemorrhages in the whites of the eyes. Severe bleeding is uncommon in Lassa fever but more frequent in the South American hemorrhagic fevers. (Argentine and Bolivian hemorrhagic fevers are known as AHF and BHF, respectively.) Patients may sustain a serious drop in blood pressure—just as victims of Ebola and Marburg do—and go into shock. At the same time, they develop severe edema (fluid) in the lungs, which causes Adult Respiratory Distress Syndrome (ARDS).

ARDS is the usual cause of death in Lassa fever, though some patients develop inflammation of the brain (encephalopathy); most of these die with convulsions and coma. Pregnant patients in the third trimester are particularly susceptible to the disease and the baby usually dies, often in the uterus. Most of the mothers die as well, particularly if they do not receive good obstetric care.

An effective treatment for Lassa exists with a drug called ribavirin. In cases of Argentine hemorrhagic fever, immune plasma has proven therapeutic (though it has not in Lassa).

Recently, new relatives have appeared in South America—Venezuelan hemorrhagic fever virus and Sabia virus from Brazil. There are undoubtedly others to be discovered. A vaccine has been produced for Argentine hemorrhagic fever, and an experimental vaccine for Lassa fever has been successfully tested in monkeys. Person-to-person transmission and hospital outbreaks of the South American diseases are almost unknown.

Lassa Fever

Lassa was first described in 1969 when an American missionary nurse died in Nigeria and another took ill and was flown back to the United States for treatment. The virus was named after the town of Lassa, where the first nurse worked. There were two laboratory infections in New York among the scientists who isolated the virus, one of whom died. This virus is found only in West Africa, in territory that extends from Nigeria to Senegal, though it has close relatives in the rest of Africa that apparently do not cause disease in man. It is spread by a common village rat, known as *Mastomys natalensis*. It has been estimated that there are about five thousand Lassa infections a year in West Africa, and that mortality ranges from 2 percent in the community to 16 percent in hospitals, and 30 percent or more in pregnant women. There have been further hospital outbreaks in West Africa, with deaths of medical staff in Nigeria, Liberia, and Sierra Leone.

3. BUNYAVIRUS FAMILY

A very large family of negative-strand RNA viruses, which usually only infect animals, birds, and their ectoparasites, particularly ticks.

Crimean Congo Hemorrhagic Fever (CCHF)

First described by the Russians at the end of World War II when large outbreaks of the disease afflicted Soviet soldiers sent to help with the harvest in the

Crimean peninsula; this is how it earned the first part of its name. The virus was originally isolated, however, in 1956 in the Congo, which is how it received the second part. It is ranked as a Level-4 organism. It usually results from a tick bite or contact with an infected animal or human blood. Human infection is very severe, but, fortunately, uncommon. Curiously, of the many animal species that can be infected, none ever develops any symptoms. The virus is carried by hard ticks and is widespread throughout Africa, the Balkans, and the Caucasus, the Middle East, Pakistan, and western China. This virus causes severe hospital outbreaks, with high mortality in infected staff.

The incubation period is two to nine days; onset is very sudden with crippling headache and body aches accompanied by high fever. There may be a petechial rash (tiny red spots on the skin) or ecchymoses (large black areas of the skin caused by bleeding). Patients drop their blood pressure and develop surgical shock, which is the usual cause of death. Bleeding from the gut may be severe. The patient may vomit blood and may actually die of blood loss.

Hemorrhagic Fever with Renal Syndrome (HFRS)

First described in the 1930s in Manchuria and in Scandinavia, HFRS probably has a history extending back to the turn of the century, when it is believed to have broken out in the Vladivostok area of Eastern Siberia and in Europe during World War I. Now it is very common in China and in Korea. Strains peculiar to urban and rural areas have been identified; the former are usually less severe in their effect than the latter. It is caused by a number of related viruses (Hantaviruses) carried in various species of mice and rats. All the Hantaviruses are Level 3. (Level 4 is used to categorize animal infections where there is a risk of the virus being spread through the air.) Hantaviruses, like the arenaviruses, infect rodents as neonates, and then persist in them for life. The rodent excretes the virus in urine, often in very large quantities. Rodent urine is probably the main source of primary human infection, most likely through scratches and abrasions. Aerosolized rodent urine in dust has been shown to be a source of human infection with Hantaviruses. Certainly, close contact with rodents is the absolute requirement for catching these viruses. Person-to-person transmission has never occurred, although thousands of patients have been hospitalized in many countries.

The disease varies markedly, depending on the virus strain. The strains vary by location. The European disease (*Nephropathia epidemica*), is the mildest strain, and is carried by bank voles. The urban disease caused by the Seoul rat virus is intermediate in its effect, while HFRS, which also has many other names, is the most severe and is caused by a virus from a common field mouse,

Apodemus agrarius. These diseases have incubation periods ranging from ten to thirty-five days; onset is slow and is marked by fever, headache, and malaise. A petechial rash (tiny red spots) may appear. Most patients recover, but some go into a state of shock for about twenty-four hours, and then after forty-eight hours, stop passing urine. Nearly all patients then recover rapidly, but some develop bleeding, mostly around the gums, which may be quite severe. Mortality ranges from zero to 15 percent, depending on the strain of the virus and the quality of medical care given to the patient. The most common cause of death is the build-up of fluid in the lungs. Death may, in fact, be precipitated by administration of too much intravenous fluid, which leaks through compromised membranes and literally drowns the patient.

In 1993, the disease appeared among humans in the United States; it was found to be caused by a relative, the *sin nombre* virus, which results in Hantavirus pulmonary syndrome. This is a virus of deer mice. Most of the cases have occurred in New Mexico and Arizona. The disease appears to be different from HFRS, since it acts much more quickly and has a severe impact on the lungs, often killing the patient. In contrast to other Hantaan viruses, it apparently causes no kidney disease.

The Hot Lab

T he spate of outbreaks in the late 1960s, Marburg disease, Lassa fever, and the the South American hemorrhagic fevers, spawned the original Level-4 laboratories at CDC. The very first laboratory was nothing more than a mobile trailer in the parking lot. By the early 1970s, a "cabinet line" laboratory was installed in its own space, with reasonably safe and efficient air handling via HEPA filters. This building was state of the art for its day. A large set of stainless-steel cabinets bolted together with flanges was interlinked with steel doors opened from the inside; any given cabinet could be closed off from the others. The entrance to this lab was keyed, so that only those authorized with the correct key (later, a magnetic-stripe card) could enter.

State of the art? Actually, this lab was more of a monster. In order to work there, it was necessary to plan any experiment very carefully. All the materials had to be gathered together at the beginning of the day. If you forgot anything, you were lost, and the day was a write off. Everything had to go in and out through an autoclave, which had doors on both ends, inside and out. The only other way to get anything in (and it had to be small, at that) was through a "dunk tank" of disinfectant with a submerged opening. It was a bit like spelunking.

To get yourself in, you had to change into a green scrub suit, a pair of sneakers, surgical gloves, and a paper cap. Then you had to remember to turn off the "OCCUPIED" light in the single change room, so that a waiting line did not develop outside. There were people (nameless) who regularly forgot this, so those waiting eventually had to beat on the door. The lab itself was entered through an air lock.

Once inside, the next trick was to rescue the materials sitting in the autoclave. This required arms with three or four telescopic universal joints to reach and open the inner door of the autoclave. You thrust your arms into nine-millimeter-thick black latex gloves, fixed to each cabinet. The gloves were designed to come up to your armpits. If you had short arms, you had problems.

To move material through the cabinet line, you hopped from glove to glove, reaching and stretching as far as each glove would go, shuffling things along a bit at a time. Inside the cabinet line was a number of pieces of simple equipment and a small section for small animals. You had to avoid escapes at all costs. It was very difficult to manipulate anything or to adjust and focus the microscope through all of these gizmos, let alone try to catch a runaway mouse.

The jokes soon wore thin. It would take at least three times as long to do an experiment as under normal conditions. Experience soon showed that the cumbersome nature of the system sometimes made it dangerous as well.

From universal dissatisfaction with the cabinet line, the "suit lab" was born. The idea was that the *researcher* would wear the protection, not the laboratory. Though tethered to an air hose, clothed in a "space suit," and wearing two pairs of lightweight rubber or latex gloves, the researcher could move much more freely and manipulate objects and animals in a more normal fashion. Arms of normal length, with standard elbows, would come back in style. By the mid 1970s, it was clear that the cabinet lab was inadequate, and Karl Johnson obtained a more modern cabinet line constructed for tumor virus work at NIH, but never used there. Then, for four years, Karl struggled to modify this into an even newer laboratory. This time it was developed as a combination of cabinet and suit laboratory: a prefabricated structure with brick and metal walls, enclosing the plywood inner walls of the laboratory itself. The suit lab portion was L shaped, accessed through an air lock, which also housed the Lysol shower for decontamination on the way out. The first arm of the L was a regular virology laboratory, with safety hoods (exhaust ventilators) and benches. At the end, another air-lock door led to the shorter arm of the L. This was an animal room, with several separate laminar-flow animal isolators. Cradled in the angle of the L was the sizable new cabinet lab, consisting of two long lines of cabinets.

The cabinet lab portion was opened well before the suit lab. At once, everyone flocked to it. When the suit laboratory was finally put into operation, a gradual migration began to this brand-new experience in a Level-4 environment. It took only a little over a year for the NIH hand-me-down cabinet line to be abandoned and used as a storage area.

Although it was an improvement, the suit lab did have drawbacks. It was very small, and it was quite common for people to be working with several different organisms and different experiments in the same lab at the same time. There were frequent problems, particularly with the holding tanks, which contained the effluent from the autoclave, showers, and other sources of fluids. There were two tanks, and when one was full it was superheated to sterilize the contents. We then had to wait several days until it cooled sufficiently to dump into the main sewer system of Atlanta without cracking all the sewers. Sometimes these tanks would fill up too fast, because of excessive use of the lab or, worse, because of a dripping faucet. That meant severe water rationing.

The lab was under substantial negative air pressure, so that if it sprung a leak, air would be sucked in, not blown out. The plywood walls bent inward, quite visibly, under the constant pressure. We improvised, patched, mended, and somehow kept it working. Other hazards included setting off the halon fire extinguishing system accidentally. This would produce a terrifying maelstrom of everyone's equipment that was stacked on top of the cabinets opposite the halon jets. Still, no one minded much; of all the things to be wary of in the suit lab, fire was the most terrifying. It would be very difficult to get out quickly, and our plastic suits would melt on our bodies.

Despite the problems, much good science was done in this lab. Though there were occasional alarms, no major accidents of any kind ever occurred in this laboratory, or in any other Level-4 facility at CDC. The only major accident in a Level-4 laboratory (at least outside the former Soviet Union, about whose program we know little) was the incident in which Geoff Platt infected himself with Ebola Zaire at Porton Down, U.K.

Still, we recognized that a much more permanent solution to the provision of Level-4 facilities was needed at CDC. Prompted partly by the emergence of HIV, funds were finally allocated, and in 1983 we started to design and build a new laboratory. The entire staff of Special Pathogens Branch now offered both good sense and some very strong views about what worked and what did not. The Special Pathogens Branch was the "end user" committee and pored over design details, making hundreds of detailed suggestions. The result was the best designed Level-4 laboratory in the world. Construction was begun in 1986, overseen by one of the Special Pathogens scientists, Dr. Mike Kiley. The design was two back-to-back suit laboratories, each one many times the size of the old suit lab. It was a palace by comparison.

What was it like actually to work in the lab? No one knows better than Sue:

If you were working alone—as I did on Sundays, when Joe dropped me off at the lab for monkey-feeding duty—you began by making an outside check of laboratory systems. You checked the Lysol shower tank, backup oxygen, and other gases. You also checked the effluent tanks. After signing off on a checklist, you put in your key card and punched in the number. Then it was clothes off, and on with a scrub suit. Step through the shower in the suit room.

Where is my suit? Buried at the back. Pull it out, identify its cord, and lower it on its pulley. Put on my first pair of gloves—regular surgeon's gloves. Check the second pair of gloves, taped securely onto the wrist bands of the suit. These are a little heavier than the surgeon's gloves, but light enough for me to perform fine tasks. Have they been recently changed? Are they undamaged? Okay.

Check the pale blue suit, and clean the visor. Slide my arms and legs in slowly to avoid the air lines that cool my feet and hands. Take a last deep gulp of fresh air, zip up the suit carefully, press the airtight seal securely shut, and then grab a breathing-air transfer unit and sling it over my shoulder.

Hop over the sill quickly into the Lysol shower, close the door, wait for it to click shut, and release the other door. Hop out over the second sill and—*ahh*—grab the first curled red air hose hanging from the ceiling. Didn't want to take time in the Lysol shower hooking and unhooking to the air. Now, if I really hold my breath and run, and can find my boots and get them on quickly, I can get right out into the main lab before grabbing another air hose. This saves a minute or two. Only problem is that, if you wait too long, your face mask steams up. Move on to the hose right by the first row of cages, let the cool, fresh air run in and clear the mask, then take a look at the "patients."

Nice long hose. Whoosh! In comes the air over my head. I can move right around the lab without changing hoses.

No one else in the lab today, so run through the checklist: incubators, freezers—check the gauges. All okay. Monkeys look fine. I'm not going to trouble them today; just see that they are active and eating and drinking. Check the water bottles and top them up, and then go to the fridge for the fresh fruit. Everybody gets half an orange and a banana and some monkey biscuits. I put in the biscuits, and start to cut the oranges. I never like cutting the oranges. You learn to dread knives and other sharp instruments in a place where all that stands between you and possibly lethal contamination is a thin plastic space suit.

After feeding, it's time to clean up and make a last check of the cages and equipment. All okay? Yes. Lights off. Out through the doors—no hose; takes too much time—boots off, and try to make it to the Lysol shower before I need more air. As the heavy inner door clicks shut, I press a button and the shower starts. Now I finally grab the hose. Rinse cycle finished, the hose is unhooked, hop out the door, and quickly unzip the suit so that I can breathe open air. Wriggle out of the suit, hook it to its cord, and pull it on its pulley, so that it drips dry, ready for tomorrow. Clothes off and into a body shower.

Although I made my Sunday rounds alone, I depended on a team for the week-day experimental work. In contrast to Level 4 at Porton Down, we were all women, with one one exception, Sam Trappier, who happened to be the only African American working in Level 4 at the time. Two others on the team, Mei Castor and Anne Conaty, had been in the Peace Corps. Mei was half Chinese, half Malaysian, petite and very pretty. She eventually left to go to veterinary school, and thence to medical school. Anne was all American, a big girl, good looking and blonde, who had been in the Peace Corps in Thailand. She developed PCR (polymerase chain reaction), a test for Lassa fever—the first PCR system actually used to diagnose a viral hemorrhagic fever. Lynette Brammer came from a clinical diagnostic laboratory, where she had worked nights for years. Cuca Perez, an elegant Puerto Rican woman, never lost the flavor of her origins and always replied, "Si" instead of "Yes."

The least likely recruit to the team was Bertha Farrar. Bertha, in her fifties, was rescued from a division of CDC that was being eliminated. She turned up for an interview, plump in a pink sweater and dyed blond hair—hardly our usual Level-4 candidate. Joe decided she had good organizational abilities and took her on. He assumed, however, that the suit lab itself was out for her. None of us thought to ask.

Bertha's organizational abilities were phenomenal, and she became a worthy successor to Helen Engleman, who had run supply operations like the ex-Marine she was. Under Bertha's management, the Sierra Leone Lassa fever project never lacked for the right supplies at the right time. She even went out to Sierra Leone and worked in the lab there. But this was not enough for Bertha. She finally told us: she wanted to work in the suit lab. At length, we relented, and she proved splendid here as well.

Teamwork in the lab was critical. It was quite difficult to communicate in the suits because of the constant rush of air in the head part of the suit. If you needed to say something, you had to cut off your air supply by pinching the hose and yell. So we all had our jobs worked out, and even though we would have six hands over each monkey—with needles—we all knew exactly what the others were doing. We depended on each other for our safety while working extensively with hot viruses and highly infectious animals, some of them quite large.

We also depended on our Safety Officer, Lee Alderman, a kind and gentle man, who was profoundly knowledgeable about the intimate workings of the suit laboratory. His detailed concern for our welfare, as well as our comfort and convenience, was a major factor in the success of a program that produced, among other things, a series of primate studies of hemorrhagic fevers comprising much of what is recorded about Ebola and Lassa virus disease in our nearest evolutionary relatives.

ACKNOWLEDGMENTS

The process of writing about one's life and work is daunting and humbling. If we're lucky, we begin to arrive at an understanding of how we developed and a deep appreciation for those who were indispensable along the way.

For Joe—

I would like first to acknowledge the support and encouragement of my mother, Jewel. Without her love and firm guidance when I was growing up on a farm in rural Indiana, I would be nowhere. I would also like to express my gratitude to my stepfather, Bob, and to my "second parents," Bun and Margaret Weathers. I owe a debt too, to Janet Kastner, and to Jim and Sue Colvert, without whom I might never have reached undergraduate school.

B. P. Reinsch, my math professor, John Sandbach, who taught me about humanity, and my college mates in the independent "Southerneer House" at Florida Southern College played key roles in my education. For my experiences in Belgium and Zaire, I must thank Luhahi Emil, Don Hill, Pete Peterson, Marijke van den Berg, and my third set of parents, M. and Mme. Noel, all of whom were wonderful teachers and worthy colleagues. Among the many who made a difference in molding my professional life were Syd Osterhout, who provided me with the chance to go to Duke Medical School; Sam Katz, who sent me to Boston; Johnny Long, who sent me back to the developing world; and Phil Lynch, my roommate at Duke Medical School. Tom Weller, the great virologist at Harvard School of Public Health, taught me and then directed me to CDC. I must credit Frank Oski, Louis Weinstein, and Stanley Plotkin, among many good teachers, for an extraordinary educational experience in pediatrics and medicine.

CDC has been exceedingly good to me. I particularly want to acknowledge the friendship and guidance of David Sencer, Bill Foege, Roger Feldman, Walter Dowdle, and Pat Webb. I am especially grateful to my virology mentor, Karl Johnson, the first American giant in hemorrhagic fever research. Thanks also to Lindsey Whitton and Mike Oldstone, who tried to teach me molecular virology, and to my colleagues David Fraser, the late Fakhry Assad, Kent Campbell, and Jim Curran. My gratitude to my faculty, research officers, and staff of the Department of Community Health Sciences at Aga Khan University. I have learned much from you. You are a great group, and you will make a difference in Pakistan. A warm thanks to Laurie Garrett, who let the light in for hemorrhagic fevers and opened doors for many of us. And, finally, I owe a very special debt of thanks for the long professional collaboration and personal friendship of Jonathan Mann.

For Sue—

My first debt of gratitude is to my father, who understood my need to go the extra distance. My greatest regret is that he died before he could see just how far that would be; the second is to my mother, who was always there for me and who helped me cope with bringing up a young child while I was going to medical school. I also wish to acknowledge my mother-in-law, who is one of my favorite people.

I especially remember Mrs. Thomas, my music teacher, who believed in me; the staff of the Farnborough Technical College; Dame Frances Gardner, Dean of the Royal Free Hospital Medical School, who thought I was worth a chance; Dame Sheila Sherlock, who assumed I could go as far as I wanted and told me to get on with it; John Tobin, who taught me laboratory virology and whose advice—"Do it as fast as you can"—I have never forgotten. I am also grateful for the contributions to my professional development of Bob Mitchell, Harold Stern, Jim Booth, Malcolm Smith, and many others in Oxford and in London. In particular, I wish to cite Chris Bartlett, who opened my eyes to epidemiology and public health; Colin Howard, who taught me molecular virology; and David and Mary Warrell, who gave me the opportunity to work overseas for the first time. Then it was Arie Zuckerman and David Simpson, who influenced me to study hemorrhagic fevers.

In Atlanta, my mentor was Joe, who has taught me most of what I know about hemorrhagic fevers. There were also others, at CDC, particularly Bobby Brown, who taught me about working with monkeys.

I would also like to thank the young Pakistanis who have worked so hard and enthusiastically with me at Aga Khan University, particularly Shaper Mirza and Tariq Malik, to whom I have passed on Dame Sheila Sherlock's message.

For both of us—

To all the folks who worked with us in the Special Pathogens Branch of CDC, many of whom are described in the book and without whom there would be much less of a story to tell; to our colleagues in so many countries, many of whom appear in this book as well, and who have become good friends, thanks.

A special thanks to Lee Alderman, who, as Biosafety Officer at CDC, constantly watched over us as we worked inside Level 4. A special recognition must go to the staff of the Lassa Fever Research Project in Sierra Leone, particularly Austin Demby. We must express our sadness over the fate that has befallen their lovely country and people.

Both of us owe a debt of gratitude to our first spouses, who supported us through many years. Of course, we want to acknowledge our children, Kit, Peter, Hannah, and Anne, who cultivate our non-scientific side, criticize us justifiably (sometimes mercilessly), and often remind us what the rest of life is about.

And when all else is said and done, we would like to acknowledge our debt to each other. We have shared so many of these experiences, and become each other's fiercest critic and fondest advocate. We have been fortunate to find both happiness and a way to enhance each other's scientific and personal lives.

Last, but not least, to Leslie Horvitz and Alan Axelrod, who patiently and professionally helped us through the process of turning our stories into this book.

JOE MCCORMICK
SUE FISHER-HOCH

I N D E X